LADIES ELECT

LADIES ELECT

Women in English
Local Government
1865–1914

PATRICIA HOLLIS

CLARENDON PRESS · OXFORD
1987

Oxford University Press, Walton Street, Oxford OX2 6DP

Oxford New York Toronto Melbourne Auckland
Delhi Bombay Calcutta Madras Karachi
Petaling Jaya Singapore Hong Kong Tokyo
Nairobi Dar es Salaam Cape Town

Associated companies in Beirut Berlin Ibadan Nicosia

OXFORD is a trade mark of Oxford University Press

Published in the United States
by Oxford University Press, New York

British Library Cataloguing in Publication Data

Hollis, Patricia
Ladies elect: women in English local
government, 1865–1914.
1. Women in local government — England
— History
I. Title
352.042   JS3078
ISBN 0–19–822699–3

Library of Congress Cataloging-in-Publication Data

Hollis, Patricia.
Ladies elect.
Bibliography: p.
Includes index.
1. Women in politics—England—History—19th century.
2. Local officials and employees—England—History—
19th century. 3. Local government—England—History—
19th century. I. Title.
HQ1236.5.G7H64 1987   320'.088042   87–5683
ISBN 0–19–822699–3

Set by Litho Link Limited, Welshpool
Printed in Great Britain by
Butler & Tanner Ltd, Frome and London

TO MABEL CLARKSON

*Norwich's first woman councillor, and
all the women councillors who have followed her.*

Preface and Acknowledgements

THIS book is unashamedly a labour of love. I stumbled upon women's local government work nearly ten years ago when compiling documents for *Women in Public*, while fitting in my own city council work around university teaching. Most friends and colleagues knew that women obtained the parliamentary vote in 1918. None knew that women had the local government vote and elected office some fifty years before that. I was later to learn that in my own county of Norfolk women were a stronger presence in local government in 1900 than in 1975.

Ladies Elect therefore started as a work of recovery, to describe and assess women's contribution as elected members of school boards, poor law boards and rural and borough councils. Many readers (like many voters) will not find the minutiae of local government business particularly appealing, and will perhaps regret that it has not been more compressed. I can only ask them to do what Victorian women did, and bear with it. Local government is properly concerned with sewers and street lighting, the politics of the gutter in an all too literal sense, as well as with the loftier language of civic service. It may be argued that I have cast women's local government work in too heroic a light. Yet knowing how much courage it takes even now for many women to stand for election, make public speeches, and chair local authority committees; and appreciating how much harder it was then for women to affront social proprieties and male sensibilities, I would claim for those elected women moral courage of a high order. It takes nerve and fortitude to smash windows and endure hunger striking as the suffragettes did; it also takes considerable courage to impose yourself on a board when your presence is deeply resented, and to offer service of a most personally demanding kind, year in and year out. Hard perhaps for men, in their overwhelming majorities and taking for granted the camaraderie of work and politics, to realize the bravery required of women who were isolated from any party or sisterly support. I would hope, however, that my admiration for their strength has not impaired a proper assessment of their work.

Have I claimed too much for their contribution, given them credit as individuals which belongs properly to their party, perhaps? In an obvious sense the book is not balanced as it does not seek to study local government as a whole but women's work within it, and therefore 'foregrounds' their efforts rather than those of other people. But I would argue that within the poor law, women did transform the administration of the workhouse; and that within education they highlighted important 'quality' questions which, without them, would have been less regarded. Their work on vestries and town councils was very much more marginal, but even there they made a contribution disproportionate to their numbers. The average councillor today is fortunate indeed if he (they are mostly he) can point to policy changes that he has personally initiated, a new community care scheme, a sheltered housing development, a framework for the arts, a swimming pool. Women councillors in the years before the First World War chalked up success after success. They made a difference.

Was that difference worth having? What, in local government eyes, seems municipal progress, can seem arbitrary and unwelcome interference to its recipients. The elected lady, imposing school attendance, clearing slums, closing pubs and brothels, reordering the workhouse, and refusing outdoor relief, was not necessarily popular, and sometimes shared the intrusive and morally authoritarian style of her philanthropic sisters. Would the poor have been better off without her? Does the book endorse her activity without sufficient awareness of what it felt like to be on the receiving end?

Two points. The first is that local government activity is not and cannot be welcome to all its constituents even most of the time. It is not about right or wrong choices (well, not usually), but about balancing sectional claims, the elderly's demand for peace and quiet against the children's desire for play: the resident's wish to limit through traffic, and that same resident's wish to drive through other people's streets. At a time when publicly financed council housing was not on the agenda, what *should* a London vestrywoman or woman councillor have done about slum housing? Let it stand, unbelievably bad though it was? Or demolish it, to be replaced with by-law housing, creating worse housing conditions for the very poorest, and better housing for the rest? What *should* a woman school board member have done about children not attending school? Was that a childish preference for play over work? A sensible rejection of irrelevant instruction? A desperate family need for children's earnings over their education? Or the sacrifice of a child's innate right to education, when she was in no position to make choices, to the parents' wishes who could?

Local government women were well aware that there were losers as well as gainers from municipal activity, and that the very poor were more often losers than anyone else. But, as best they could, women sought consent to what they were doing. On the whole, elected local government women were less authoritarian, less censorious, and more respectful of the views of the poor than their male colleagues, both elected members and officers alike. The exception, and it is a large one, was the early generation of poor law women on the subject of outdoor relief, where their temperance and philanthropic training made them more judgemental and case-hardened than the men. That aside, in newspaper report after report, in election after election, and in town after town, women stood for the poorest wards, were run by working men's associations for office, fought hard to distribute municipal resources towards the neediest, and tried to soften the harsher aspects of local government intervention. They believed in what they were doing, and they did their best.

The book is not a study of women employed in local government (teachers, clerks, poor law nurses, sanitary inspectors, or rate collectors), except incidentally, though that book requires writing; nor of women co-opted to local government work, as members of care committees, distress committees, or the local education authority itself, without passing through the process of election. It is a study of the three thousand or so women who between 1870 and 1914 sought votes and won seats, often in the face of considerable male hostility, to the parishes, vestries, boards, and councils of English local government.

A word about methodology. The book is based on personal research in some twenty towns and cities; London, Liverpool, Manchester, Birmingham, Leeds, Bradford, Oldham, Bristol, Sheffield, Nottingham, Norwich, Oxford, Cambridge, Reading, Great Yarmouth, Brighton, Eastbourne, Bolton, and Southport. I was also sent marvellously full material on Tynemouth (through the good offices of Mr Eric Hollerton, district librarian), on Staleybridge, Guildford, Richmond on Thames, and Godalming. Towns chosen were simply a mixture of the largest, and of those spa/sand/spire towns where women were particularly prominent. Almost entirely missing are the great port and dockyard towns of Plymouth, Portsmouth, Southampton, Hull, and Newcastle where women for long remained politically invisible. Ten Norfolk rural district councils were also studied in some detail, but it is not easy to recover the work of rural local government women as the sources are somewhat sparse. Scattered material was acquired on a further hundred or so authorities, from Falmouth to Westmorland, and this offered biographical material of varying fullness on over two thousand women. Readers may rightly complain that chapters could have been slimmer had some of the references to local women been discarded. I retained them, on reflection, because I hope that others may be sufficiently intrigued to recover women's work in their locality for themselves. By working across such a large canvas I will, I fear, have sometimes misjudged a situation or made mistakes which will be painfully obvious to the local historian. I would welcome any correction they care to make. The detail would have been more secure, but perhaps the findings less so, had the book confined itself to just three or four towns.

Each chapter of the book is organized somewhat differently, according to the nature of the material. The London school board, finite in its existence, has been treated narratively. Provincial school boards, again because they too faced finite problems over a defined time-scale, have been presented as a dozen local case-studies. Women's poor law work has been described thematically, client group by client group; while women's work in council has been discussed in more general terms. There is more 'setting' in the chapter on rural local government than the others, but as far as I am aware, nothing has been written on the Parish and District Councils Act of 1894 since 1895, to which one can refer the reader.

Working in so many towns has meant that I have incurred a host of debts which it is my pleasure to acknowledge. At their head must come my thanks to the Economic and Social Research Council for funding a year's research without which I could not have completed the book. Of the many archivists and local librarians I have consulted, I would especially like to thank David Doughan of the City of London Polytechnic Fawcett Library; Mr Eric Hollerton, North Shields district librarian; Mr David James, Bradford city archivist, who kindly shared his own research findings with me; and the staff of Staleybridge, Guildford, Southport, West Sussex Record Office, Godalming, and Richmond on Thames libraries, all of whom have been helpful far beyond reasonable expectation. I would like to thank the librarian of Loughborough College for permission to borrow, and use, their material on the Women's Co-operative Guild; the librarian, Girton College, for access to the Women Guardians Society minute books as well as the Blackburn and Davies collection; the secretary of the National Council of Women for access to the papers of the National Union of Women Workers as well as their friendly cups of tea. Among colleagues, my thanks to John Garrard for his remarks on northern politics, to Jill Liddington for references to Lady O'Hagan, to Martin Pugh for an advance sight of his *The Tory Party and the People*, to Professor Rosemary Van Ardel for comment on Florence Fenwick Miller, to Linda Walker for her thesis on women's party political activity in the period, to Brian Harrison and David Rubinstein for helpful comment, and John Gyford for recent work on women in local government. Alex Alexander gave me help on Edith Sutton of Reading; my sister, Rosalyn Clarke, worked with me through the back numbers of Nottingham newspapers. Professor Bryan Keith-Lucas kindly gave me permission to reproduce from *The English Local Government Franchise*, 1952, his summary of local government legislation, which is featured in Appendix D; Dr Alan Crosby has allowed me to reprint his exquisitely drawn map of Norfolk, and Dr Paul Thompson to reproduce his maps of London, first published in his *Socialists, Liberals and Labour: The Struggle for London 1885–1914,* 1967. Finally, my thanks to the members and officers of Norwich City Council, past and present, who have taught me in the past twenty years to value local government.

September 1986   PATRICIA HOLLIS

Contents

Contents xv

List of Illustrations

duced with kind permission from the London School of Economics); Miss Mary Clifford of Bristol (reproduced with kind permission of Bristol Central Library); Miss Eleanor Rathbone of Liverpool (reproduced with kind permission of the University of Liverpool).

9. Mrs Ethel Leach, canvassing at Great Yarmouth, *The Daily Mirror*, 3 November 1908.

10. Miss Margaret Ashton, canvassing at Manchester 1909 (reproduced with kind permission from the Ashton collection, Manchester Central Library).

11. Mrs Lees, Mayor of Oldham, and her work. *The Designer*, February 1912. (Reproduced with kind permission from Oldham Central Library).

12. Miss Mabel Clarkson, of Norwich, and her election address of 1913, (reproduced with kind permission from the Colman and Rye Library, Norwich).

List of Maps

Abbreviations

BWTA	British Women's Temperance Association
CAB	Citizens' Advice Bureau
COR	*Charity Organisation Review*
COS	Charity Organisation Society
EDP	*Eastern Daily Press*
EWR	*Englishwoman's Review*
GFS	Girls Friendly Society
GLC	Greater London Council
HMI	Her Majesty's Inspector
ILP	Independent Labour Party
LCC	London County Council
LEA	Local Education Authority
LGB	Local Government Board
LNA	Ladies' National Association
LSB	London School Board
LSE	London School of Economics
MABYS	Metropolitan Association for Befriending Young Servants
MBW	Metropolitan Board of Works
MOH	Medical Officer of Health
NSPCC	National Society for the Prevention of Cruelty to Children
NUT	National Union of Teachers
NUWSS	National Union of Women's Suffrage Societies
NUWW	National Union of Women Workers
NVA	National Vigilance Association
RC	Royal Commission
RDC	Rural District Council
RSPCA	Royal Society for the Prevention of Cruelty to Animals
SB	School Board
SBC	*School Board Chronicle*
SSA	Social Science Association
u.d.	undated
UDC	Urban District Council
WCG	Women's Co-operative Guild
WDP	*Western Daily Press*
WGS	Women Guardians Society
WH	*Women's Herald*
WLA	Women's Liberal Association
WLF	Women's Liberal Federation
WLGS	Women's Local Government Society
WLL	Women's Labour League
WLUA	Women's Liberal Unionist Association
WPP	*Women's Penny Paper*

WS	*Women's Signal*
WSJ	*Women's Suffrage Journal*
WSPU	Women's Social and Political Union
WVS	Workhouse Visiting Society
YWCA	Young Women's Christian Association

I

Philanthropists, Electors, and Party Activists

M ISS H ENRY was elected in 1894 as a rural district councillor
for Newbury; that meant she was also a poor law guardian
and served on a school attendance committee. Some fifteen
years later she was speaking to the Anglo-Japanese Congress
of 1910.

One day in a school I was told, 'We had three inspectors yesterday,
the Diocesan Inspector, the Inspector of the Education Department,
and the County Council Inspector'. I replied, 'Well, now you have
one more!' I went into the infants' room and at once said 'What is
the matter with that child?' The child came to me, and was standing
on her toes; she could not put her feet to the ground. Now that child
was crippled and yet had to walk more than a mile to school every
day. I felt that we could not let that go on, or she would be in the
workhouse by the time she was sixteen. So I went to our Board of
Guardians and they agreed to give me a ticket for a London hospital.
The child went up to the hospital; and her legs and feet were made
perfectly straight. I had forgotten all about it, but when I went to the
school afterwards the teacher called the child out, and she ran to me
and said, 'Look at my straight legs'. Now that child has as good a
chance in life as any other. Where you get one chance of doing good
in connection with a church or parish, you get fifty as a district
councillor.[1]

It was a small yet revealing incident that could have been
repeated by countless women in local government before the
First World War. There was Miss Henry, conscientiously
inspecting schools many miles from home, undeterred by the
information that the school had been thoroughly inspected the
day before; noticing as a woman would but three professional
male inspectors had not, that a child was crippled; concerned
both at the cost to the child in suffering but also at the long-
term cost of that child to the rates; taking swift and effective

---

[1] *Report of the Anglo-Japanese Congress*, 1910, p. 5.

action as a guardian; and fully aware of the privileges and power for good that she possessed as an elected member which far exceeded what might be done through church or charity. As Mrs Lees, who became one of the first woman mayors in the country, had said in July 1903 when pressing the claims of Oldham women to be poor law guardians, 'a member of a Public Board has a position and power for usefulness that no other person can possibly have'.[2]

By the late 1890s, when Miss Henry was serving as a rural councillor, some fifteen hundred women were holding elected local office. They were members of London vestries, school and poor law boards, parish, rural, and urban district councils, and they would have agreed with her. Half a century before women obtained the parliamentary vote and held parliamentary seats, they had obtained the borough vote and been elected to local government office. From the 1860s, developments in local government and in the women's movement had converged to bring forward 'ladies elect', who were to exercise an influence and offer a service, that, as was said of Miss Sturge of Bristol, 'may deliberately be called a splendid service'.[3]

It is perhaps worth reminding ourselves of the tangled nature of nineteenth-century local government. The administrative geography of England had for centuries been rooted in the parish and in the municipal borough, old towns like Norwich and Bristol. The new industrial towns of the North, such as Manchester, which had grown out of villages, remained parishes in name, without MPs or borough government, and ill-equipped to deal with problems of overcrowding, disturbances, or disease. In 1832 such towns acquired MPs of their own, and in 1835 the Municipal Corporations Act allowed them to apply for borough status. All boroughs, old and new, were now to have a town council, elected by ratepayers, able to address problems of public order and public health. Rural areas slumbered on unchanged—county councils were half a century away. In the countryside, law and order was dispensed by JPs, public health and highways handled by a miscellaneous array of parishes and boards. One

---

[2] Mrs Lees, letter to Liberal and Conservative associations, Lees collection, 133.
[3] See below, p. 157.

piece of legislation had, however, cut through their torpor. The 1834 Poor Law Amendment Act had grouped parishes into unions, the better to build workhouses, and these unions were to become the rural sanitary districts of 1872 and the rural district councils of 1894. Within the towns, poor law responsibilities remained with urban poor law boards (and their weighted franchise) and did not pass over to the new town councils. In the absence of a common and coherent structure of local government across town and country, government continued during the nineteenth century to permit or invent *ad hoc* boards for particular tasks, such as improvement works or education.

The result was a tangle of authorities and agencies, each with different boundaries, different electorates, different qualifications for office, different voting procedures, and different rating powers. The president of the Local Government Board told the House of Commons in 1893 that England and Wales had by now 62 county councils, 302 municipal boroughs, 31 improvement act districts, 688 local government act districts, 574 rural sanitary districts, 58 port sanitary districts, 2,302 school board districts, 362 highway districts, 6,477 highway parishes, 1,052 burial board districts, 648 poor law unions, 13,755 ecclesiastical and nearly 15,000 civil parishes, all criss-crossing and cross-cutting each other; and he suspected he had mislaid some in the count.[4] Over the years, however, most of the *ad hoc* bodies were absorbed by the borough councils—sanitary improvements by the 1870s, education in 1902, poor law not until 1929. When, following the Third Reform Act of 1884, county councils and then parish and rural district councils were invented to provide rural administration, the new map of English local government was essentially complete. Slowly, elected councils acquired a general competance for their communities.[5]

Parliamentary reform had knock-on implications for local

---

[4] Fowler, Hansard, House of Commons, 21 Mar. 1893, col. 681.

[5] For studies of local government, see D. Fraser, *Urban Politics in Victorian England*, 1976; *Power and Authority in the Victorian City*, 1979; E. P. Hennock, *Fit and Proper Persons*, 1973; J. Dunbabbin, 'British Local Government Reform: The 19th Century and After', *EHR*, Oct. 1977, pp. 777–803; J. Garrard, *Leadership and Power in Victorian Industrial Towns, 1830–1880*, 1983; P. J. Waller, *Town, City and Nation: England 1850–1914*, 1983. See Appendix A for the structure of local government functions.

government, its boundaries and its electoral registers. It also helped to focus the emerging women's movement. Nineteenth-century feminism did not start with the demand for the vote, but from the need of 'surplus' middle-class single women to find respectable paid work apart from governessing; from the demand of married middle-class women to acquire legal rights within marriage; and from the hunger of leisured middle-class women to engage in worthwhile public service.

Given the poverty of girls' education, single women unwilling or unable to marry were ill-equipped to enter the labour market. As Emily Davies tartly remarked, 'It is indeed no wonder that people who have not learnt to do anything cannot find anything to do.' She began in the 1860s to promote secondary and higher education for girls. Her local committees, bringing girls forward for public examinations, were later to run their mothers and their sisters and their aunts for local school boards. Around her was clustered a powerful circle of women, Elizabeth Garrett, Barbara Leigh Smith Bodichon, Bessie Rayner Parkes, Maria Rye, and Jessie Boucheret, who were carving out other areas of work for women, such as clerical work, printing, law copying. Their periodical, *The English Woman's Journal*, started in 1858 from their Langham Place offices, was the first recognizably feminist publication in the field. It began to network women's efforts. Meanwhile Florence Nightingale was cleaning up nursing, Elizabeth Garrett was forcing open medical doors. It was not by accident that Emily Davies and Elizabeth Garrett were to be the first two women to hold elected office in 1870.[6]

The second cluster of feminist concerns, at a time of wider legal reforms, was the legal status of married women. Barbara Leigh Smith's *Summary of the Laws of England* (1854) had listed the legal and civil rights lost by women on marriage, when all

---

[6] Thirty per cent of all women aged 24 to 35 years were single in 1871, but M. Vicinus argues that the plight and numbers of single 'surplus' middle-class women were overstated. M. Vicinus, *Independent Women: Work and Community for Single Women 1850–1920*, 1985, p. 26. For Emily Davies, see B. Stephen, *Emily Davies and Girton College*, 1927. She was building on the pioneering work of the Governesses' Benevolent Institution which started Queen's College in London in 1848, followed by Bedford College, Miss Buss's North Collegiate School (1850), and Miss Beale's Cheltenham Ladies College (1858). More generally, see J. Burstyn, *Victorian Education and the Ideal of Womanhood*, 1980; R. J. Evans, *The Feminists*, 1979; L. Holcombe, *Victorian Ladies at Work 1850–1914*, 1973; J. Lewis, *Women in England, 1870–1950*, 1984.

their rights were subsumed into those of their husbands. Husband and wife were one, said Blackstone, and 'that is he'. Women now tried to disaggregate their rights from those of their husbands, by seeking independent control of their property within marriage and independent custody of their children outside it. The 1839 and 1873 Infant Custody Acts, the 1857 Divorce Act (despite its double standards), and the 1870 and 1882 Married Women's Property Acts, marked some of the steps along the way.[7]

A third source of pressure came from those single and married women with leisure bought by industrial prosperity and a growing servant class, searching for something useful, good, and godly to do. Florence Nightingale had found it by training at Kaiserworth. Anna Jameson's call to a communion of labour had stirred the conscience of many others. As Catherine Winkworth of Bristol wrote to her sister, Mrs Shaen, in 1874, 'I am often tossed in my mind whether a life in which I neither write books nor visit the poor is rightfully arranged for someone who is not married nor an invalid'.[8] Many women turned to philanthropy, where they were organized into voluntary societies and district visiting; others devoted their energy to mission work; some turned to public service. Miss Winkworth and Mrs Shaen both became poor law guardians.

Many of these threads came together in the Social Science Association formed in 1857. Its members, lawyers, MPs, philanthropists, and women voluntary workers, met in congress to encourage best practice at the interface of official and voluntary public work, promoting sanitary reform, workhouse visiting, rescue work among women and children, legal reform, and employment opportunities for women. Its sectional committees often flowered into free-standing voluntary organizations of their own. Said Emily Davies, 'The Association was of immense use to the women's movement in

---

[7] H. Burton, *Barbara Bodichon*, 1949; R. Graveson and F. Crane, eds., *1857–1957: A Century of Family Law*, 1958. L. Holcombe, *Wives and Property*, 1983, describes the two conflicting legal systems of common law and equity, which were fused in the great Judicature Act of 1873. *The Summary* is reprinted in P. Hollis, ed., *Women in Public 1850–1900: Documents of the Women's Movement*, 1979, pp. 171–8.

[8] C. Winkwork, *Memorials of Two Sisters*, 1908, p. 308. More generally, Vicinus, op. cit.

giving us a platform from which we could bring our views before the sort of people who were likely to be disposed to help in carrying them out'.[9]

As the parliamentary reform movement of the 1860s grew, however, the demand for the vote for women on the same terms as men became the overarching issue, their 'common principle of action'.[10] The circle around the Langham Place women now included Helen Taylor and had constituted itself the suffragist Kensington Society. When in 1867 John Stuart Mill presented their suffrage petition in a magnificent Commons speech, the women's movement came out in public and on to the parliamentary agenda. When in 1869 Mill published his *Subjection of Women*, he gave them their testament.

With the Second Reform Act of 1867, parliament was again committed to consequential amendments of the local government system. No better campaign ground could have been devised for the women's movement. Local government seemed to offer a sphere of work and a way of going about it particularly suited to women. It embraced 'the domestic work of the nation', said the *Anti-Suffrage Review*, a belated convert to its value. When women stood for election so that they might inspect workhouse beds and childrens' bodies for vermin, bring the damaged and delinquent child to school, campaign for clean milk, pure water, and no drink, they were reading local government, in the phrases of time, as compulsory philanthropy, as municipal housekeeping, as domestic politics. 'When the community undertook more and more the administrations of these home duties', wrote the *Englishwoman* in 1914, 'many women followed their natural work into the larger world.'[11] As local government enlarged its responsibilities, as care of the young and old, poor and sick moved from the private to the public domain, so did women.

Yet local government was more than just practical rate-funded philanthropy. It called for service within the map of

[9] Quoted Stephen, op. cit., p. 75.

[10] G. Hill, *Women in English Life*, 1896, pp. 346–7. More generally, see O. Banks, *Faces of Feminism*, 1981; C. Rover, *Women's Suffrage and Party Politics 1866–1914*, 1967; B. H. Harrison, *Separate Spheres*, 1978; L. Hume, *The National Union of Women's Suffrage Societies, 1897–1914*, 1982.

[11] *Anti-Suffrage Review*, Dec. 1908, p. 2; *The Englishwoman*, 16, 1914, p. 16.

local politics. It required women to seek election; mobilize an electorate; gain the endorsement of political parties and the confidence of local interest groups; accept the drudgery of canvassing and committee work, the exposure in the press and on the public platform, and the discipline of accountability. In every way, women believed, it was an apprenticeship for parliament. Local government, in other words, caught up notions of suffrage and service, philanthropy and practical Christianity; and permitted the pursuit of good causes, temperance and liberalism, moral and social purity. It offered multiple languages within which women could advance their rights, plead their cause, pursue their duties, fulfil their mission, and lay claim to full citizenship. Women coming from the field of philanthropy might cast their service in the language of separate spheres; they were different from men, an argument which men would find hard to dispute. Women coming out of the suffrage movement might lay greater stress on the equal needs, equal rights of women. They were not essentially different from men . . . 'the same origin . . . the same chequered and difficult path through life . . . and the same final destiny'.[12] Most women most of the time were happy to use both arguments at the same time without embarrassment. And just as the changes within local government presented opportunities to the women's movement, so the entry of women into local government helped to mould the nature of local government service itself.

Those opportunities opened almost immediately. Some five thousand women ratepayers claimed the parliamentary vote in 1867, but this was dismissed by the courts a year later. However, in 1869 when government tidied up the compounding question that had hung over from the 1867 Reform Act, Jacob Bright slipped through an amendment late at night giving women ratepayers the borough vote. This was narrowed by the courts in 1872 following the case of *Regina* versus *Harrold* to unmarried women ratepayers. None the less, women soon formed up to a fifth of the borough electorate. By 1900, over a million women had the local government vote.

When in 1870 the government devised school boards to supplement voluntary education, women ratepayers could

[12] Jacob Bright, Hansard, House of Commons, 6 June 1877, col. 1410.

vote for and any woman could stand for the new boards. Elizabeth Garrett headed the London school board results with 48,000 votes, three times as many as the next highest candidate, T. H. Huxley. By 1879, there were some seventy women on school boards, many of them drawn from Emily Davies's 'ladies' educational societies. By 1900 all the major towns had women members, sought along with working men to balance the political slate. They were strikingly child-centred, insisting, in Mrs Homan's words, that education was not about dry facts but about 'making the children something better'.[13] They introduced kindergartens, PE, and pianos, to the mingled irritation and enthusiasm of their male colleagues.

It was not clear whether women were eligible to serve on poor law boards, but in 1875 Miss Martha Merrington quietly stood for a London union and was elected. Twenty years on, there were over eight hundred of them. Many of these women guardians were experienced workhouse visitors; and had been trained, in London at least, by the Charity Organisation Society. Confident that they could discriminate between the deserving and the undeserving poor, they sought to repress outdoor relief, while at the same time seeking to improve workhouse life for women and children, the old and the sick. Miss Brodie Hall of Eastbourne took 'barracked and warehoused' children out into foster care. Mrs Evans of the Strand found one hundred and twenty girls sharing five brushes, eight towels, and ringworm. Mrs Leach learnt that Yarmouth's sick wards were headed by a pauper nurse of eighty-one, assisted by another of eighty-five and an inebriate midwife of seventy-one. None of them could read the medicine labels. She introduced trained nurses, tender loving care, and sacked the misogynist clerk to the board, no mean feat as he was a deputy lord-lieutenant of the county, and moved in powerful circles.

The Third Reform Act of 1884, which extended the house-hold vote to rural England, did not extend it to women, to their chagrin and dismay; but the Act was followed by the 1888 County Councils Act (for which women were not eligible) and the 1894 Parish and Rural Districts Act (for which

---

[13] Mrs Homan, see below, pp. 119 f.

they were). At local level, women and working men swept to power in what was considered 'a rural revolution'.[14] They closed unfit houses, sank wells, cleaned ponds, walked footpaths, bought burial grounds, acquired allotments, and regulated nuisances. A few of them joined London vestries where they grappled with the problems of the built environment of the East End, its overcrowding, squalor, and disease. From 1907 women were finally admitted to borough and county councils, and though they found it hard to win seats, successful women city councillors did much to bring down infant mortality and slum property. From their ranks came women like Eleanor Rathbone of Liverpool, Susan Lawrence of the LCC, Marion Phillips and Ethel Bentham of Kensington Borough, who after the War were to serve as the first generation of women MPs. Margaret Ashton of Manchester found herself in 1910 with a 'watching brief' (her words) for all women's questions, from municipal milk and midwifery to public lavatories and lodging-houses. Such women came from a philanthropic and usually liberal and nonconformist constituency. They were the women of whom Josephine Butler had presciently written back in 1869:

There are few, scarcely any women in these Northern towns who live a simply fashionable life. They are often humble middle classes chiefly, and are for the most part sensible earnest working women. A very considerable number are trading in their own names. The *men* of the towns are apt to be a little too much absorbed in business and money-making, and the women who have somewhat more of leisure are frequently as it is, called upon to supply to members of town councils opinions on all the matters with which town councils deal. They are matters which concern women and children equally with men, and in which the advice and practical help of women will be needed more and more. The question of pauperism alone, including that of nursing out the pauper infants, is one in which men will require all the help they can get from women of insight, and delicacy of perception, for it is a question which has to deal with human beings of all ages and both sexes. Again on all such questions as those of public health, prison government, the details of expenditure in large institutions etc. are questions in which the united common sense of men and women is required, and in which women could be

[14] See ch. 7.

granted a greater share of responsibility with advantage to the whole community.[15]

Some now exercised that greater responsibility on parish or rural councils where they sat, the wives and daughters of the vicar and the squire, using their social status to intimidate parsimonious farmers into spending the rates. In the larger towns, women issued their manifestos alongside men, canvassed vigorously, addressed meetings at length, chaired committees, headed deputations, gave evidence to Royal Commissions, and served on government committees. They mobilized behind them the women's vote, the philanthropic vote, and they could (except in poor law work) count on the working-class vote as well. They built networks of mutual support, and circulated secretaries, societies, letters, journals, conferences, pamphlets, address books, brothers, and husbands amongst themselves. They included motherly, middle-aged conservative churchwomen, hard-edged non-conformist Liberals, undogmatic and pragmatic party workers, interventionist Progressives, and a few visionary and argumentative Socialists. Divided though they often were by political, religious, and class loyalties, they none the less shared a feminist perspective, a 'trade union spirit' as a sour member of the London school board put it, which they deployed, quite explicitly, on behalf of women and children, the old and the sick, the morally, mentally, and physically deformed. They spoke to the moral community; they left men to address civic pride. As Caroline Ashurst Biggs, editor of the *Englishwoman's Review* wrote in 1881, 'It is the most Christian work a woman can take up. The poor we have always with us . . . Indiscriminate alms giving is not Christianity, neither is paying your rates. But giving time and trouble to see that the rates are applied properly, for the greatest benefit, morally as well as physically, is the noblest Christianity of all'.[16]

Philanthropists

Women came into local government from the world of philanthropy. Since the days of Hannah More, personal

---

[15] Josephine Butler, 1 July 1869, quoted in R. Walton, *Women in Social Work*, 1975, p. 68.

[16] C. A. Biggs, 'Women as Poor Law Guardians', *EWR*, 15 Sept. 1881.

philanthropy was considered a woman's profession, requiring tact and moral taste, an attention to detail and a sympathy with the domestic scene, for which men had neither the time nor the aptitude. In rural society, with parishes small, relatively stable, and knowable, it was thought easy enough for the charitable lady to assess 'the worth and the wants' of parishioners (Hannah More's phrase), and provide them with coals and baby clothes accordingly. The recipients no doubt displayed their gratitude and concealed their resentment as best they could.[17] However, in large and residentially segregated cities, the quarters of the urban poor were out of sight, threatening public health with cholera and public order with Chartism. The poor needed faith, hope, and charity. From the late 1840s, newly formed voluntary societies sent waves of women district visitors into the slums, visiting house to house, bearing with them the religious tract, the sisterly touch, moral certitude and carbolic soap; inspired by Josephine Butler's imperative, 'the extension beyond the home of the home influence'.[18]

In the 1880s as depression deepened, the housing crisis worsened, and political demonstrations again took to the streets, such work seemed even more urgent. By the late nineteenth century, Louisa Hubbard estimated that at least twenty thousand salaried, half a million voluntary, and many thousands of church women were at work befriending the homeless, rootless, and handicapped.[19] It was a hugely impressive effort in time, money, imagination, and benevolence. The self-denial was genuine: their recent historian quotes a survey of forty-two families in the 1890s, which found that they spent more on charity than on any other item in their budget, including rent, clothes, and servants' wages, except food.[20] So was the compassion. When a speaker casually referred to fallen women, she was speedily corrected by a rescue worker. 'Call them "knocked down women" if you will, but not fallen'.[21] More problematic was the amateurism. J. R.

---

[17] See Joseph Arch's autobiography, and M. Ashby, *Joseph Asby of Tysoe, 1854–1919*, 1961, for descriptions of rural charity.

[18] J. Butler, *Women's Work and Women's Culture*, 1869, p. xxxvi. More generally, F. Prochaska, *Women and Philanthropy in 19th Century England*, 1980.

[19] A. Burdett-Coutts, ed., *Women's Mission*, 1893, p. 363.

[20] Prochaska, op. cit., p. 228.   [21] Burdett-Coutts, op. cit., p. 157.

Green in 1867 could talk dismissively of the lady-brigades from the West End descending on the East in gift-bearing rivalry. Miss Emily Janes, organizer of the National Union of Women Workers (NUWW) reminded her conference in 1891 of some of the ways of charitable women:

Miss A., for instance will teach at the Girls club, but you must not depend upon her if it is wet or cold, or if some unusual attraction in the way of entertainment turns up; Mrs B. will be delighted to look after a little maiden in her place, so long as the child gives no trouble, otherwise 'she really can't undertake the responsibility'; Miss C. will take the Bible class (the odd thing is that everyone supposes herself capable of explaining the mysteries of the Faith) but you cannot expect her either to attend weekly lectures, or to make up for the want of definite preparation by home study; Mrs D., poor lady, thinks it would be an easy and pleasant way of getting a living to take up the superintendence of a Girls Friendly Society Lodge, or a training school for young servants, and is surprised to find that it is not sufficient to walk around the house in the morning to see that it has been set in order, and then to devote the remainder of the day in mild gossip and the study of three volume novels from the circulating library around the corner; Mrs E. credits herself with being philanthropic on the score of a few guineas given in answer to emotional appeals, and a few tears shed as she reads by her comfortable fireside, of the woes of little children and of despairing women; Miss F. turns hospital nurse for a few weeks, because her imagination is fired by the idea of wearing a becoming dress and sitting in interesting attitudes by her patient's bedside; Miss G. dear soul, hopes to fit herself for mission work in China by the study of Hebrew and the theory of music; Miss H. takes up slumming in a violent hurry, because it is the fashion, and drops it again as quickly; and Mrs J. whose feeble goodygoody fiction does not alas! all find its way into the wastepaper basket—are not all these types you and I have often met with, camp followers as it were, hanging on to the rear of those fighting under strain and stress, and sorely hindering real progress?[22]

Many of the larger charities made creditable efforts to train their workers; and the promise of disciplined endeavour underlay much of the appeal of religious sisterhoods such as the Kaiserworth deaconesses, and the Nightingale nursing

[22] Miss Janes, NUWW Annual Report, Liverpool 1891, p. 203; J. R. Green, quoted in H. Bosanquet, *Social Work in London 1869–1912*, 1914, p. 12.

school. Georgina Hill commented in 1896 that 'women have developed an unexpected capacity for organization, an enterprise in arduous undertakings, and an enthusiasm for difficult, disagreeable and unpromising work'. From their philanthropic work, many women acquired self-assurance, committee discipline, organizational skills, the courage to canvass, the patience to fund-raise, the ability to read accounts, the fluency to address public meetings, and the confidence to chair functions. All of them skills equally valuable in local government.[23] Women for their part sternly lectured each other on the need to exhibit self-discipline and staying power, to regard philanthropic service as work. Mrs Sheldon Amos could talk of 'the arduous unpaid work of the national housekeeping' to the NUWW conference in Leeds in 1893. One reason why the much-maligned Charity Organisation Society (COS) possessed the influence and retained the loyalty it did, even when its moralism was thought outdated, was its investment in training. Octavia Hill had moulded an entire generation of able women in professional 'scientific' philanthropy, based on careful enquiry, detailed observation, and personal casework. Her students went on to become district visitors, rent collectors, sanitary inspectors, and poor law guardians: women like Eva Muller, the Lambeth guardian who later founded the Women's Liberal Federation, the Sturge sisters of Bristol, and Emma Cons, the first LCC woman alderman. When Miss Mallory was appointed a guardian in Hackney by the Local Government Board she promptly went out and bought half a street full of property to practise housing management and learn her trade along best COS lines. The COS taught its women to have the confidence of their training, to be tough—perhaps too much so, for they often became case-hardened and, in Sophia Lonsdale's phrase, 'quite fearless of unpopularity'. Miss Lonsdale, coming from a jolly, horsey, rectory background on to Lichfield's poor law board, 'was of opinion that nothing was so calculated as to give a young woman security as to herself, her circumstances, the world, her future work, and the dominant ideas of the day, as a COS training . . . A young woman ought not to leave the

---

[23] Prochaska, op. cit., p. 228

COS until she has found herself'. And she added, 'I am body and soul with the COS.'[24]

Other women from the late 1880s found their way to the new settlement houses in Bermondsey, Lambeth, and the East End where, inspired by the example of Canon Barnett's Toynbee Hall, they befriended women who were imaginatively as well as financially deprived. Mornings were spent in district visiting, afternoons studying, holding clinics and mothers meetings, evenings running girls clubs, clothing clubs, and endless (though not always thankless) meetings and societies. To their modest terraced houses came residential community workers as well as a floating population of the more casually charitable. It was noticeable that later London school board women (Honnor Morten and Mary Bridges Adams included), as well as women standing for the London boroughs after 1907, had done their stint in a settlement where they had attempted to experience working-class life from the inside out. They were no doubt motivated by class guilt, as Beatrice Webb suggested, but they were also seeking training in social administration. Many of the settlements, as at Liverpool, had close links with the newly formed university social science departments. Jane Brownlow, producing the Women's Local Government Society handbook in 1911, insisted that any woman seeking elected office needed 'a severe training', either by serving on a 'well ordered committee' such as one of the London care committees, working for the COS, or living six months in a university settlement.[25]

Although women tried to confront or allay men's fears that they would be soft, silly, and sentimental in their dealings with the poor, men found it hard to take their contribution seriously or value it properly. This was as true for philanthropy as it was for public service. The street workers were women; the decision makers and committee members were men, even within those societies, like animal and child welfare, with which women had a particular affinity. Women found it just as hard to gain the citadels of philanthropy as to

[24] Mrs Sheldon Amos, NUWW conference report, Leeds 1893, p. 108; S. Lonsdale in V. Martineau, *Recollections of Miss Sophia Lonsdale*, 1936, pp. 217–18.
[25] J. Brownlow, *Women's Work in Local Government*, 1911, p. 3; M. Vicinus, 'Settlement Houses: A Community Ideal for the Poor', in *Independent Women*.

enter the town halls of local government. Indeed, women were already well represented on school and poor law boards before, for example, they obtained committee seats in 1896 on the RSPCA. NSPCC literature had quite clear views on women's role:

Much is said and written today concerning women's rights. Zealous upon the Platform and in the Press, uttering their furious indignation against being deprived of votes at Parliamentary elections, and of standing for election to public bodies, all, too, of the pride and malice of a man. But, happily, a far larger number of women are capable of indignation at the masses of helpless children [and at the] wrongs which inflict on their tender years fearful sufferings and life-long injury to their health. This is the sphere in which we should ask women to labour . . . Of their own citizenship and rights true women can think little and say nothing till the citizenship and rights of children are established.[26]

Emily Davies had written ruefully to Barbara Leigh Smith Bodichon back in the 1860s, 'There is nothing at all new in women working together. All over the country there are Ladies Associations, Ladies Committees, Schools managed by ladies, Magazines conducted by ladies etc. etc., which get on well enough. The new and difficult thing is for men and women to work together on equal terms.' Forty years later Mrs Louise Creighton, retiring president of the NUWW, noted sadly that 'men have not learnt to work with women. Some women have learnt to work with men; and I think just a few men have learnt to appreciate having them working with them'.[27]

The embarrassment, indifference, marginalizing, or outright hostility that some women faced in local government, therefore, was not peculiar to elected office. Women faced much the same problems and met them in much the same ways in all dimensions of public life including philanthropy. They persisted. When voluntary schools were being supplemented by board schools; when the workhouse found itself holding more deserted children than any orphanage and

[26] *Child's Guardian*, Oct. 1895, quoted in G. Behlmer, *Child Abuse and Moral Reform in England, 1870–1908*, 1982, p. 144.
[27] E. Davies to Barbara Leigh Smith Bodichon, Bodichon papers, u.d. [14 Jan.] Cambridge, Mrs Creighton, NUWW Conference, York 1904, p. 226.

its sick wards nursing more invalids than any voluntary hospital; when town councils were not only promoting bills to sewer and sanitize but to provide baths, books, and parks for their urban poor, many women believed that their mission of service was more suitably fulfilled in local government than in philanthropy or church work. Dr Ethel Williams supported Miss Maud Burnett of Tyneside on her election platform in 1910, saying,

Under the changed conditions of modern life, much of the work that was especially women's work, and which had hitherto been done by women, in missions and charitable work, was now being taken over by Town councils. There was work to be done on these public authorities which, unless a woman did it, was left undone. She hoped that Miss Burnett would not go on to the Council in too quiet a spirit. Fighting was good, and the interests of women and children, housing reform and the care of the tuberculous, were things that were worth fighting about.[28]

The National Union of Women Workers called on its members to offer this service. Said the Hon. Mrs Lyttleton at Brighton in 1900:

A great many very good women still believe that what I call directly philanthropic work is more suitable and more womanly and perhaps, more religious, than is the work of social reforms which they would have to undertake on public bodies. Until we get that idea rather broken I do not think we shall have as many women coming forward as we ought to have. I have never been able to understand why it is thought quite right to wipe up the milk when it is spilt and unwomanly or aggressive to see that the jug which holds the milk is not first cracked and then broken. Women think it is meritorious to visit hospitals, to support homes and to visit the poor, but they do not realize that better even than to visit hospitals is to see that the sanitary arrangements and the water supply, both in town and country, are put right. These are often the direct or indirect cause of much indifferent health among the poor . . . Women think it right to start Homes, and do not see the necessity of preventing overcrowding which, although it may not lead directly to immorality, leads to a want of that sense of decency and that lack of personal refinement which is one of the great causes of immorality.

[28] *Shields Daily News*, 22 Oct. 1910.

They will visit the poor most kindly and sympathetically, but they do not think it necessary to ascertain whether the daughters in factories and workshops are working under proper conditions. Women at present do not see the connection between these two things, and that it is more important to stop an evil at its beginning than to ameliorate or cure it when it has grown. That is what we want to bring home to the women of England.

She herself had three years experience as a guardian, and found it most valuable.

The value of knowing that you are working so to speak, before the public, and that if you make a mistake it will be known, is immense. I would like to run every woman through a course of such public work before allowing her to do ordinary philanthropic work. It seems to me so far from being in any way unwomanly or aggressive, social and public work is most valuable as an educational factor . . .[29]

It was a call to which hundreds of women responded. Two may perhaps stand for many. Miss Jenny Laura Foster-Newton, the daughter of a city merchant, was elected to the Richmond guardians at the age of thirty-six. She started infant nurseries, opened a dispensary, organized a penny savings bank, ran a mothers' home, rounded up and rescued the fallen. She stood duty at the late night railway stations to prevent girls being accosted and led astray; and if they were, accompanied them to court as a voluntary worker, and received them into her charge as a self-appointed probation officer. She collected a clutch of committee posts—hon. sec. of the local Metropolitan Association for Befriending Young Servants, hon. treasurer of the NSPCC, committee member of the RSPCA, vice-president of the Women's Liberal Federation—to add to her activity in the London suffrage movement. She founded a sanitary association to improve the town's water-supply and a coffee stall to remove the temptation of liquor. She herself took tea with the Duchess of Teck in Richmond Lodge. Banned from workhouse visiting by irate male staff and barred from entering less reputable pubs by their owners, she managed to open the one and close the other

[29] NUWW Conference, Brighton 1900, pp. 74–6.

after well-publicized battles in the press. A poor law guardian up to 1925, she was, as her obituary said, a 'bonnie fighter until her death'.[30]

Twenty years later, and Miss Maud Burnett, daughter of a Tyneside merchant, stood for the Tynemouth council at the age of forty-seven. For nearly thirty years she had run Bible classes, an invalid children's aid society, play schemes, children's holidays, a school for the blind, penny banks, and girl guides. Formal committee responsibility came in 1902 when she was co-opted to the local education authority. She was also a committee member of the YWCA, hon. sec. of the Women's Liberal Federation, in 1907 founded her own branch of the Women's Local Government Society, and in 1908 promoted both a mothers' guild and a day nursery. Standing for the council she laid claim to a mixture of motives: women's rights, women's duties, women's needs. But above all she, like Miss Henry, saw local government as the most effective way of promoting philanthropic concerns. When campaigning in 1910 she made an impassioned plea for 'poor crushed little souls', defective children:

There were women who saw all the sorrow around them, who heard the cry of the children, who knew there was sorrow and misery in the world, and who said, 'You must let us help'. Was it right to answer, 'No, stand aside, because you are a woman?' She did not seek a seat on the council for her own sake, but because she cared for the weak and oppressed, and because she wished to show that women could help. The whole effort to try and keep women out of it seemed too small and petty when one considered these things.

When one did, there was not a dry eye in the place.[31]

Such women moved freely between charitable and civic work. Maud Burnett had many years of voluntary work behind her before winning her council seat in 1910. She there served on education, health, infant welfare, housing, watch, insurance, local pensions, and mental deficiency committees. She became a JP in 1920, and then retired back into voluntary service in 1921, 'in order that she could do better work among the people of the borough'. In 1926, however, she

---

[30] *Thames Valley Times*, 22 Sept. 1937; *Richmond Herald*, 25 Oct. 1937.

[31] *Shields Daily News*, 20 Oct. 1910, 8 Jan., 9 Nov. 1928.

returned to the council in a by-election and two years later was made mayor.[32]

When such women found that a client group needed special care they were as likely to mobilize their charitable network as they were to move a motion in council. As guardians, for example, when horrified at the effect of institutional life on children, they brought them out into private hands to be fostered and mothered. Upset at the salacious treatment of pregnant girls or the unpleasant bullying of the feeble-minded, they helped find and found voluntary homes and colonies. Louisa Twining, who more than anyone else opened poor law work to women, offered *Suggestions to Women Guardians* in 1885 which counselled them to seek voluntary help—Girls Friendly Society, bands of hope, training homes, Sunday schools, the Workhouse Nursing Association, girls aid societies—at every step of their work. Miss Elizabeth Lidgett, an experienced London guardian, said to the NUWW conference in Glasgow in 1894, 'Women guardians serve as the chief link between the poor law authorities and charitable work outside. Some people tell us that charitable work is out of place. However this may be in some places, it is most valuable in connection with the poor law. I know well that the Board to which I belong would be certainly maimed in bringing some of the best work to good effect if we were to lose the co-operation of the bands of visitors in our infirmary and workhouse'.[33]

Women joining school boards trod the same path. They found children listless from malnutrition; they enlisted the charitable world to feed, clothe, and care for them. Socialists like Annie Besant, who were somewhat short of official charitable connections, launched feeding societies of their own. Years later, the women-led LCC child care committees brought together school managers, health visitors, charity and settlement workers, and medical staff, to supervise the welfare of the whole child.[34] Likewise, as town councillors, women made infant mortality the focus of their work and so started

[32] *Shields Daily News*, 28 Sept. 1921.
[33] Miss Lidgett, NUWW Conference, Glasgow 1894. Miss Twining's *Suggestions* are reprinted in Hollis, *Women in Public*, pp. 243–7.
[34] Miss M. Frere, *NUWW Occasional papers*, Feb. 1911; M. Stocks, *My Commonplace Book*, 1970, p. 57.

dozens of voluntary mother and baby clinics which were subsequently absorbed by the local authority.

As Lord Salisbury told the Lords in the 1899 debates on the London Boroughs bill:

Women are in closer touch with [the poor] than any man can be. What touch there is, what contact there is, between the working classes and the classes that are above them—apart from matters of business—passes almost entirely through the hands of women. All the charitable work . . . all the sympathetic work, and the knowledge of their daily life, is, to a very large extent, the possession and the privilege of women and not of men . . . They are able to guide the bodies on which they sit in the administration of those parts of the law which most closely concern the social life and the moral existence of the working classes . . .[35]

Late Victorian and Edwardian philanthropy, and local government (revealingly described as 'compulsory philanthropy') developed alongside and enlarged each other. The one did not displace the other. As certain needs were met, others were revealed. Where local authority work required personal service, befriending, counselling, it was usually devolved to voluntary hands and ladies committees; where a voluntary group operating on the fringe of public work was strapped for cash (as with Manchester's subsidized milk) or short of coercive power (as with children at risk), they pressed tasks on to local authorities. Typically, guardians devolved fostering on to voluntary ladies committees; the committees in turn sought official inspection.

Three organizations in particular—the Charity Organisation Society, the Guilds of Help, and the National Union of Women Workers—bridged women's voluntary and municipal work in the years before the First World War.

The COS was the most controversial of the three. It was founded in 1869 to repress mendicancy and to rationalize charity.[36] Firm supporters of the principles of the 1834 poor law, its members believed that pauperism was a state of mind, a feckless willingness to be dependent on others. Reward that fecklessness by alms or outdoor relief, and it spread. 'The poor

[35] Lord Salisbury, Hansard, House of Lords, 26 June 1899, col. 550.
[36] C. L. Mowat, *The Charity Organisation Society*, 1961; M. Roofe, *A Hundred Years of Family Welfare, 1869–1969*, 1972.

starve because of the alms they receive', said Canon Barnett in a famous utterance. The only help worth giving would permanently reform the character, would force families to face up to their responsibilities. One exchange that sums it up came when the COS attacked the Salvation Army's 'General' Booth (the COS always put the General into quotation marks) for distributing food 'without enquiry' in the severe winter of 1888–9. Booth accepted that 'there were other plans very much better than his own, but these better plans were a long way off, and meanwhile the people were hungry'. Charles Loch, secretary of the COS, could not restrain himself. 'Why should Mr Booth forsake the better plan because it is a long way off? The people he says are hungry; they will be hungry if he doubled his depots and quadrupled his meals. He is adding to the number of the hungry. The hunger of the mass of people can only be met by their own exertions'.[37] Naturally, the COS opposed school meals, as they would reward mothers who were indolent. When Francis Peek of the London school board set up a boot fund to entice barefoot truant children to school, the COS complained bitterly that the school board 'cared little about restoring families of the children to independence and much about the increase of attendance through the immediate and easy supply of clothing'. Quite so. School superintendents quickly learned to bypass the COS with its inquisitorial investigations, and appealed directly through the press for cast-off clothing.[38]

The COS worked hardest to establish sound principles in the field of poor relief. They believed, falsely, that pauperism was on the increase (it was rising, but by less than population growth), and that this was due to indulgent outdoor relief. Who would bother to work or save if public relief came easy; if, said Mrs Haycraft, a Brighton guardian, 'the honest and hardworking were forced to provide out of their penury for the support of the dissolute and worthless?' Only the strict application of the principles of 1834 would cut the rates and save

---

[37] *The Times*, 17 Jan., 21 Jan. 1889; H. Barnett, *Canon Barnett, His Life, Work and Friends, by his Wife*, 1921. COS district committees would boast of how few they had felt obliged to help. A third of their cases were rejected, a third passed on, and only 10 per cent of the rest merited help. Roofe, op. cit., p. 69.

[38] Bosanquet, *Social Work in London*, pp. 237–9.

the poor, and where small shopkeepers might care about the first, women guardians were obsessed by the second. Women guardians referred to COS handbooks in a not entirely casual way as their bibles. Mary Clifford, an outstanding Bristol guardian, struck the missionary note: 'The old view that the province of the poor law was simply to keep destitute people if they wished it from starving, has given way to an intense desire to deal with pauperism aggressively, to check it, to separate the hopeful elements from the practically hopeless, to rescue and restore; to fight against hereditary misery'.[39]

Accordingly, COS women workers sometimes sat in on guardians' meetings to pluck the rescuable from the pauper abyss. The COS in its turn head-hunted guardians on to its local committees. By 1900 fifty-nine COS workers were sitting on LCC committees. Likewise, the forty-nine members of Paddington's COS committee in 1909–10 included six guardians, thirteen LCC care committee members, two from the Women's Local Government Society, one from the Paddington School for Mothers, and others from insurance and pension committees, charities and churches. Mary Stocks accompanied her aunt on poor law work, her mother on COS work; her uncle was chairman of the Kensington guardians. As they staffed their country house with broken-down and unemployable Irish domestics met in poor law work, their COS principles, however, were not unrelenting.[40]

The COS, one should add, was by no means *laissez-faire* in its attitude towards problems beyond the capacity of moralized families to remedy. It called for stringent enforcement of sanitary legislation—Octavia Hill wondered whether to join a vestry to show them how to do it—and had helped to sponsor the 1875 Artisans Dwelling Act. The COS also showed imagination and compassion when in 1875, long before the language of eugenics, it identified a class of 'improvable idiots', the feeble-minded. COS members of the London school board pioneered special schools. Mary Dendy,

---

[39] Mrs Haycraft, *EWR* 15 July 1885; Miss Clifford, NUWW Conference, Liverpool 1891, p. 124. See also comments of Mr Chance, hon. sec. of the poor law conferences, North district Poor Law conference, 1896, p. 133; women were 'the most pronounced opponents of outdoor relief'.

[40] Roofe, op. cit., p. 278; Bosanquet, op. cit., p. 103; Stocks, op. cit., pp. 31, 47, 52.

sister of Helen Bosanquet of the COS and a member of Manchester school board, was joined by Ellen Pinsent, later a Birmingham councillor, to promote permanent care for the retarded. A Royal Commission later, COS proposals were enshrined in the Mental Deficiency Act of 1913.

The COS was extraordinarily powerful among a small, highly influential group. Its thinking dominated the poor law section of the Local Government Board; and it represented advanced thought within official poor law conference circles. Few women guardians were untouched by it. But it was never popular. Its censorious style ('Cringe Or Starve') was loathed by the urban poor, and must have deterred the deserving at least as much as the undeserving. Canon Barnett, when he broke from them, denounced their 'thin and narrow, timid and hard' cast of mind. It was also perceived to be less and less relevant. The reports of social investigators, Andrew Mearns's *Bitter Cry of Outcast London* in 1883, General Booth's *In Darkest England* of 1890, Charles Booth's seventeen-volume *Life and Labour* of the 1890s, the work of Rowntree in York and Bowley in other provincial cities, all revealed that destitution was less a problem of the inadequate mother, work-shy father, and demoralized family, as of the under-paid father, the large family, and the exploitative workplace. To its critics were also added the 'new liberals', focused around the writings of Masterman, the Fabians, and the socialists who naïvely wanted to know why unearned income demoralized the poor but not apparently the rich; and who posited structural rather than personal explanations for great poverty amid great wealth. The COS did not flinch. 'Flagrant poverty' had little to do with 'insufficient earnings', it said. 'A wise economy is all that is needed.' Anyone who thought otherwise was an imperfect and untrained observer.[41]

Enough did, and the COS, to its chagrin, found itself displaced in many large provincial cities during Edwardian years by Guilds of Help, no more free with their money but less free with their moral judgements. These Guilds were civic committees, headed by the mayor and councillors, who went out to enrol women and working men behind concepts of

---

[41] Helen Bosanquet, in C. S. Loch, ed., *Methods of Social Advance*, 1904, p. 33.

neighbourliness.[42] Like the COS, the Guilds of Help sought to co-ordinate voluntary effort; unlike the COS they made little distinction between private and public welfare, and considered that unemployment, sweating, and overcrowding were as relevant to family well-being as drink, thrift, and moral health. The first, the Bradford Guild of Help, was launched in 1904 by progressive liberals to mitigate trade depression and to moderate the demands of the strengthening Labour movement. By 1911 some seventy cities had a similar Guild each with hundreds of members enrolled. As the City of Birmingham Aid Society expressed it, 'Under present circumstances, individuals and families are liable through no fault of their own, to be exposed, to circumstances which, except for timely aid, are likely to lead to destitution'.[43] In Bradford, Alice Priestman of the school board and Florence Moser of the guardians were to the fore. In Liverpool, Eleanor Rathbone was a pivotal figure in the town's Council of Voluntary Aid, which brought together official bodies, sixty-five voluntary agencies, and three hundred and forty-one churches. Miss Rathbone sat on its social improvement and education committee along with chairmen of council committees, dispensing grants from both the city council and the guardians. Further south, in Oxford, the COS retained its name and organizational form, but Councillor Miss Merivale helped to shift its attention away from investigating families, towards sanitary work, health, housing, and infant mortality.[44] In Reading, the COS Poverty Committee was relaunched as Reading's Guild of Help in April 1910 by Councillor Edith Sutton and the mayor, after they attended the Bradford Conference on Guilds of Help. They offered a

---

[42] M. J. Moore, 'Social Work and Social Welfare: The Organization of Philanthropic Resources in Britain 1900–14', *Journal of British Studies*, 1977, pp. 85–104; M. Cahill & T. Jowitt, 'The New Philanthropy: The Emergence of the Bradford City Guild of Help', *Journal of Social Policy*, 1980, pp. 359–82. For rivalry with the COS see *COR* 24, 1908, pp. 40–59.

[43] *City of Birmingham Aid Society, Objects and Organization*, 1911, *City Aid Review*, 1909–11.

[44] For Bradford, see Cahill & Jowitt, op. cit.; for Liverpool, see H. Poole, *The Council of Social Service, 1910–1959*, 1960; *University of Liverpool Recorder*, Jan. 1955; *The Sphinx*, 24 Jan. 1906; Liverpool Central Relief Society, Annual Reports. For Oxford, COS Oxford Annual Reports, 1909, 1913 (incomplete); Reading Guild of Help Annual Reports, 1909–14.

poor man's lawyer service, as well as exhibitions on TB, and clothing depots. In 1913 they were shocked by Bowley's study of poverty in Reading published in the Journal of the Royal Statistical Society of June of that year, which revealed that 25 per cent to 30 per cent of Reading families were at or below the efficiency line. Guild members recognized that they needed to depart from family casework entirely and become a Council of Social Welfare.[45]

Both the COS and the Guilds of Help interlocked with public provision at almost every point. Women guardians, at least until 1894, were moulded by the COS. London vestry-women and borough councillors had strong roots in the settlement movement. Many of the provincial guardians elected after 1894, LEA members, and women councillors were active in Guilds of Help. Their more generous sympathies, their refusal to censure the recipients of aid and advice, their community work base, their openness to research and social science, were reflected in women members' own attitudes to local government.

Women's public work was also mobilized by a third and exclusively female body, the National Union of Women Workers (later to become the National Council of Women. It had nothing to do with trade-unionism.) The NUWW sprang from meetings of women's associations in the 1880s, for the care of friendless girls. It met more formally in conference and in council from 1888 where a thousand ladies and a few men from church and charities sought to 'link all women workers for women'. Thereafter, its annual conference circuiting provincial cities, together with its annual reports, journals, and handbooks, networked thousands of women engaged in philanthropic, religious, and public work. The Society was determinedly non-party, non-sectarian, and until 1902, non-suffrage. By 1914 it had 6,500 individual women and 1,450 societies affiliated to it, from women's trade unions, colleges and settlements, health and temperance bodies, to rescue and refuge groups, and local government societies. It brought together the roll call of the women's movement: suffragists (Mrs Fawcett) and anti-suffragists (Mrs Humphrey Ward); Dames of the Primrose League and socialists like Mrs Ramsay

[45] Reading Guild of Help Annual Reports, 1909–14.

MacDonald; Girton graduates, and women trade-unionists such as Mrs Amie Hicks; Bible women; and Lady Frances Balfour, Henrietta Barnett, and Beatrice Webb. Emily Janes, their organizer, caught its spirit in 1890 when she said of its work for women, 'We have props for the feeble ones, crutches for the stumbling feet, leading strings for those as yet unable to run alone. We care for the weak-hearted and weak-willed; for young girls in great poverty, perplexity or danger; for the children of the state; for the ignorant; for the morally poor'.[46] Beatrice Webb noted that they kept in 'touch with women all over the country, the silent, good and narrow women who do so much to form the undercurrent of public opinion'. The Nottingham conference she attended in June 1895 consisted of 'about six hundred, mostly middle aged well-to-do, but a good many hard-working professional philanthropists, guardians of the poor etc. A very fair assembly well meaning, with a slight tendency to "cant" but sober and on the whole openminded, thoroughly typical of provincial English middle class'.[47] The NUWW network was much valued, especially by more isolated provincial women, though, as one delegate reported, packing an audience of women into a conference room presented problems as 'their garments take up so much room'.[48]

The NUWW started as a somewhat conservative, philanthropic, and religious union, bringing women together to share experiences and offer support. By the turn of the century it saw one of its prime tasks as placing women on public bodies and in local government. Though not a feminist organization in itself, it did much to dispel the notion that public work was unfeminine at best and polluting at worst. The most effective Good Samaritan must stand for election before crossing the street. Every locality was encouraged either to form a local government committee or to work with the Women's Local Government Society (itself affiliated to the NUWW); and to select, place, and elect women to office. Branches canvassed the party caucuses and political agents

[46] Miss Janes, NUWW Conference, Birmingham 1890. More generally, I. Grant, *The National Council of Women: The First 60 Years, 1895–1955*, 1955; and S. Kelly, 'A Sisterhood of Service', Univ. of London MA thesis, 1985.

[47] B. Webb, *Diaries*, vol. ii, entry for 18 Oct. 1895.

[48] Miss Tillard, *COR* Dec. 1892, p. 413.

for vacancies; campaigned for their candidates on the street and in the press; and, with their candidate elected, acted as her link between public and philanthropic worlds. Miss Janes noted with satisfaction in the NUWW handbook of 1912–13, that 'one function of a branch is to recommend from intimate knowledge new candidates for Public Service whom we believe to deserve public confidence; and this has been done over and over again by us'.[49] Each year the NUWW discussed women's public service, reminding its members that local government work was more truly religious than mere philanthropy could ever be.

By the end of the nineteenth century, a host of women's organizations was encouraging women to stand for elected office; most obviously the Women's Local Government Society and Women Guardians Society,[50] but also women's temperance, social purity, and party political associations. The COS, the Guilds of Help, and the NUWW offered the most acceptable route of all, that from voluntary service to elected service. Philanthropy's language of succour to the needy, its commitment to the domestic and the local, its training and networking, were perhaps more useful to women than other ways in, and certainly less threatening to men.

But in the later 1890s women were coming into local government who had consciously rejected the philanthropic perspective, who found it intrusive and imperialist, however careful they were to knock at the door, who did not believe that personal wealth endowed moral worth, and who wanted to change the economic and legal frameworks of society and not merely offer personal casework to its more damaged members. Miss Elizabeth Lidgett, a staunch old-style COS guardian, could say to the NUWW in 1909, 'We may devise

[49] NUWW Handbook, 1912–13, p. 116. In April 1910, for example, to take a snapshot, the Bristol branch had held a public meeting with Cllr Margaret Ashton, Cheltenham was seeking out women for the poor law elections, Gloucester was running five women, Hull had found two. Portsmouth members were pressing their LEA to start school meals, Scarborough was threatening to run women as spoiling candidates for the council if the main parties did not make space for them, Torquay was looking for women to supervise boarding out of poor law children, while at Tunbridge Wells women guardians had got additional staff appointed by their board. They had also led a deputation to the mayor to reduce public house licenses. NUWW *Occasional Papers*, April 1910.

[50] For the WLGS see below, pp. 317–36; for the WGS, pp. 231–7.

schemes, we may appoint officials, but it is only the living human soul that can save a soul alive.'[51] The radical Countess of Carlisle would have none of it, announcing back in 1891, 'Almsgiving is such a paltry thing ... We want no Lady Bountifuls in this last end of our nineteenth century; we want Radical women in whatever class they may chance to live giving to the people their legitimate rights', rights to decent housing, free education, and better working conditions. Another radical countess, Daisy of Warwick, herself for nine years a poor law guardian, commenting on the hungry children, the tramp of the unemployed, the aged behind workhouse walls, asked 'Of what possible use is it to plaster this state of things with philanthropy?' At a NUWW conference in 1905, Miss Honnor Morten, a former London school board and settlement worker, and somehow also an organizing secretary of the COS, argued for free school meals. They were no more pauperizing than free education, she said. Miss Margaret Horn, a guardian, was most indignant. By careful investigation for the COS, she had reduced the number of families claiming free school meals from sixty down to three, and Miss Horn had her doubts even about those. Left without cooking to do, mothers spent their time gossiping in the streets or drinking in the pub. Mrs Margaret MacDonald, wife of Ramsay MacDonald, retorted angrily that 'she thought state interference was preferable to COS, or district visitor enquiries'. And Ada Chew, a former factory hand, remarked that she had heard a lot of talk about the poor needing the help of people better off. What they wanted was the vote to help themselves. Anything else was merely 'methods of alleviation'.[52]

Women who were by no means socialist, like the wealthy Mrs Summers of Staleybridge, the even wealthier Mrs Lees of Oldham, and powerful Eleanor Rathbone of Liverpool, were renouncing the self-indulgence of personal philanthropy for the discipline of elected and accountable public office. The

[51] Miss Lidgett, NUWW Conference, Portsmouth, 1909, p. 63.
[52] Countess of Carlisle, quoted C. Roberts, *The Radical Countess: Rosalind, Countess of Carlisle*, 1962, pp. 169 ff.; Countess of Warwick, *Life's Ebb and Flow*, 1929, p. 93; Miss Morten, NUWW Conference, Birmingham, 1905, p. 167; Ada Chew, NUWW *Handbook*, 1911–12, p. 215.

mark of the distance travelled came when Lady Trevelyan penned a leaflet for the Women's Local Government Society in 1904 fearing that opportunities for women in public life were narrowing. 'Why does the public reject these good services offered to them, and drive their clever, capable women back into the paths of individual charity, and philanthropy, where so much effort is wasted?'[53]

Electors and the Local Suffrage

Philanthropy, women's duties, and an impatience with its limitations, had brought many women into local government work. A second impulse was the suffrage movement, women's rights, and the claim to political citizenship.

Local and parliamentary politics were linked at every step and in every mind. The reasons were obvious. For men, the local and parliamentary franchises were broadly coextensive by the late nineteenth century.[54] Party caucuses in most cities selected their candidates, canvassed their electorates and ran their machines for local as for parliamentary elections. The drink and the temperance vote, the church and chapel vote, the Irish and the Labour vote were alike wooed. Local government power was worth having for its own sake, as the great cities were proud and powerful bodies; but local elections were also increasingly regarded as straw polls of national political opinion.

The links were at policy level as well. Growing parliamentary time was spent on local government. For its part local government was created and amended by statute, some of its finance[55] and all of its power to raise loans came from central government, and even the qualifications of certain staff and their hiring and firing were determined by the Local Government Board. It was not all one way. Then, as now, local councils speaking for their community freely passed motions or sent petitions on matters of parliamentary policy to Westminster, which they expected their MPs to endorse.

Men's parliamentary vote rested on the same property

[53] Lady Trevelyan, *Women on County Councils*, 1904.
[54] See N. Blewett, 'The Franchise in the U.K. 1885–1918', *Past & Present*, 32, 1965; B. Keith-Lucas, *The English Local Government Franchise*, 1952. Some parliamentary qualifications, such as the university and freemen vote, were not local qualifications.

qualification as gave them the local vote. 'Men in this country', wrote Lydia Becker in 1879,

obtained parliamentary representation in and through local government. They used the power they had, and they obtained more extended power. We urge women to follow their example—to take an interest in the local affairs in which they have a legal right to be represented, to make their votes felt as a power which must be recognized by all who would govern such affairs, and be ready to fill personally such offices as they are liable to be nominated for, and to seek those positions to which they are eligible for election . . . Political freedom begins for women, as it began for men, with freedom in local government. It rests with women to pursue the advantage that has been won, and to advance from the position that has been conceded to them in local representation to that which is the goal of our efforts—the concession of the right to a share in the representation of our common country.[56]

The experience gained on boards, said Honnor Morten, 'may be regarded as preparation for Parliament . . . Let women crowd on to all bodies and committees that will receive them, and there learn the country's laws, administer the public funds, experience the discipline of associated action, and so broaden their judgement . . . that they are capable of imperial government; they have got to use their municipal vote so as to gain the parliamentary vote.[57]

Women suffrage workers were quick to come forward. Mill had presented the Kensington Society's suffrage petition in May 1867. Local suffrage societies swiftly formed in Manchester, Birmingham, Edinburgh, and Bristol, federating with the London societies later that year to form the National Society for Women's Suffrage. Lydia Becker, the first provincial school board woman, became their organizing secretary until on her death in 1890 the leadership passed to Elizabeth Garrett's younger sister, Millicent Fawcett. Suffrage lecturers criss-crossed the country, organizing meetings, forming women's groups, urging them to take part in local govern-

---

[55] Central government grants to local government rose from around £1m. early 1870s, to £12m. 1900, and £21m. 1910, representing 11 per cent of Exchequer expenditure.

[56] L. Becker, *The Rights and Duties of Women in Local Government*, 1879.

[57] H. Morten, *Questions for Women*, 1899, pp. 63, 68; see also NUWSS 18th Annual Report, 1890, p. 10.

ment. Miss Becker's *Women's Suffrage Journal*, which she started in 1870, held the early movement together.

Their first local government success came in 1869. Women property owners had from time immemorial played their part in parish life, sometimes personally undertaking parochial office where such responsibilities attached to their land and their families. Crabbe's verse on the woman farmer was quoted with relish:

> No parish business in the place could stir
> Without direction or assent from her
> By turns she took each duty as it fell—
> Knew all their duties and discharged them well.[58]

Women parish ratepayers had taken their vote, and presumably a right to stand as guardians, into the enlarged poor law unions formed in 1834. The Municipal Corporations Act of 1835, however, had followed the precedent of the 1832 Reform Act and referred to 'male persons' when it made local ratepayers the local electors. Those few women freemen with a local corporation vote, lost it. The rather greater number of women who paid rates, did not acquire it. In 1869 Jacob Bright quietly and deftly amended the borough electoral qualifications along lines suggested by Lydia Becker, giving women ratepayers the local vote on the same terms as men. It went through the Commons late at night without debate. In the Lords, the Earl of Kimberley spoke to it sympathetically, noting that this was 'not a proposition giving to women the municipal franchise for the first time' but restored to women rights they had possessed before 1835; adding, inevitably, that it had nothing to do with the parliamentary franchise.[59]

Women ratepayers were now admitted to all local government electoral registers, forming about 17 per cent of the electorate overall, a figure rising to some 25 per cent in spa and spire cities.[60] By 1889 there were 412,000 women on the county and 239,000 women on the borough registers. By 1900

---

[58] *WSJ* 1 June 1876.

[59] Hansard, House of Lords, 19 July 1869, col. 145.

[60] Bath had 1,660 women to 5,900 men voters, Bristol 3,500 to 24,000, Brighton 3,000 to 11,900; Birmingham and Liverpool each had some 9,300 to 63,900; Norwich 2,600 to 14,000; Manchester 9,400 to 45,000; Leeds 8,700 to 52,700; Nottingham 5,400 to 34,000; Oxford 1,100 to 5,000; Sheffield 6,000 to 44,900. *WSJ* 1 Sept. 1885.

over a million women had the local vote. They included spinster ladies, young widows struggling to bring up a family, and working women with shops, lodging-houses, laundries, and small businesses of their own. The Liberal leadership (and some historians too)[61] concluded that these women were middle class, Tory, and not to be trusted. The evidence is scattered, but it suggests that women voted from much the same background and with much the same convictions as men. The Labour Party calculated from the returns of fifty towns in 1905 that 82 per cent of women local voters were working class, as were 95 per cent of the women voters in Nelson, 90 per cent in Bolton, and 60 per cent in St Pancras. In two wards in Leeds, 532 of 536 and 1,100 of 1,190 women on the local register were regarded as working class. Clara Collett's more detailed study in 1908 estimated that 51 per cent of women ratepayers were in gainful employment (shopkeeping, clothing, medicine, and education); 38 per cent were housewives without servants; 5 per cent had one servant; and just 6 per cent had two servants or more.[62] Suffragists argued from this that enfranchising women ratepayers would not skew the political system.[63]

Women local voters would have been more numerous still if all women who occupied property, and not just the single and widowed, were allowed the ratepayer vote. The question had been left open in 1869. The 1870 Married Women's Property Act encouraged some married women to place themselves on the register. Shortly thereafter, a man standing for Sunderland borough council was beaten by just one vote. He went to court on the grounds that two of the voters were married women and therefore not entitled to vote. By *Regina* v. *Harrold*, and in a decision that was to darken the rights of married women, the courts in 1872 agreed with him. The Act of 1869 was not to override the common law principle of coverture

---

[61] See e.g. M. Sheppard, *Bulletin of the Society for the Study of Labour History*, 45, Autumn 1982, p. 22.

[62] J. Keir Hardie, *Votes for Women*, 1905; Mr Slack, Hansard, House of Commons, 12 May 1905, col. 219; C. Collet, *Journal of Royal Statistical Society*, 30 Sept. 1908.

[63] They were somewhat disconcerted by the anti-suffragists' claim that they had canvassed women municipal voters and found that most did not want the parliamentary vote. The suffragists complained that the canvass was suspect and the questions loaded. See *Anti-Suffrage Review*, Apr., June 1911, Apr. 1912.

which subsumed a married woman's rights in those of her husband. The courts held that, on marriage, the woman ratepayer voluntarily opted out of the 1869 Act. The two votes were void, the claimant obtained his seat. Over the years, the Women's Local Government Society sought amending legislation but got little help from the women's suffrage movement as they were divided by the selfsame question. Mrs Fawcett, for one, would not separate husband and wife; the Liberals were probably right to suspect that married women ratepayers would be both wealthier and more tory than their single or widowed counterparts, especially as well-to-do couples could arrange their affairs to give their households two votes. It was left to the breakaway Women's Franchise League, formed in 1889 by Mrs Wolstenholme Elmy, Mrs Fenwick Miller, and Mrs M'Ilquham (who as a married woman forced her nomination papers on county returning officers in 1889), to claim citizen rights for married women at both local and national level.

Women voters also turned out to vote in much the same proportion as men. Early municipal returns for 1869 to 1871 suggested that about half of qualified women, and of men, were voting, though this concealed wide variations. In Durham in 1870, 98 out of 128 women voted, in Lyme Regis only 11 of 69. Guardian and vestry elections produced polls of 10 per cent or less, lower in London than outside. School boards ran rather higher and reflected party energy at the polls. Town council elections could in marginal wards generate polls almost as high as parliamentary contests. Again, the evidence is scattered, but in York in 1889 Liberal canvassers found that of nearly 300 female voters, 19 did not care to vote, 18 were undecided, while 199 were fully determined to vote. In Tunbridge Wells where there was an active women's movement, 114 of 165 women voted in 1896, 514 of 976 men. The Women's Local Government Society analysed some 1912 results, and found almost identical proportions of men and women voting, 48 per cent both of men and women in a Kensington ward, and between 75 per cent and 80 per cent of both men and women in Huddersfield and Wolverhampton. Women usually polled strong support from other women, from clergy, doctors, and philanthropists; and poor law apart, from working men. London always polled poorly.[64]

Conflicting reports came in from the early municipal elections. In Leeds, Liberal canvassers were so delighted with the way women voted that one hundred and fifty working men of Holbeck 'spontaneously' subscribed towards a timepiece to present to the first woman voter, Mrs Simpson, herself the widow of a man who had, it just so happened, worked for the suffrage movement. In Manchester's Ardwick ward, however, the returning officer was upset by the unseemly sight of women taken to the public house and plied with drink before voting. There were a few reports of drunken prostitutes and illiterate Irish coming to the polls but these were rare after the secret ballot was introduced in 1872. More typical was the account from Bury where seven thousand voters formed immense crowds around the solitary polling booth to elect local commissioners. Women voters were given a separate entrance and 'shown every consideration'. Policemen proffered their umbrellas, and the roughest of working men made way for the ladies to cast their votes.[65]

Miss Sturge, a member of Birmingham's school board, told a Huddersfield audience in 1875 of the experiences of a wealthy woman Friend, who did not vote because she thought it was unwomanly.

The morning after the election, she met one of the prominent gentlemen in the political party with which she sympathized (it chanced that the Liberals had won by a very small majority), and he said to her, 'We did not receive any help from you yesterday, did we?' She told her how she replied with great pride, 'No, I don't wish to vote'. She went a little further and met with one of her Friends, a man in whom she had great confidence, and who had won his seat in the town council by a majority of one, and he said to her, 'I was not indebted for my seat in the Town council for thy vote was I?' She told her that she hesitated a little, and then she replied, 'No, and I am ashamed of it'. Realizing that her vote could have affected the outcome, that he was a most valuable man of whose principles she approved, she not only subsequently voted, but persuaded other

---

[64] *WSJ* 1 Oct. 1872; reported to NUWW Conference, Tunbridge Wells 1896, p. 12; WLGS Annual Reports. For York, see L. Walker, 'The Women's Movement in England in the late 19th and 20th centuries', Manchester Ph.D. thesis, 1984, p. 129.

[65] *WSJ* 1 May 1871, 1 July 1870.

women to cast their votes, if not for themselves, then for the good of the town and the welfare of poorer women.[66]

Women school board candidates, like Eleanor Smith of Oxford, naturally appealed to women voters, encouraging them to attend political meetings alongside men, putting out posters and election material for their eyes only. Their vote was sufficiently substantial for many caucuses to include a woman on their slate. When they did not, would-be women candidates threatened to run as spoiling independents unless space was made for them. They did and it was. From this it was a short step to mobilizing the women's vote as an electoral bloc in other elections. Miss Becker of Manchester and Mrs Scatcherd of Leeds held dozens of meetings in Yorkshire and Lancashire during the later 1870s, begging women to use their vote; reminding them of the social wage a town council could offer women in the shape of baths and washhouses, open space, street lighting, and cleared slums; and inviting them to appraise and catechize male candidates for their views on temperance and suffrage. Such meetings were always well attended. On one occasion at least, they were gatecrashed by men supporting their party favourites, and Miss Becker was so heckled that she could not complete her speech. Undaunted and unperturbed, she called a meeting for women only the following night (mainly shopkeepers and working women, she noted) and though political feeling ran high among the seven to eight hundred members of the audience, it was a calm and successful meeting. Local newspapers compared the 'superior intelligence and self control of the women' to the rowdiness of the men the night before.[67] Similar work was being done in Birmingham and Bristol. Mrs Scatcherd, an experienced suffrage worker, started cottage meetings where she talked with eight or ten working-class women in their own homes. More ambitiously, she was employed by the Liberal caucus to hold women electors' meetings, which the party would finance and arrange. As she reported to Manchester's suffrage society,

It marked a great step in their cause when a political association of great standing invited a woman to give addresses to the women electors of a large manufacturing town, and she thought it meant success. (applause). The result of the meetings had been that a large number of women ratepayers recorded their votes, and they had a distinct contradiction of the often-repeated statement that women would always vote one way. In the North ward the women rate-payers openly declared themselves in favour of the conservative candidate who was successful; in the adjoining North-East ward women were strongly in favour of the Liberal candidate, who was carried. Again some of the canvassers had told her that women would not go to the polling booths in cabs saying it was a sort of bribe, and that they would walk to the polling station. She was further glad to hear that some women had said, 'We will neither go in a blue coach, nor a yellow coach, because they stop too often at the public houses for us'. (applause). In one of the wards by half past twelve, only women had voted, and they had voted in large numbers; and she knew of two women who sent word to their employers that they would be rather late in getting to work that morning, because they intended to call and give their votes on the way (hear, hear).[68]

Naturally she sought payment in kind—the adoption of a second woman on the Liberal school board slate. In the ward she herself organized, she brought out 413 of the 451 women voters (an amazing 92 per cent poll), despite sickness, absence, and removals. All reports suggest that, outside London, and where work had been done on the women's vote, between two-thirds and three-quarters of registered women could be expected to poll.

Male candidates soon graced women's platforms, offering their pledge on sanitary matters, both moral and physical, their life's loyalty to women's suffrage, and spaces on their nomination forms for women to sign. One Midland ward notorious for its rough elements could find no respectable man to contest it. When a lady asked an acquaintance to stand, she was told 'utterly impossible'. 'Do you know there are 450 women electors in the ward?' 'No.' 'Let me call a meeting of the women electors, and talk to them.' She brought them together, he stood and was returned by a majority of over 400.[69] In Liverpool in 1889, the ladies committee decided they

[68] *WSJ* 1 Dec. 1879.
[69] *WSJ* 1 May. 1883; Mr Courtney, Hansard, House of Commons, 7 Mar. 1879, col. 411.

must fight a great brewer who had been nominated for the county council. Risking physical assault from his heavies and much verbal abuse, 'brave and enthusiastical ladies [faced] blinding snow on polling day so that Dr Wood might score a victory for temperance and purity'. As Mr Courtney told the House, 'Women take a greater interest in municipal matters; they readily come to the polls in about as great a proportion as men; that they are very careful in their choice of persons to represent them; that they are known for a strict regard for character; and that the one thing they desire to secure is the presence of a respectable person on the board which is to administer the affairs of the town or school district'. Women voters were known to show more independence than men by preferring moral character over party loyalty in their candidates.[70] The *Women's Suffrage Journal* now confidently targeted some two dozen towns and cities where the women's vote was large enough and the seat marginal enough to repay effort. At the same time Lydia Becker pressed home her bloc vote by sending women's suffrage petitions to every borough council in the country. Those who opposed it were marked men. Alderman Woodhead seconded the motion at Huddersfield by suggesting in not entirely jocular fashion that as so many men 'were indebted to ladies for the seats they occupied at the Board, he conceived that they would not be unwilling to extend to their female constituents the privilege of voting for MPs especially as they had displayed such excellent taste in voting them into the council'.[71]

Over the years, municipal petitions for women's suffrage flowed in from towns across the country. Only very occasionally did town councils refuse. As the *Women's Signal* put it, women must make their influence felt 'at every stage in the election. Whether it is in choosing candidates, in canvassing or in voting, we should be to the front. Men have to learn that they can reckon on women turning up everywhere, and that they can no longer afford to ignore them either in framing their programmes, making their speeches or pledging their

[70] Mr Courtney, Hansard, House of Commons, 19 June 1878, col. 1809; NUWW Conference, Manchester 1869, p. 12.

[71] *WSJ* 1 June 1880, *EWR* 15 Mar. 1880, *WSJ* 1 June 1881; Alderman Woodhead, *WSJ* 1 Apr. 1876.

votes ... What is necessary above all is to convince every-body, especially the candidates on both sides, that without women's support they are undone'.[72]

London was more problematic. Sidney Webb complained bitterly that at the 1895 LCC elections, barely 20,000 of London's 100,000 women electors bothered to vote. At recent vestry elections, 360 publicans won seats, he said, because women stayed at home. Half of the LCC divisions were held with majorities of less than 500; each had between 500 and 2,500 women electors. If women wanted open space and public health, purity and temperance, and a 'moral' rather than a market wage for women staff, then they must come out and vote. Mrs Elmy scathingly replied that no man was going to tell women if and how they should vote. Women were answerable only to their own conscience and not to Mr Webb's. What about male apathy? Men after all, outnum-bered women voters six to one. 'It is always "the woman" who is blamed for every social grievance, always "the woman" whom from the earliest times till now man has made the scapegoat of his own crimes and vices.'[73] Feminist or not, Sidney Webb was right to be worried. Experienced women canvassers working in his Deptford division found that women voters, unlike men, neither knew nor cared about the LCC. Perhaps this was not surprising in the large and anonymous tracts of poorer working-class London: but it mattered precisely because women rested their parliamentary hopes on their local government record. If women who had the 'domes-tic vote', as MPs were fond of calling it, did not use it, why believe that they wanted, needed, or would use the 'imperial' parliamentary vote? The anti-suffragist Violet Markham was not the only one to say that if women were so indifferent to their existing powers, it was 'humbug' to seek a wider franch-ise. 'Let them get on with the rights they had'.[74]

London was, of course, not just a problem for women but for radical and liberal causes more generally. Elsewhere, women voted in much the same strength and much the same

---

[72] See e.g. *WSJ* 1 Mar. 1879, 1 Apr. 1880, 1 July 1889; *WS* 14 Feb. 1895. For a petition signed by 634 county councillors, *EWR* 15 Apr. 1893.

[73] *WS* 21, 28 Feb., 7, 21 Mar. 1895.

[74] V. Markham, *Return Passage: The Autobiography of Violet Markham*, 1953, p. 55.

way as men. After 1870 there was never any serious proposal to limit the local government franchise. That achievement was safe.

Following the granting of the borough vote in 1869 had come the Elementary Education Act of 1870 which offered women not just the opportunity to vote but also to stand in a very public way for school boards. Towns with the most developed suffrage groups—London, Edinburgh, Manchester, Bristol, Birmingham—were the first to come forward with women candidates, committees, and cash. As Mrs Hallett declared to the AGM of the West of England Suffrage Society in 1883, without the suffrage society 'in all probability no lady would now be a member of the board of guardians or of the school board . . . It was from the suffrage office that committees had emanated for the organization of these elections'. It was the suffrage societies that had undertaken the educational work which had made women welcome.[75]

The London school board was regarded as the most prestigious and important of all late nineteenth-century local government bodies and could count peers and MPs among its members. Women like Elizabeth Garrett and Emily Davies stood not from any enthusiasm for elementary education (they were somewhat bored by it), but from a determination to stake out women's claims to public life from the first. Elizabeth Garrett found it all great fun, as she spoke from orange boxes and beer barrels, and in cavernous public halls. Emily Davies found it most painful but gritted her teeth, hoping against all the evidence that the next speech would be easier than the last. The first handful of elected women sought out and brought on other women for public life, trained them in the small print of their work (literally so, when they provided book lists), supported them on the platform, arranged committees of Friends, were impatient when their protégées' health broke down, as it frequently did, and indignant if women proved unreliable or unworthy, which was fortunately seldom. These first 'ladies elect' knew they were setting themselves desperately high standards, far higher than any men, and to that degree were colluding with the same double standard in public life that they were fighting in

[75] *WSJ* 1 Mar. 1883.

private life. In practice it was not a problem, as most towns had a wealth of leisured women formidably well qualified for public office, born into powerful, political families, trained in voluntary work, feminist, cultivated, self-possessed, and competent. As Miss Martyn said, 'the majority of well educated women secretly yearn for some sphere in which they can make themselves useful to the community at large'.[76] Local government gave them precisely the outlet they sought.

Women's local government work was recognized and respected in the House, particularly by those MPs who served alongside women on the London school board or on the LCC. Tribute was paid them from the earliest suffrage debates and more than one MP commented on the absurdity of women holding their own on school boards with the most able and eminent men of their generation, yet unable to cast a discreet vote by secret ballot for the MP of St Ives. Mr Henly, indeed, a traditional anti-suffragist, confessed that he had changed his mind and come to support women's parliamentary vote on the basis of their local government work. Every success, every set-back, was quoted in the House. As Parnell teasingly put it in 1879, 'If social life has not been disturbed, if the sanctity of the heart has not been invaded, if all the finer feelings which women ought to be possessed of have not been destroyed or vitiated by the participation of women in municipal and school board elections, then womanliness would surely not be destroyed by casting a parliamentary vote by secret ballot once every five years.'[77]

MPs persistently made the connection, as women suffragists knew they would, between local and parliamentary politics, though they made it in many different ways. A sizeable number of MPs wanted women to have full political rights at national as well as local level; a few 'antis' wanted them to have none of either. Most MPs most of the time fell between the two and the problem for them was where to draw a line that had any logic or plausibility. As Jacob Bright warned them, members of the House of Commons 'either have gone too far or else they have not gone far enough'.[78]

[76] Miss Martyn, 'Women in Public Life', *National Review*, Sept. 1889, p. 279.
[77] Parnell, Hansard, House of Commons, 7 Mar. 1879, col. 481.
[78] Bright, Hansard, House of Commons, 26 Apr. 1876, col. 1701.

For feminists like Walter McClaren, whose wife was a guardian, or Leonard Courtney, brother-in-law of Beatrice Webb, it was simple. They wanted women to vote and stand at local elections; they hoped to see them with votes and seats in parliament. Senior members of both front benches, and the majority of Liberal back-benchers, were more equivocal. They were much in favour of women's local government work, and they supported women's claim to the parliamentary franchise. As Balfour said in 1892, 'There is no fundamental distinction between giving women the right to vote in municipal affairs and giving them the right to vote in imperial affairs'. The Tory Baron Henry de Worms stated that voting for a councillor and voting for an MP were equally 'a function of citizenship'.[79] But neither they nor, for example, Fowler of the Local Government Board, wished women to become MPs. They were rather uncomfortable if anyone deduced from women's local government work that women would be useful members of Westminster government. Instead they preferred to argue that as women voted in local elections and worked for political parties, they were already *de facto* members of the political community; common decency meant that women should have the vote as well.

Beyond them was a large group of anti-suffrage MPs, mainly Tory, but ready enough to support the right of single women ratepayers to protect their property with a local vote. Once they thought about it, they did not object either to women serving on school and poor law boards (that was work with women and children), or on rural district councils (that was local work). The group included James Bryce, the educationalist brother-in-law of Margaret Ashton, Manchester's first woman councillor, and Asquith himself. They drew the line at women becoming borough and county councillors, however, precisely because such councils were less local and more parliamentary in style. As they ruefully admitted, if they agreed to that, how could they then resist the demand for the parliamentary vote? And, more horrifying still, parliamentary seats? Lady Frances Balfour, dining at Westminster, reported back to Mrs Fawcett after MPs refused to reinstate women on

[79] Balfour, Hansard, House of Commons, 27 Apr. 1892, col. 1528; De Worms, ibid., 12 June 1884.

to London boroughs in 1899: 'What interested me was the feeling that many expressed, namely that at the bottom of it lay the deepest hostility to the Suffrage.'

Anti-suffrage women around Mrs Humphrey Ward took the battle into the enemy camp. Her Anti-Suffrage Local Government Advancement Committee of 1911 argued that as women had the domestic vote in the fields of education, poor law, and public health, this meant not that women had earned the parliamentary vote but that on the contrary, they no longer needed it. Everything that properly concerned women now lay within their competence. In her own words, 'We have not only to show the risks and perils of the suffrage; we have also to show the young, the generous, the public spirited, that all they desire in the way of reform and beneficial change, is already within their power, if they will but use what they possess'.[80]

The final group of MPs, small in number, were the authentic 'antis', mostly of the far Right, such as the high-church Tory Beresford Hope (founder of the *Saturday Review*, who in 1889 had Lady Sandhurst's election to the LCC declared invalid), Boulnois, and Bouverie, but including so-called radicals like Labouchere, who fought women's claims every inch of the way. Women, they said, were inferior in reason and superior in moral texture to men. On the first count, women would damage public life; on the second count, public life would damage them. Either way, there was no place for women in public. Labouchere's curious contribution was that all domestic bliss would end 'if a man was perpetually leaving his own wife and visiting another man's wife on the plea that he wanted to be a local councillor'.[81]

MPs found themselves in some silly positions at times. Certain members would give women at local level votes but not seats; others, welcoming the expertise of the few but fearing the enthusiasm of the many, would give women seats but not votes. Some MPs argued that women had so little sense of public responsibility that they should not have votes;

[80] Lady F. Balfour to Mrs Fawcett, July 1899, Fawcett autograph collection. *Anti-Suffrage Review*, 1 Apr. 1911. More generally, see B. H. Harrison, *Separate Spheres*, pp. 133–6.

[81] Labouchere, Hansard, House of Commons, 16 Mar. 1904, col. 1314.

others, that they had so much of it they should only be indirectly elected as aldermen and not be directly elected as councillors. Behind every Commons debate was a coded party calculation. Liberal suffragists continued to argue, as Labour did after them, that women voters would come from the working and not the idle classes; and that they would be political (i.e. Liberal), rather than independent (i.e. Tory), in the same way as men. Their front bench was unpersuaded. Both sides of the House tacitly agreed not to press the right of married women with property to have a ratepayer vote of their own. The Tories would not challenge husbands' hegemony, the Liberals feared that such a vote would enfranchise the more affluent and tory. This left anti-suffragists, from Gladstone to Asquith, free to argue that they could support no proposals which would reward single propertied women and punish married propertied women, and therefore left them free to do nothing at all.[82]

On the whole, parliament was not unhelpful towards local government women, but it refused to make any declaratory law, preferring to retain discretion and deal pragmatically with each board, body, or council as it was thought up. The result was confusion compounded. From 1869 single women ratepayers always had the vote for everything; and a few married women ratepayers had the vote for some things, like parishes and poor law boards. But each type of local authority had different qualifications for its candidates, and the result was a bemusing tangle of electoral law. In some cases it was easier for a woman to be a candidate than a voter, in other cases it was not possible to be a candidate at all. On school boards, any woman, with or without house or husband, could be a candidate, but only single women ratepayers could vote for her. In poor law work before 1894, women needed a broadly similar property qualification both to vote and to stand, though to be a candidate required more of it. For borough councils, however, single women ratepayers could vote but not stand, married women ratepayers could do neither, and parliament resisted enabling legislation until after the Liberal landslide of 1906. When in 1888 and 1894

---

[82] For Gladstone, see his letter to Samuel Smith, 11 Apr. 1892, reprinted in Hollis, *Women in Public*, p. 319.

rural local government structures were devised, parish and district councils more or less followed the precedent of school boards, the new county councils the precedent of the borough councils. However, as the new rural district councils were also the rural poor law bodies, they in turn set a precedent for existing urban poor law boards; so from 1894 the property qualification for poor law work was abolished. Residence was enough. Married women now became guardians; but as they had to be ratepayer occupiers to be electors, married women unless they farmed land or ran a business could seldom vote. London had somewhat different qualifications from the rest of the country, and the City of London somewhat different qualifications from the rest of London. If this were not complicated enough, it was simply unclear even to the President of the Local Government Board himself whether women could stand for the LCC in 1889, whether women could stand for London vestries before 1894, whether married women were eligible to vote and stand for poor law boards before 1894, and whether married women could become borough councillors after 1907.[83] The law was ambiguous. The test case of *Regina* v. *Harrold* in 1872 had disenfranchised married women ratepayers in the boroughs. Did it also apply to school boards? Poor law boards? The presumption was yes to the first and no to the second, but no one was sure. The Local Government Board always refused to give a ruling, so it was left to local revising barristers who drew up the electoral register, defeated candidates, party agents, aggrieved ratepayers, and the stoic Women's Local Government Society, to fight it out in the courts. As the courts were always more impressed by women's common-law disabilities than parliament had ever intended,

[83] The burgess register, used for borough and county councils and school boards, by R. v. *Harrold* excluded all married women ratepayers from voting, until amended in 1914. The parish register until 1894 included married women who either owned property, such as cottages on which they paid rates, or occupied it, perhaps as shopkeepers. The 1894 Act ended the right of both owner and tenant to acquire a vote from the same property, and confined it to occupiers. As the husband occupied and acquired his vote from the matrimonial home, the married woman who had to be separately qualified, would normally only have a vote if she occupied a property for business purposes. The parish register was used for parish, district, and poor law elections. Hansard, House of Commons, 21 Nov. 1893, cols. 1388–98; 1 Jan. 1894, cols. 975–80. See also Mrs Elmy, 'Women in Local Administration', *Westminster Review*, July 1898, p. 39.

and not at all by women's offer of service for the common good, this usually worked to women's disadvantage. There was one final unfairness: when, as a result of population growth, a rural district council became an urban district council, or an urban district council became a borough council, then all the rules and electoral arrangements changed as well. Whenever local councils aspired to grander things, women lost out.

Let us look at one moment, 1899, when these issues were aired, as MPs debated whether to turn London's vestries into London boroughs.

The LCC in 1888 had been ushered into life over the top of London's existing parish vestries. As a by-product of the 1894 Parish and District Councils Act, women found themselves clearly entitled to serve on these vestries. Some fifteen ladies promptly won seats, and tried to bring down infant mortality and slum property. Few people denied that they had done valuable work. But now in 1899 it was proposed to reshape the lower tier of London local government by amalgamating vestries and creating boroughs. These would possess essentially the same functions as the old vestries though with some modest additional powers in the field of housing. Did that mean the new boroughs were still parishes or possibly urban district councils, on both of which women could serve? Or were they like provincial borough councils, on which women could not? Fine calculations were made in the Commons as to which was the lesser evil, debarring women from a field where most people thought they had worked well, or setting a possible precedent for their admission to borough councils which were considered the last bastion against the parliamentary vote.

Mr Boulnois, the Marylebone MP, put the 'antis' case. He understood that some women served on vestries but these could be dismissed as 'failures' and 'practical nonentities'. With that out of the way, he advanced his argument. What did women know about sewers, building regulations, and property assessment? 'We do not want these ladies on the borough councils' but only men 'who have business capabilities, who have been trained to public life and . . . [possess] practical experience'. He admitted that women had done good work as guardians where they worked with women and children but that was no reason for admitting them 'to other spheres' where they would be out of place. And, he pronounced, if the House agreed to this measure, 'there is no logical reason whatever, for excluding them hereafter from the House of Commons itself'.[84]

John Burns made one of the most effective speeches on behalf of women.

He appealed to the House of Commons to recognize that they were shutting the door to some of the best citizens that they had in that vast city. In many districts in the East End where one-room tenements prevailed, it was not only not proper, but it was not decent at certain times for male councillors to enter. In regard to disorderly houses, and in regard to many questions of public health administration, women were better suited than men, in many cases they were the only persons qualified to carry on the work . . . Women were already in the turmoil of political life. What was the Primrose League? Women were now used as auxiliaries of men in political elections in a way that did them more harm than if they were on local vestries. At the last General Election one would have thought that Battersea was the Court of King Arthur and that he was Sir Galahad with all the noble dames anxious to help or oppose him. Women were useful to grind the axes of political mediocrities or to help titled nobles to oust men who devoted their time and interests to municipal life; but when they asked that women should take their proper place in local municipal life they were told that they were degrading English womanhood. That was not true. The curse of English municipal life was that women did not take that interest in local life that they ought, but if they were given an opportunity for voting for the ablest of their sisters, their apathy and indifference would be broken down. Women had never been identified with jobbery or maladministration, and he asked the House of Commons, on behalf of hundreds of wealthy women with leisure, means and inclination, who only wanted to help their poorer sisters, to rise to its true level, and to put underneath the young bloods who, to the permanent belittlement of women, wished to prevent them from taking a part in local municipal life.

Leonard Courtney added, 'They need not go into any *a priori* arguments, nor need they talk about sentiments, or what women were fated for according to the conception obtained

[84] Boulnois, Hansard, House of Commons, 6 June 1899, cols. 468–9.

from books. They had got experience of women in actual life
and they proved their own use.'[85]

The Commons agreed with him; the Lords did not, and
amended the bill. When it was recommitted to the Commons,
the Tory majority agreed not to press the point and
acquiesced in the Lords' amendment. Women lost their seats.
As the suffrage societies noted, men would not have dared do
that to women if women had possessed the parliamentary
vote. As late as 1907 when Burns piloted his bill through the
House qualifying women to stand for borough and county
councils, some MPs were still insisting that women would be
placed 'in a sphere for which they were not fitted', 'the rough
and tumble of politics'. Yet, as Mr Woodall had said back in
the suffrage debate of 1892, 'The irony of the situation is that
while so many members hesitate to confer the vote upon
women they are willing to have them associated with them in
the rough work of political contests, acting on electoral
committees, undertaking the work of canvassing, and
persuading men how they are to vote, and in the course of
which women have displayed the very highest kind of political
acumen, almost amounting to that of professional experts'.[86]

Party Activists

Many women, it has been suggested so far, came into local
government from philanthropy, as their caring work moved
from the private to the public realm. They were often deeply
religious, usually apolitical, and indifferent to the claims of
party. By choice, they found their way on to poor law
boards, where they could efface themselves in gruelling
personal service. For others, whose commitment was essentially to the feminist and suffrage movement, local government offered a splendid way to stake out their larger political
claims, to show in Mr Woodall's phrase that they were a
trained electorate and ready for parliament. These women
were instinctively liberal, out to remove any impediment to
personal freedom and social progress. They were by no means

---

[85] Burns, Hansard, House of Commons, 27 Apr. 1899, cols. 776–7; Courtney, ibid.
[86] NUWSS 28th Annual Report, 1898, p. 6; Helmsley, Hansard, House of
Commons, 12 Aug. 1907, col. 933; Woodall, ibid., 27 Apr. 1892, col. 1492.

apolitical, but they wanted as far as possible, to foreground the claims of women as women. If they could stand as independents rather than carry a party label, they would.

There was a third impulse taking women into local government. The town hall and the elected board possessed corridors of power for which church struggled with chapel, temperance workers confronted the drink trade, sanitarians took on slum landlords, promoters of trams competed with owners of omnibuses, social purity moralists waged war on purveyors of unsavoury theatre and dirty books; and for which, above all, Liberal fought Tory. A powerful presence in the town hall could deliver success to many a local pressure group seeking to free, limit, license, ban, or defeat the activities of others. In certain of these pressure groups, women were increasingly prominent. They had their pet causes, and like men, looked increasingly to local government to do something about it.

In some cases, this grew out of their charitable work as it developed a political cutting edge. Miss Foster Newton patrolled railway stations at midnight to stop pimps picking up young girls fresh from the country. She also lobbied to have parks policed and strict licensing laws. The streets, the parks, the music halls, the pubs, the common lodging-houses, and the brothels were the terrain over which virtue warred with vice, very often in the committee room of the council.

One of the most influential pressure groups was the temperance movement. It was very much a women's issue, since they, their children, and their homes were at the receiving end of the domestic violence and destitution that went with men's heavy drinking. The British Women's Temperance Association, the leading women's temperance group, built an extensive network of branches, affiliates, and auxiliaries, claiming 570 branches and 50,000 members by 1892. Steered by Lady Henry Somerset, the BWTA was not content with mere rescue work but from its inception in 1876 tackled the problems which tempted men to drink. Women were urged to use their local government vote, and to catechize male candidates, asking them if, when elected, they would allow pub licences to lapse, would ban drinking from public halls and institutions, would support labour legislation, better housing, and munici-

pal water-supplies for the urban poor.[87] Local temperance
branches worked with the National Union of Women Workers
to select and support women willing to stand for local office,
and with the moral and physical courage to take on the drink
trade. Even one or two women on a board could, they hoped,
elevate the tone of committee meetings, get temperance taught
in cookery and Bible lessons, start Bands of Hope, exclude
drink from the workhouse and the drunkard from outdoor
relief. In their simplistic and tough-minded way, COS women
held drink to be responsible for 90 per cent of workhouse
inmates, 70 per cent of hospital patients, and 60 per cent of
lunatics.[88] Almost without exception, women guardians and
many school board members were active temperance workers
whatever their politics. Many of those who tried to avoid
party labels counted instead on the temperance societies as
well as the NUWW to canvass for them to keep them in office.
They needed all the help they could get. Overzealous temperance pursuits made many women guardians unpopular with
the workhouse poor and uncomfortably priggish colleagues on
the board. Eva McClaren found herself run out of Bradford on
the issue. Male guardians and parliamentary candidates, who
took a more relaxed view of the matter, were regarded by
these women as thoroughly unsatisfactory public representatives. The BWTA concluded from this that women needed the
parliamentary vote. 'Without the vote', said Miss Morgan,
lecturing on women's civic responsibilities, 'much of our work
in temperance reform remains ineffectual.' Another corres-
pondent urged the BWTA to 'join hands with those Societies
which are working for the political enfranchisement of
women'. Though in 1893 the women's temperance movement
divided over how much effort should be invested in wider
questions of social reform, both wings continued to urge
temperance women into local government as one of the most
fruitful ways of delivering sobriety.[89]

[87] *British Women's Temperance Journal*, 1 Jan. 1889, 1 Feb. 1892. More generally, see
B. H. Harrison, *Drink and the Victorians*, 1971; L. L. Shiman, ' "Changes are dangerous": Women and Temperance in Victorian England', in G. Malmgreen, ed., *Religion
in the Lives of English Women 1760–1930*, 1986.
[88] 27 Sept. 1894; *British Women's Temperance Journal*, 1 Nov. 1891, pp. 123–4.
[89] Miss Morgan, *British Women's Temperance Journal*, 1 July 1890, p. 78, 1 Oct. 1891,
p. 114; *Wings*, Feb. 1893, Feb. 1895. For Eva McClaren see below, pp. 61, 217.

A host of other moral and social purity organizations, such as the Moral Reform Union and the Social Purity Alliance, turned for leverage to local government in the late nineteenth century. Their work had its roots in Josephine Butler's campaign against the Contagious Diseases Acts in the 1870s, where, with a sharp feminist edge, she denounced legislation which forcibly sanitized poor women so that they provided safe sex for rich men. In the early 1880s she urged women to use the municipal vote in the garrison towns 'to purify the governing bodies of men tainted with the evil system around them'. This was not just prudery or moral authoritarianism on her part: women and children were at risk. When in 1885 W. T. Stead, editor of the *Pall Mall Gazette*, bought thirteen-year-old Eliza Armstrong, his revelations of child prostitution and white slavery generated the National Vigilance Association 'to repress criminal vice and public immorality'. Members included temperance school board women, such as Henrietta Muller, sister of Eva McClaren and editor of the *Women's Penny Paper*, and Florence Fenwick Miller, who edited the paper in its later manifestation as the *Women's Signal*. Also prominent were COS-trained guardians, Miss Elizabeth Lidgett, Miss Whitehead, Miss Donkin; the founding members of the National Union of Women Workers, Emily Janes and Mrs Louise Creighton, whose husband as Bishop of London chaired the Public Morality Council, as well as Miss Ellice Hopkins, who for ten years swore thousands of men to chastity in her White Cross Army. They were joined by suffrage workers, such as Mrs Sheldon Amos, of whom it was said that the two questions inspiring her as a woman 'interested in all that concerned women' were suffrage and the National Vigilance Association, as well as by Mrs Fawcett herself.

They scrutinized plays, books, posters, newspapers, picture postcards, music halls, waxworks, and tableaux vivants. They put pressure on police and public bodies to suppress brothels, massage parlours, and pornographic picture galleries. They encouraged the London school board to ban young children from appearing on the stage. They urged the LCC to revoke the public entertainment licences of bawdy theatres and to patrol the parks; and called for decent municipal lodging-houses for men, safe shelters for girls, sound and sanitary homes for working people in well-lit and sanitary streets. In 1890, the LCC responded by appointing twenty-three inspectors to scrutinize places of entertainment; the leading provincial towns were at the same time clearing prostitutes off the street and closing down their brothels (thus forcing them into the arms of pimps for protection). Westminster's borough council watch committee boasted that it had closed down five hundred brothels by 1914. One reason why newly elected women councillors were so anxious to serve on watch committees was because they wished to reclaim the prostitute, and not merely to render her homeless. It was largely National Vigilance Women who came together in 1888 to found the Women's Local Government Society. They wanted to place women on the new LCC not only to advance women's rights and educate women ratepayers, but also to clean up London's moral sewers. In their crusades they were joined by churchmen denouncing lax moral standards, by doctors anxious to suppress quack medicine and contraception, by headmasters protecting the purity of their pupils, by newspaper men making money out of hysterical white slavery headlines, by the prurient and by the decidedly odd.[90]

It is not easy to sympathize with much of what they did, when Manchester's Vigilance Association burnt 25,000 copies of Zola and when Henry Vizetelly, one of the heroic free-thought publishers of the century, was as an old man broken by penal imprisonment. Nor when Hull's library committee banned Wells's *Anne Veronica*, when homosexuals were flogged for their offence, brothel keepers' children removed from parents' care without any evidence of abuse, and when popular working-class music halls were hounded into closure. Yet the British Women's Temperance Association and the National Vigilance Association always linked moral reform and social purity to feminism. Men who move freely around their cities have not always understood the anger, fear, and

[90] W. Coote, ed., *The Romance of Philanthropy*, 1916, p. 9, for the NVA. Mrs Sheldon Amos, *NUWW Occasional Papers*, obit. Feb. 1908, pp. 30–1. More generally, see B. H. Harrison, 'State Intervention and Moral Reform' in P. Hollis, ed., *Pressure from Without*, 1974; P. McHugh, *Prostitution and Victorian Social Reform*, 1980, ch. 7 and p. 244; E. J. Bristow, *Vice and Vigilance*, 1977.

indignation of women who could not. Only when the streets were safe for respectable women, literally and metaphorically, could women come out of the home into the public domain. That meant not just repressing the outward and visible signs of brothels, gin palaces, and obscene window displays. It also meant challenging assumptions about men and women's sexual and social nature. Women faced a public world which had two spheres and double standards; they would if they could raise women to the public standing of men, and men to the moral standards of women. As Josephine Butler's crusade against the Contagious Diseases Acts had understood, this meant inhibiting, as well as prohibiting, men's baser urges, in order to expand the public space available to women.

It was precisely because a woman's vote would be on the side of cleaning up politics, corruption, drink, crime, and vice that many American and Australian states enfranchised women. They would tame and domesticate the frontier.[91] Given that the 'antis' case was largely based on the polluting nature of public work, the rough and tumble of electioneering, the soliciting of votes, the degraded clientele of poor law work, and the vulgar male club atmosphere of elected boards, women insisted that they needed to raise male standards of public life for the sake of the clients, women and children. That left Liberals torn between respect for women's moral and social conservatism and fear that it might translate into political conservatism. It was an abyss that the Gladstone who rescued prostitutes was uniquely qualified to negotiate. One reason why Gladstone, who after all thwarted women's suffrage, was none the less revered by so many public women, was precisely because he cast their social and moral puritanism as a liberating and liberal ethic. Hugh Price Hugh's rallying cry, that whatever is morally wrong cannot be politically right, applied not just to Ireland and foreign policy, but to the daily minutiae of local authority work.

Other bodies, like the Anglican Church and Nonconformist chapels, also demanded access to local authority decision-making, especially in the field of education. They courted the women's vote though seldom ran women candidates of their own. In practice, most women Liberals were evangelical and

[91] R. Evans, *The Feminists*, 1979, pp. 58–60.

Nonconformists, a few such as Lydia Becker were church-women, and during the 1890s some professed no faith at all. Moderate (Tory) women were almost always Anglican. The organizations with the most obvious interest in local government, however—and which had the most ambivalent attitude to women's work within it—were the political parties themselves.

Between the two Reform Acts of 1832 and 1867, political activity was very often pressure-group activity and the parties were to a large degree umbrellas sheltering interlocking groups and causes. Women played a modest role in certain of these. They raised funds and distributed literature for the anti-slavery and anti-corn law campaigns in the 1830s and 1840s; worked with the peace movement, for Italian unity, on behalf of foreign refugees in the 1850s; strengthened the Nonconformist conscience with temperance and social purity campaigns in the 1870s. As H. J. Wilson, Sheffield's MP was to say in 1882, 'The back bone of the Liberal party is more and more teetotal and philanthropic in its sympathies.'[92] The wives, daughters, and daughters-in-law of the Cobden and Bright families, the McClarens, and the Shaen Ashurst women, to name but the most prominent clans, provided an informal feminist network that was progressive, provincial though with London connections, intermarried, Nonconformist, and suffragist.

In the years following the 1867 Reform Act, women came into parliamentary politics encouraged both by the moral hegemony of Gladstone and then by the style of party built by Chamberlain and Churchill. Gladstone's Midlothian Campaign of the late 1870s, which demanded morality in foreign policy, and justice for Ireland, generated a heartfelt response in many women who instinctively identified with oppressed minorities denied their civil and civic rights. *The Women's Suffrage Journal* noted that women were flocking to political meetings for the first time and subscribing their pence to general election funds. At the same time, the new party machinery—the caucus, the ward committee, the delegate general council—was being developed in

---

[92] Quoted in P. McHugh, op. cit., p. 239. Women seem largely to have avoided pressure groups connected with economic, administrative, church, and land reform.

Birmingham by Joseph Chamberlain and Schnadhorst to cope with the enlarged electorate and multiple voting of the 1867 Reform Act. Sometimes the caucus made space for active women. A handful of women's Liberal associations worked in Bristol, Darlington, London, and York at the 1880 general election. The Bristol Women's Liberal Association was invited to run two wards in 1883 and brought them home as Liberal gains. In Leeds, Mrs Scatcherd was working as a covert Liberal party agent. In Birmingham women delegates on the general council were helping to select Bright as their parliamentary candidate in 1885. By 1888, twenty-three women sat on Birmingham's General Council of two thousand and four were on its Executive of four hundred.[93] The Women's Political League from 1885 tried to place more women on party executives.

The founding of the Primrose League in 1883 formalized women's party political work. The Tories had lost the 1880 general election, Randolph Churchill believed, because the Liberals had the more efficient machine. The Tories responded with the Primrose League, wishful-medieval in structure, with its dames, knights, and habitations, Disraelian in its message, social in its style. Teas, lantern slide shows, and country outings were to mobilize popular toryism behind Empire, Church, and Union. Women were members virtually from the beginning, sometimes in women-only habitations (Croydon had nearly two thousand women members in 1896) but more often in mixed branches. By 1891 the League had a million members and women may well have comprised half of them. It was not a feminist organization and its dames refused to support women-only societies such as the WLGS. Its aim was 'unselfish': to build popular conservatism between elections, and return fathers, husbands, and brothers at elections. It asked nothing for women themselves, and certainly not the vote. Many women rose to be ruling councillors of mixed habitations (if their social status was superior enough) and sustained much of the League's activity.[94]

---

[93] E. Orme, *Lady Fry of Darlington*, 1898, p. 11; *EWR* 15 Mar. 1883; *Birmingham Daily Post*, 16 Apr. 1885; Birmingham Women's Suffrage Society 18th Annual Report, 18 Mar. 1888.

[94] Lady Llangattock, 'The Primrose League: Why Women should Support it', *The Lady's Realm,* June 1898; anon., 'Women in Politics', *The Gentlewoman*, 28 Apr., 12 May

The League was given greater saliency with the passing of the Corrupt Practices Act of 1883 which limited the number of paid cavassers a party could employ, as well as forbidding treating, bribing, and the hire of vehicles. At the general election of 1885 and faced with Chamberlain's Unauthorized Programme, an enlarged county electorate enfranchising the agricultural labourer, new constituency boundaries, and a limit on official expenditure, the Tories needed all the help that Primrose dames could give them, particularly in the countryside. Tory women canvassed villagers, addressed campaign literature, entertained political meetings with puppet shows, used their charitable donations and tradesmens' accounts with discrimination; and on polling day brought voters to the polls. Reports from London and Birmingham confirmed 'the very active part taken on both sides by women'. Sir Charles Dilke found himself beset by dames in Chelsea and threatened to report two of them for electioneering offences. Meanwhile women Liberals had helped to select several of the London candidates, and worked for the election of many others. York enjoyed the attentions of Miss Milner's habitation on one platform and the Pease Liberal ladies on another. When two years later a Tory candidate reported from South Kensington that at the last moment a radical was standing against him, he had nearly a hundred dames working for him within the day.[95]

At every subsequent election, Primrose dames and women Liberals were to be found building up the register and chasing removals. In a key by-election in Darlington in 1898, for example, Primrose dames traced a remarkable 690 of 720 removals. They canvassed assiduously. Meresia Nevill, daughter of Lady Dorothy Nevill, considered women made the best canvassers because they were experienced district visitors. Lady Llangattock made the same point. 'Who is a better canvasser than a woman? . . . By this I do not mean the hurried visit with the canvass book during the few weeks that elapse immediately before an election; but rather that

1900; more generally, *Primrose Record, Primrose League Gazette*; J. Robb, *The Primrose League 1883–1906*, 1942; M. Pugh, *The Tories and the People 1880–1935*, 1986.

[95] *EWR* 15 Dec. 1885; Birmingham Women's Suffrage Society AGM Report, 19 May 1886, p. 5.

systematic continuous friendly visiting, with the kindly chat for the believer, the cogent reason for the objector, and the ready information for the inquirer'. With such work women would, as in philanthropy, help weld all social classes together. Far from protecting women from the rough and tumble of political life, anti-suffrage MPs and their agents shrewdly sent dames into the roughest areas, such as the East End and South Wales, knowing that their courage, style, social class, and sex would command civility rather than the customary rotten eggs.[96]

Anti-suffrage MPs rightly faced constant jibes in the Commons for refusing women the heavy burden of a secret vote every five years while dragging them through the mud and mire of electioneering. Mrs Fawcett got equally cross as she addressed the inaugural meeting of the Women's Liberal Unionist Association in 1888.

> Sometimes you may hear it said that it is a very undesirable thing for women to take part in political organization; but at the same time, that very same person may ask women to take part in politics on behalf of the party in which he is interested himself . . . We do not ask you to believe that women's share in politics is a wrong thing, and that we are to act wrongly because it will help the right side but . . . because we think it a right thing.[97]

The general election of 1892 generated still more demands for women's help. Lord Roseberry, it was gleefully reported in women's Liberal circles, had, despite his hostility to women's suffrage, felt obliged to contribute £100 to women's Liberal organizations; Sir William Harcourt, known to be impatient with women's claims, 'has been compelled by his local agents to seek the aid of Liberal women in order to keep his seat at Derby'. Mr Rollit's women's suffrage bill was only narrowly defeated in the House that summer. Mrs Fawcett attributed 'the unexpected smallness of the hostile majority . . . to the increased activity of women in political affairs'; MPs knowing

---

[96] *Primrose League Gazette*, 1 Oct. 1898 (I owe this reference to Dr Pugh). *Lady's Realm*, June 1898, p. 183. For some sound canvassing advice, see 'The Worker's Catechism', *WS* 14 Feb. 1895; for some decidedly naive advice, *WPP* 9 Aug. 1890. For a mocking account, E. Banks, 'Electioneering Women', *Nineteenth Century*, Nov. 1900, pp. 791–800.

[97] Inaugural speech to WLUA, 5 July 1888, pamphlet. See also Mr Grant Lawson, Hansard, House of Commons, 31 Mar. 1905, col. 75.

that they would need 'to appeal to women in their constituencies to help them retain their seats', hesitated to make their usual rabid and ribald remarks. During the winter of 1900 when many men were away fighting the Boers, husbands returned from South Africa to find that their wives had canvassed for them, issued their election address, and seen them safely elected in their absence and somewhat to their surprise.[98]

Liberal women were contemptuous of the fripperies, vanities, and perversion of charity indulged in by the League. It was 'everything except politics, everything except political teaching, and everything except that which elevates and educates the people', said Lady Sandhurst bitingly. But they were also respectful of the League's money and membership. So in 1886 seventeen local women's Liberal associations met at the house of Mrs Sophia Fry to form the Women's Liberal Federation. Their tasks, they decided, were to advance the principles of liberalism which, they added, were larger than the views of Liberal men within the Liberal party; to promote political education and protective legislation for women; and to engage in 'pure' election work (unlike the impure sort pursued by the dames). Mrs Fry urged those present to co-opt temperance, COS, and professional women into political work. She firmly believed that party politics at both national and local level was a healthy development—it aroused interest and increased voting turn-out.[99] Eva McClaren became effectively their national organizer.

Over the next decade WLF membership rose to some 80,000 in 470 branches. It was more urban, middle class, and high-minded than the Primrose League, given to talks on sanitation rather than shadow shows, preferring to circulate

[98] *WH* 25 June, 1892; *WS* 7 Feb. 1895; Mrs Fawcett, 'The Women's Suffrage Question', *Contemporary Review*, June 1892; *EWR* 15 June 1901. Similarly, the antisuffragist Asquith had to be warned by his chief whip in 1910 not 'to drive the whole Women's Movement into the most bitter opposition: nor to weaken and in many cases, alienate the support of the most active Liberal women workers'. Quoted D. Morgan, *Suffragists and Liberals*, 1975, p. 70.

[99] This account is drawn from the Annual Reports of the WLF, quarterly leaflets of the WNLA, contemporary biographies, pamphlets, and the women's press. There is as yet no published study of women's political liberalism, but see L. Walker, 'The Women's Movement'. For Lady Sandhurst, *WH* 9 Jan. 1892; 'Women Workers in the Liberal Cause', *Westminster Review*, June 1887; Orme, op. cit., pp. 120–39.

political literature rather than sing patriotic songs. At their conferences they discussed women's employment and education, trade and taxation, temperance and moral reform. Above all, the WLF was always a woman's organization, standing apart and sometimes aloof from male Liberalism, insisting on its independence and autonomy. Miss Clephan, a member of Leicester school board, rejoiced in 'the feeling of sisterhood and fellowship which the women's Liberal associations are calculated to foster among women'. Rosalind, the somewhat tempestuous Countess of Carlisle, told local women in 1890, 'Sisters, it is a great comfort to work with one another, linked together in our common womanhood.' Separate associations would train women to be self-reliant; to be free to hold fast to the principles of liberalism. 'It was right, it was best' to stay apart, said Miss Conybeare at Redruth, because women would never enjoy equal rights with men in any organization until they had the vote. Whereas Primrose League dames suspended their habitations at election time to act as Tory party workers, Liberal women always retained the right to say no, to refuse to work for men whose private life or political views brought women into disrepute.[1]

Primrose dames were most active at parliamentary elections. Popular patriotism and imperialism lent itself less easily to the sewers and street lighting of local government, though dames could on occasion be found working in county council and LCC elections, which were more parliamentary in style. Not until 1907 did Mrs Humphrey Ward encourage Tory women to work in and stand for local government when she formed her Local Government Advancement Committee. Surprisingly, given the 'domestic' quality of local government work, Tory women remained largely indifferent to it. Women's Liberal associations, on the other hand, grasped every occasion to advance women's local government work. As they built up the register, methodically and efficiently, and canvassed from house to house, they chivvied women voters

---

[1] Miss Clephan, *WPP* 20 Dec. 1890; Lady Carlisle, Roberts, *The Radical Countess*, p. 183. (Lady Carlisle was sister of the Hon. Lyulph Stanley, leader of the LSB, and daughter of Lady Stanley of Alderley, a stalwart of Girton College.) Miss Conybeare, *WPP* 13 Dec. 1890; *WH* 4 June 1892. Sir Charles Dilke was to be hounded by the moral purists within the WLF when the personal scandals surrounding him became public.

into political self-consciousness. Every bad house, every bad school, said Lady Carlisle, was the fault of women who failed to vote. Liberal women workers held around a thousand local election meetings every year. They negotiated for one or two seats on the school board to pass down the female line; sought out women with the necessary property qualification to become guardians; where they were admitted to the caucus executive, helped to select men for town and county councils; drafted and delivered manifestos on their bicycles, chased the absent, arranged transport for the invalid, ran ward committee rooms, and scoured the streets for votes, returning sometimes to find their tyres punctured by the Tory faithful.[2] Because Liberal women shared interlocking membership with temperance, suffrage, social purity, and philanthropic groups, they were an important organizational asset for grass-roots liberalism. As, unlike the dames, they kept their autonomy, linked their work to suffrage claims, and asked for a share of local government's glittering prizes at the expense of male candidates, they were also resented. Many Liberal men, while publicly praising the moral virtue, intelligence, and self-respect of women's Liberalism, clearly wondered in private why a woman Liberal could not be more like a dame.

Some women seeking election insisted on standing as independents, to emphasize their claims as women rather than as Liberals. This was encouraged by the Women's Local Government Society and by the National Union of Women Workers. In the smaller towns which lacked elaborate caucus or party machinery women could often collect more votes that way. Keen party women in the larger towns, however, counted it a greater achievement to be integrated into the official Liberal slate, on equal terms with men. Miss Emily Sturge at Bristol, Mrs Wilson and Mrs Ripper at Sheffield, Mrs Green at Norwich, Mrs Cowen at Nottingham, and Miss Eliza Sturge, Miss Dale, and Miss Kenrick at Birmingham, joined school boards proud to bear a party label. They could all count on a male caucus that was suffragist. All had behind them a strong women's Liberal association which they had often founded themselves; but then so did Mrs Byles of

[2] Roberts, op. cit., p. 190; WLF 10th Annual Report 1897; WNLA *Quarterly Leaflet*, Feb. 1906, pp. 10, 12.

Bradford and Mrs Connon of Leeds, and neither could persuade the men to move over. Where the Liberal caucus refused to accept a woman, Liberal women at Oxford, Ipswich, Bradford, and Swindon had the courage to run as independents. Making full use of the cumulative vote, they forced their way on to school boards, often displacing an official male Liberal in the process, and insultingly often coming in head of the poll as well. Some women worked their way in, others bought their way in, raising funds to pay off the debts incurred by business men's Liberal clubs, and still others were sought out to help revive Liberal fortunes. It was said of Mrs Cowen of Nottingham, that she was 'a winning card'. Miss Sturge of Bristol was run by the party to displace an independent woman who was holding the balance of power and who was a temperance fanatic to boot. Mrs Lees as mayor-elect brought victory to the Liberals at Oldham.[3]

School board seats generated most tension. The average school board had less than a quarter of the members of the average town council. Seats were few in number and prestigious in standing. As towns were not warded but each voter had as many votes as there were seats, school board elections required tight central caucus control if they were to win a working majority. As voters could plump all their votes for one candidate, women candidates and electors had a countervailing weapon of their own. They could force the caucus to negotiate. Poor law seats were far more numerous and less contentious. They attracted little publicity, less prestige, and were not part of the local political career ladder for men. Once the property qualification disappeared in 1894, women eased their way on to the boards in considerable numbers. On borough councils, however, women found they were in the familiar position of a poor relation, without the dowry of the cumulative vote, and with only their service to offer. And not much was thought of that. Local government was—is—about local difference. Nineteenth-century women led in Leeds, offered solid achievement in Bristol, were sought in Sheffield, were loathed in Bradford, were repressively tolerated in

[3] Mrs Cowen, *Nottingham Daily Express*, 19 Nov. 1883; for Miss Sturge and Mrs Lees see below, pp. 156–9, 403–4.

Birmingham, were an afterthought in Norwich, and were ignored entirely in Plymouth.

Eva McClaren's career is illuminating. As Eva Muller, wealthy daughter of an *émigré* merchant, she was a Lambeth guardian. Her sister was on the London school board and founder-editor of the *Women's Penny Paper*, the main voice of women's liberalism and temperance in the late 1880s. Eva Muller had been trained by Octavia Hill in housing management, by Agnes Jones in nursing, and by Josephine Butler in pressure-group politics. On marriage she moved to Bradford where she again became a poor law guardian, but alienated the local male Liberals when she called in the Local Government Board to discipline the kindly, mournful, intemperate workhouse master. After vitriolic controversy, she lost her seat and abandoned Bradford to drink. Her husband became MP for Crewe. She then devoted herself to women's organizations, on the Executive of the Suffrage Society, hon. treasurer and pivotal figure in the Women's Liberal Federation, vice-president of the British Women's Temperance Association, and committee member of the Women's Local Government Society. Her husband was one of the most faithful suffrage MPs (at the age of sixteen he had escorted women medical students safely to their lectures at Edinburgh), and her mother-in-law was Priscilla Bright McClaren, one of the dowagers of the women's movement.[4]

Women Liberals worked to heighten women's political awareness, to elect Liberal women to local government, and to aid the male Liberal party at local and national level. They also operated as a women's lobby, radical and feminist, delighting in the power, however modest, that their municipal vote gave them. They pressed male colleagues on town councils to introduce free libraries, baths, and washhouses, to improve street lighting and sanitation, and to feed hungry schoolchildren. Recalcitrant men were censured, the supportive were permitted to address women Liberals on their work and to count on their canvassing and electoral support. Women presented petitions to their council in favour of women's suffrage and local government work, and lobbied MPs and ministers. Here they often acted as foot-soldiers for

[4] For Eva McClaren, see below, p. 217.

the WLGS, ensuring in 1893, for example, that Fowler amended the Parish and District Councils Bill along WLGS lines so that its provisions would not subsequently be nullified by the courts.[5] Women joined the Liberal outcry against the Tory education bills of the late 1890s, and when women were no longer eligible to stand for election to LEAs they helped to ensure that women were at least statutorily co-opted. Led by the WLGS, they tried in vain to retain women's seats on the new London boroughs in 1899; and they provided much of the local leverage which gave the WLGS bill of 1907 its majority, allowing women at long last to stand for borough and county councils.

Women's liberalism was a parallel liberalism, and was therefore exposed to the same divisions over home rule as the parliamentary party. In 1888, women Liberal Unionists, including Lady Stanley, Mrs Fawcett, and Lady Frances Balfour, withdrew from the WLF to form the Women's Liberal Unionist Association, though not without some male opposition. By the 1890s they claimed 15,000 members and thirty branches.[6] Such was the hegemony of the Irish issue that gradually their women drifted into collaboration with the Tories. The alliance was clearly uncomfortable at local government level since Liberal Unionists remained distinguished by their progressivism, their suffragism, and their national-efficiency imperialism, which led women like Mrs Pinsent of Birmingham to devote her life to retarded children. Chamberlain's betrayal of free trade in 1903–4, however, brought many of them back into the Liberal camp; others withdrew from politics altogether.

An equally troubling split occurred in 1892 over the suffrage question. Almost all active Liberal women were suffragist; but

[5] See below, pp. 325–6.
[6] Kate Courtney, wife of the suffragist MP Leonard Courtney and sister of Beatrice Webb, was one of the founding members of the WLUA. After their first tentative meeting she noted in her diary, 'Some foolish people get up a scare among the LU men who are so alarmed at some of the women . . . because they have worked in other causes etc—that when Thursday comes [the next meeting] most of my ladies come to say their husbands won't let them join'. The dowager Lady Stanley took upon herself the task of turning the husbands around. Courtney diaries, vol. 23, Friday, 11 May 1888. See also Lady F. Balfour, *Ne Obliviscaris*, 1930, vol. ii, pp. 114 ff. For a fascinating recent account of the Potter sisters' public, philanthropic and political work, see B. Caine, *Destined to be Wives. The Sisters of Beatrice Webb*, 1986, ch. ix.

though they believed that votes for women were properly part of Liberal principles, female suffrage was unfortunately not a plank of the Liberal party programme. When Gladstone wrote his open letter to Mr Samuel Smith, MP, in April 1892, rejecting women's suffrage, he divided women's Liberalism into those who would make it a test question, those who would keep it an open question, and those who would lay it aside. There were many women Liberals who objected on principle to single plank Liberalism, others who feared that they would become yet another suffrage society, and as many again, like Arabella Shore, who pondered why it was acceptable for Gladstone to make home rule a test question, giving the Irish the right to govern themselves, but not women's suffrage, giving the same right to half the nation. Others believed that men would never concede women's suffrage if they could get women's work and women's service without it. They would not in conscience work for any anti-suffrage MP however sound he might be on other matters. Said Mrs Brownlow, 'I would not lift a finger to help any man who would not help my sisters'.[7]

The WLF had for many years contained both tendencies, 'suffrage' and 'neutral', each meeting in caucus before executive committees and each with one of the WLF's hon. secs., Mrs Broadly Reid and Mrs Sophia Fry. Following Gladstone's letter of 1892, they battled it out in three days of conference. The neutrals, who were mainly official ladies and the wives of prominent Liberal men, were no match for the committed suffragists, who contained almost all of the tough and experienced local government women; the neutrals withdrew, taking some fifty to sixty branches and seven to ten thousand members with them into a newly formed Women's National Liberal Association. The larger towns, with the exception of Bradford and Bolton, remained with the WLF. The WLF became increasingly a women's organization first, and a Liberal body only second; the WNLA, a ladies auxiliary of the Liberal party, much like the Primrose dames. In practice however, branches retained considerable local autonomy and many of those formally affiliated to the WLF

[7] A. Shore, *WPP* 20 Nov. 1890; Mrs Brownlow, *WS* 2 Dec. 1896. See also the cartoon among the illustrations.

continued to work for anti-suffrage men (they could not bear as an election approached to remain on the sidelines), and many of those within the WNLA declared over the years for women's suffrage.[8] Relations remained cordial and they were formally reunited in 1919.

Not surprisingly, WLF branches had a more difficult time with their male Liberal clubs than did the WNLA or more acquiescent women Liberals. Miss Kate Ryley was the strong-minded and strong-willed leader of Southport women's Liberalism. She had come on to Birkdale's school board,[9] where she was embroiled in constant controversy as she tried to raise standards of teaching, in the process harrying an incompetent (and rather unpleasant) headmaster into an early grave, for which she was not forgiven. She would not support any anti-suffrage Liberals.

The position of a political woman without the vote seems to me the most abject that can be conceived. She is called upon to use her time, her influence, and her money in the furtherance of a political platform, in the drawing up of which she has no say, and in the choice of whose candidate she has not been called upon to express her opinion. The influence of political women of this anomalous description must become more and more dangerous, for they will acquire a large amount of political knowledge, which they will be free to use as personal or private motives dictate, and without the reponsibility which justly attaches to the possession of the vote.[10]

Politics without the vote was 'only an idle game'. When her branch refused to work for anti-suffrage men at elections, it found to its fury that Liberal women from outside the constituency came in as blackleg labour, canvassed hard, and destroyed their protest. Bitterness and bad blood increased between Liberal men and Liberal women in Southport. Official male Liberals determined to oust Miss Ryley from the school board; meanwhile, official male Tories were muzzling her right to speak at the board. She had a rough few years, and resigned from the WLF. For the same reasons, Margaret Ashton, a Withington urban district councillor, left the Liberal party and resigned her presidency of the North of

---

[8] *WPP* 13 Dec. 1890, *WH* 14 May 1892.
[9] Miss Ryley, see below, pp. 171–5.
[10] *Southport Guardian*, 7 Dec. 1889.

England women's Liberal movement, a few years later. Such women, said Lady Florence Dixie, would no longer 'be used to do the *dirty work* of politics'.[11]

As with philanthropy, there was no simple one-way movement between women's party political activism and their local government work. Women were well established on school boards and had a recognizable presence on poor law boards before the Primrose League or the WLF were formally founded. The first cohort of poor law women were mostly Liberal, when they thought about it, but came out of philanthropy and in particular the COS. Indeed, it was just because women had made their mark in local government as elected members as well as electors, and had in the process created and mobilized a female constituency, that male politicians made space for them, though whether grudgingly or gracefully depended on the local scene. The women's political organizations did give women their staying power. After that initial heady enthusiasm, with local newspapers profiling their first elected lady, local government work required steady and careful attention to detail, a reliable flow of women candidates, systematic canvassing and fund raising, and a political capital of charitable contacts and sisterly support to offset men's clubland of caucus, business, and commerce. The party political organizations sustained the second and subsequent generations of elected ladies.

From the later 1890s, ladies were joined by women as the small socialist parties became a political force. The Social Democratic Federation was somewhat dismissive of feminism, despite numbering Helen Taylor and Annie Besant among its adherents, believing it to be a bourgeois diversion. Unlike Liberal and Tory organizations, Labour women were full members of their local branches (forming perhaps 10 per cent of the membership) and had in many towns the option of women's sections as well. The ILP freely included women on its slate, alongside working men, but then most of their candidates were forlorn hopes. Some women, a few of them working class, became Labour poor law guardians, especially in the

<hr />

[11] Letter of Miss Ryley to Mrs McClaren, Minute-book of the Southport WLA, 21 Oct. 1899; Lady Florence Dixie, *Women's Position: Social, Physical and Political*, Women's Emancipation Union Paper, Oct. 1892, p. 5.

North, after the 1894 Act removed the property qualification. Mrs Pankhurst in Manchester, Isabella Ford in Leeds, Sarah Reddish in Bolton, Selina Cooper in Nelson, Ada Chew in Crewe, Mrs Reeves in Norwich, Hannah Mitchell in Ashton, and Mrs Despard in London were among the better known. Often they were the candidates of the Women's Co-operative Guild as well.[12] George Edwards, who refounded agricultural trade-unionism in Norfolk, and his wife were both guardians in Erpingham, where they fought, successfully, for more generous outdoor relief. School boards were abolished before many Labour women had worked the cumulative vote to their advantage, though Margaret McMillan at Bradford and Mrs Bridges Adams on the London school board pressed forward the ILP programme. As Hannah Mitchell and countless guildswomen wryly noted, 'Socialists were not necessarily feminists', despite the party commitment to full equality, and especially when it came to sharing domestic responsibilities and demanding home-made pies and cakes for tea.[13] Working-class women activists had a hard time of it.

The Women's Labour League, founded by Margaret MacDonald in 1906, reaffirmed the value of separate women's political organizations, if hesitant, inarticulate, and politically inhibited women were to stop making tea and start making policy. As Marion Phillips, national woman's organizer of the Labour Party, was to say in 1918, '. . . women are so newly come into political life that their development will be hindered and forced along the ordinary lines of political thought amongst men, thus losing the value of women's rich experience, if the whole of their work is conducted in organizations including both sexes'.[14]

The League's twenty branches in 1907 had become seventy branches by 1910, with a membership more genuinely comprehensive than the older political associations. 'We have many members', wrote Margaret MacDonald,

---

[12] For women's work in the early socialist parties, see Walker, op. cit., p. 52. For the Women's Co-operative Guild, see below, pp. 242–6. The WLL, the WCG, and Mary Macarthur's Women's Trade Union League interlocked, seeking to heighten the consciousness of working-class women voters, mothers, and workers.

[13] H. Mitchell, *The Hard Way Up*, 1968 edn., p. 99.

[14] Marion Phillips, ed., *Women and the Labour Party*, 1918, quoted Walker, op. cit., p. 82.

whose chief work is the care of their homes and families. Some of these are very poor in this world's goods, hardly knowing where the next meal for their families will come from, but they recognize that the best way to give their children a better chance in life than themselves is to work for the recognition of the rights of labour and the responsibility of the community for every individual citizen. The League makes a special effort to enrol wives and daughters of Trade Unionists and Socialists, since it recognizes the weakness and danger where the wife is not in sympathy with her husband's enthusiasm for Labour politics. Then we have amongst us many professional women, teachers, nurses, doctors, inspectors, post office clerks etc. The facts of life have driven them to make common cause with the wage-earners, and they see in our movement the only hope of real social reform. Side by side with toilers who can hardly read or write, we have young women from the Universities giving their powers of writing and speaking, and organizing to help their less educated comrades, who, on their side, contribute the practical knowledge, which the middle-class girl so often lacks. We try to get as many working girls as possible enrolled, and to cooperate with trade unions for women . . . Many of our branches are in towns, but we are attacking less thickly populated areas too, and in some of the northern colliery villages our members walk miles by dark country roads to their fortnightly meetings.[15]

Along with the Women's Co-operative Guild, the League educated its members in political and social questions, with meetings, discussions, leafleting, petitioning, canvassing, and lobbying. When issues like school meals and municipal milk came to the town hall, Labour women presented themselves as a visible and vociferous constituency, filling the galleries, lobbying councillors, and helping Labour members press their motions and amendments with unladylike cheers and groans. The League was less successful in its other ambition, to place Labour women in elected office. As the Tories strengthened their grip on local government, there were few winnable seats for any Labour candidate, male or female, in the Edwardian years. By 1913 the Women's Labour League had some thirty women serving as poor law guardians, mainly in the North, but only in London did a handful of socialist women win borough council seats after 1907, and their feminism was inevitably ambiguous. Socialism, into which they embedded

---

[15] Mrs J. R. MacDonald, *Women and the Labour Party*, 1909, pp. 5–6.

the needs of working-class women, always took priority over 'middle-class' women's rights, and adult suffrage precedence over women's suffrage. However, with a parliamentary party committed to full and equal civic and political rights for both men and women, Labour women were spared that desperately painful clash of loyalties which bedevilled women's Liberalism for over twenty years. They lost some women to the Pankhursts' Women's Social and Political Union; but from 1910 Mrs Fawcett's suffragists, many of them former Liberals, were working the constituencies on behalf of Labour as the only party committed to women's suffrage.[16]

The heroic days of Labour local government were to come in the late 1920s and 1930s when cities like Sheffield and Norwich fell to fifty years of Labour control. Only then did Labour women enter local government in any number. But certainly one reason why so many Progressive women in local government drifted towards the Labour party[17] was precisely because it offered a less grubby accommodation between practical policies and political principles for women than they had experienced in the late Victorian and Edwardian Liberal party.

[16] The WLL left 'a free hand' to members on the question of women's suffrage. 'Some of our most active members are advocates of the immediate enfranchisement of women on the same terms as men, others think it best to work for adult suffrage only. Of the former, a few have left us in order to work with the "militant" suffrage societies irrespective of any Party at all, and these Societies have perhaps delayed the growth of the League by absorbing the energies of many women who would otherwise naturally have worked in its ranks.' MacDonald, op. cit., p. 7. See also L. Hume, *The National Union of Women's Suffrage Societies, 1897–1914*, 1982.

[17] Miss Clarkson of Norwich, Miss Ashton of Manchester, and Miss Sutton of Reading were three of these. This point is confirmed by recent research of Prof. Banks, *Becoming a Feminist: The Social Origins of 'First Wave' Feminism*, 1986: 'The decline of liberal feminism therefore occurred at a time when the Liberal party was still the main alternative government'. Most of her sample feminists born after 1850 became Socialists (p. 22).

EDUCATION

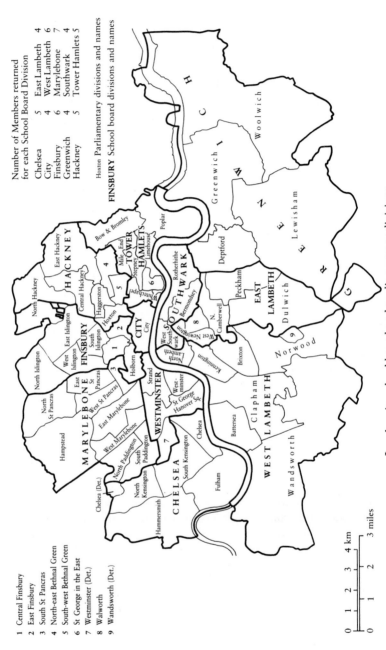

Number of Members returned
for each School Board Division

Chelsea	5	East Lambeth	4
City	4	West Lambeth	6
Finsbury	6	Marylebone	7
Greenwich	4	Southwark	4
Hackney	5	Tower Hamlets	5

Hoxton Parliamentary divisions and names

FINSBURY School board divisions and names

1 Central Finsbury
2 East Finsbury
3 South St Pancras
4 North-east Bethnal Green
5 South-west Bethnal Green
6 St George in the East
7 Westminster (Det.)
8 Walworth
9 Wandsworth (Det.)

1. London, its school board and parliamentary divisions

2

The London School Board 1870–1904

The Triumph of Elizabeth Garrett: 1870–1876

IN the autumn of 1870, London was electing its first school
board. The Newcastle Commission of 1861 had revealed what
many parliamentary Liberals, provincial radicals, and
Nonconformists had long suspected: that, despite half a
century of voluntary religious effort and considerable
parliamentary grants, the leading educational societies,
headed by the Anglican National Society, were not reaching
those most in need, the destitute and delinquent children of
the city slums. Already Birmingham's radical Education
Society, led by Dixon and Chamberlain, was outlining the
progressive educational agenda of the next thirty years by
demanding local schools that would be free, compulsory,
unsectarian and comprehensive, financed and inspected by
the State but managed by the town hall. So the incoming
Liberal ministry of 1868 looked first to local government to
make good the shortfall. Town councils had the money,
power, and authority to reach educationally deprived
children, as voluntary societies never could. But government
bills to make education the responsibility of town councils
foundered on the lack of any equivalent rural authorities.
County and district councils were some twenty years away. In
any case, even within the towns, it was not entirely obvious
that municipal councils of business men, concerned with the
'built environment' of streets and sewers, were the most suitable people to care for the educational, moral, and religious
welfare of the young. Accordingly, Forster agreed to adopt the
American pattern of directly elected school boards in both
town and country, wherever voluntary provision was found to
be inadequate. Board schools were not to displace voluntary
schools, but to make good any gaps in the network of

voluntary provision. Any adult, male or female, cleric or lay, was eligible to stand for election to the boards; and the rights of minorities (such as Catholics) would be protected by the cumulative vote.[1] Led by London, the great cities went to the poll.

Educational destitution was at its most severe in London. Men and women went into the autumn elections with a sense that there was everything to play for. As *The Times* put it, 'No equally powerful body will exist in England outside Parliament, if power be measured by influence for good or evil over masses of human beings'.[2] Hopes were high, the candidates impressive. Among them were three women. The *Lancet*, writing of a fellow doctor, commented:

We are extremely glad to see that Miss Garrett has permitted herself to be nominated as a candidate . . . her sex, if it influence her at all, will do so by giving her a warmer sympathy with everything that can help to elevate the child. It is a good augury for the working of the Act that Miss Garrett should even seek to be one of its administrators . . .[3]

Elizabeth Garrett at thirty-four had behind her five years of local medical practice, a triumphantly acquired MD from Paris just three months before, a national name, and a circle of staunch feminist friends, men and women, who were anxious to lever open public life for women. In June 1866, she and her close friend Emily Davies, had presented J. S. Mill with the first women's suffrage petition. In September, 1870, the prestigious Social Science Association, meeting at Newcastle, had heard Mrs Wolstenholme Elmy urge women to stand for and vote in the forthcoming school board elections. So when

[1] Forster was not only pressured by the Birmingham Education Society (later the National Educational League) but was also impressed by Melly's official inquiry into educational provision in Manchester, Birmingham, Liverpool, and Leeds, which revealed horrifying shortfalls in voluntary school places. Clergy were to join the new school boards as much in an effort to protect voluntary schools from competition, as to enhance popular education. The Birmingham League remained unhappy about much of Forster's Act: its directly elected boards, its optional sectarianism, its fee paying, and its neglect of teacher training. The League became the leading progressive pressure group in the educational field. See A. Taylor, 'The History of the Birmingham School Board, 1870–1903', Univ. of Birmingham MA 1955.

[2] *The Times*, 29 Nov. 1870. Marylebone was one of ten electoral divisions each with between four and seven members.

[3] *Lancet*, 29 Oct. 1870.

Elizabeth Garrett was approached by a deputation of Marylebone working men (whose wives she treated at her dispensary) to join their slate alongside Thomas Huxley, President of the British Association, and Randal Cremer, a leading trade-unionist, she was surprised but promptly agreed.

She faced a formidable fight, a vast division which included the parliamentary seats of Marylebone, Paddington, St Pancras, and Hampstead, an unreliable electoral register, and no party machine. The contest itself was slippery with church, chapel, voluntarist, and vestry rivalries; the prize, a newly invented authority whose powers were undefined and whose task was awesome. She was also the first woman ever to seek public election, and although education was traditionally women's work, no one knew whether this would count for or against her. In her favour, however, was the electoral system for the boards (which allowed voters in a multi-member division to plump for just one candidate who could therefore be elected by a determined minority); her 'name'; and her own superb talents as a candidate.

She had first to construct a large campaign committee of influential and respectable residents to give weight to her candidature. She spent October 'very busy writing to ask people to come upon my committee', and recruited, among others, the Revd Llewelyn Davies (Christian Socialist brother of Emily and father of Margaret), John Westlake QC and his wife Alice, Canon Barnett, Miss Octavia Hill, her brother-in-law Henry Fawcett, and the business man J. S. Anderson, who became first her chairman and then her husband. Only Shaftesbury refused to lend his name.[4] She had next to find workers to canvass and distribute literature. Within a week, three hundred women, largely unknown to her, poured in from all over London, Kent, and Sussex, to offer help. Anderson organized them into parish teams, she said,

'one band for each of the four parishes into which the division of Marylebone is divided, and had subdivided each parish on the ordnance map, each woman taking so many streets and courts and mews, and so thoroughly worked the district—worked it in a way

---

[4] L. G. Anderson, *Elizabeth Garrett Anderson,* 1939, p. 148.

that no other district in London was worked . . . [with] an amount of electioneering talent and skill that surprised not only myself but all the committee who were behind me'.[5]

Miss Buss wrote and distributed five thousand leaflets to women ratepayers, Anderson drafted her election address, and issued fulsome press releases. No expense was spared, and the candidate was happy to pick up the final deficit of £100 from her own pocket, thinking it 'a very moderate price to pay for such a pinnacle'.[6]

With an influential list of supporters behind her and a thorough canvass around her, she had now to seek the endorsement of as many churches, chapels, charities, vestries, and associations as possible. The all-important church and chapel vote she bought by distancing herself from Huxley's atheism and by opposing secular education; and she resigned herself in the process to losing the advanced Radical vote.[7] The final task was to take to the public platform. At a time when conventions of womanliness required even well-established ladies like Mary Carpenter to have their professional papers read for them by men, it took courage to address large mixed audiences, risk hecklers, and offend proprieties. In some trepidation, Elizabeth Garrett wrote to Emily Davies:

They [the working men's association] think there must be meetings to teach people to be interested . . . They propose holding one in each of the four divisions of the district and they wish me to attend and speak. Huxley has agreed to do so. I suppose it is part of the whole thing and ought not to be refused, though I am sorry it is so. I dare say when it has to be done I can do it and it is no use asking for women to be taken into public work and yet to wish them to avoid publicity. We must be ready to go into the thing as men do if we go at all, and in time there will be no more awkwardness on our side than there now is on theirs. Still I am very sorry it is necessary, especially as I can't think of anything to say for four speeches! and after Huxley too, who speaks in epigrams! However I shall hope to avoid bad taste even if I am commonplace . . .

And the eternal plea from despairing candidates, 'Quotations

    [5] *WSJ* 1 May 1871.
    [6] E. Garrett to J. Anderson, 8 Nov. 1870, u.d. Nov. 1870, 16 Dec. 1870, Fawcett Autograph Collection, vol. x.
    [7] E. Garrett to E. Davies, 24 Oct. 1870, Davies papers.

either from the Bible or Milton especially one to wind up with would be very precious. Bless us! It's a tough and toilsome business'.[8]

She turned out to be a gifted candidate, buoyant, breezy, quick-witted, confident, hardworking, and fluent. She quickly divided up her day: morning surgeries, lunch-time factory meetings, afternoons dispensing medicine to patients and encouragement to canvassers, early evening committee meetings, a working dinner followed by one or two evening speeches, at the end of which she often dashed across London to aid another woman candidate. By mid-November, she cheerily claimed to have 'knocked under' sceptical vestrymen, undertakers, and piano workers. Her speeches were all very general and inoffensive—religious difficulties she thought were exaggerated, compulsory school attendance she thought essential, PE and women board members she thought desirable. She was roundly told off by Anderson whenever she said something definite which could lose her support. In the absence of party slates and any prior experience, she had to construct what coalition she could, without much sense of how she was doing for size:[9] and that meant being all things to as many voters as possible. Given that she had few views on any aspect of education, this was less of a problem for her than it might have been.

Her spare energy, time, and supporters she brought to the aid of the other two London women candidates, Emily Davies and Maria Grey, both of them wracked with self-doubt, and loathing the whole business. The widowed Maria Grey had, with her sister Emily Shirreff, been writing on girls' education since the 1840s, and the sisters went on to found the Girls' Public Day School Trust. Elizabeth Garrett persuaded Mrs Grey to stand for Chelsea, helped plan her campaign and construct her committee, but despite zealous friends and painful public meetings, the campaign was bedevilled by a large number of contestants, a shortage of workers, and a candidate who clearly wanted to retire from the fight.[10]

Emily Davies, busy with her women's college at Hitchin,

[8] E. Garrett to Davies, 24 Oct. 1870, Davies papers.
[9] E. Garrett to J. Anderson, Fawcett Autograph Collection, vol. x.
[10] E. Garrett to E. Davies, 27 Oct. 1870, Davies papers; *EWR*, 15 Jan. 71.

was also pressed to stand, in this case for the even more unlikely seat of the City. 'I am torn in two between various advisers', she confided to a friend, 'My mother and Miss Garrett urge me that if a woman is wanted to try it, I ought not to lose time by holding back. My brother insists that it would be too audacious to offer to stand for "the greatest constituency in the world" without *more* invitation . . . [as] an ignominious and ridiculous defeat would do harm'.[11] Mr Tidman, the uncle of one of her students, now suggested that she should instead consider Greenwich, making out 'such a moving case of the anxiety of the electors of Greenwich to return a woman (*what* woman being a secondary consideration) that I thought I ought not to refuse . . . It is a coalition of very respectable ladies, "Low Church parsons" (perhaps High Church too) and Mr Mill'.[12] Mr Tidman became her full-time agent, arranging her timetable, briefing her on audiences and issues, overcoming her self-doubts—and casting a somewhat envious eye at the panache displayed by Elizabeth Garrett. 'Don't be afraid of having to say the same thing over and over again', he reassured his candidate. 'You will have very different audiences from night to night.' That evening she faced artisans and shopkeepers, and so should talk about buying sugar; the moral would be that in education as in trade you got what you paid for. The following night, with a better class of audience, she could talk about the industrial training of girls.[13]

Emily Davies gritted her teeth and embarked on nightly meetings.

'I am a little disappointed at finding that speaking does not get easier as one goes on. It was quite easy at Greenwich but at Blackheath I found it hard and had an uncomfortable sense of failure when it was over. I felt that I have been nervously hurrying on to get it over, and I am afraid the audience must have felt the same. The chilling reception to the opening speeches was depressing, but of course one ought to be able to resist such influence . . . There was more opposition near the end, of a more vulgar sort, and I could not judge how far it called out sympathy on the other side.

[11] E. Davies to Mr Tomlinson, 7 Nov. 1870, Davies papers. Unlikely, because the City had no women voters on its register.
[12] E. Davies to Mr Tomlinson, 14 Nov. 1870, Davies papers.
[13] Mr Tidman to E. Davies, 17 Nov. 1870, Davies papers.

The chief feeling I have through it all is a kind of sense of being half asleep and having nothing to do with it. But I cannot help wishing you had been at Greenwich instead of Blackheath. You would have liked Miss Garrett's speech—it was only too generous—and the meeting was enthusiastic. The Hall was fuller than it would hold (it holds 1,000) and the women came crowding into the committee room at the end to shake hands and promise their votes. Shaking hands seems to be a chief part of the candidate's business. There are to be meetings at Lewisham tonight, Woolwich tomorrow, Sydenham on Wednesday, Eltham on Friday, and a 3 o'clock meeting at the Crystal Palace is talked of for Saturday. I am afraid that Sydenham and the C.P. may be like Blackheath. Those rows of ladies with apathetic faces, from which one cannot guess whether they are agreeing or contradicting are hard to encounter'.[14]

At her adoption meeting, she was told by hecklers to keep to her proper sphere of school visiting, and not get 'mixed up with men' on the LSB. She was also forced to concede, with painful honesty, that elementary education had never been an interest of hers, but thought she could still offer honest service, if the electors so decided.[15]

London went to the polls at the end of November. Most of the press supported the women. The *Spectator* thought Elizabeth Garrett had more ability than most MPs, the *Saturday Review* thought she had 'a capacity for doing men's work without ceasing to be womanly'.[16] Emily Davies came in head of the poll in Greenwich, with over 12,000 votes; the next three successful candidates followed 2,000 votes behind; and eight were defeated. Maria Grey was less fortunate. Here eleven candidates competed for four seats, and she came in fifth, trailing by just 108 votes. Elizabeth Garrett was one of twenty-two candidates fighting for seven seats in Marylebone. When the results came in at midnight, she slipped out to send Anderson a postcard with the figures, 'Garrett 47,858; Huxley 13,494; Thorold 12,181 . . . '. She not only headed the poll; had not only helped to carry the other women to creditable results; but had more votes cast for her than any candidate in any election anyone could remember. She was front-page news; and to her amusement heard herself talked about in

---

[14] E. Davies to Mr Tomlinson, 21 Nov. 1870, Davies papers.
[15] *EWR*. 15 Jan. 1871.
[16] Cuttings u.d. in Davies papers, cuttings collection.

railway carriages. She had settled definitively the right of women to stand for school boards. When a couple of months later she married her chairman, J. S. Anderson, and had neither the grace to offer her resignation nor any board member the courage to ask for it, she settled the right of married women board members to serve as well.[17]

The *Telegraph* coyly noted:

Of course they may break down. Miss Garrett if a motion of hers is rejected may burst into tears, though those who know the lady deny the probability; Miss Emily Davies if called to order, may faint in the arms of Mr. Hepworth Davies, though why she should do so nobody could possibly say. But if these things do not occur—if the ladies are calm, businesslike and useful, and keep to the point in public debate—what is to become of us men? . . . If education is a women's subject because there are girl pupils, so is crime because there are women criminals; so is public health because there are women sick; so is the poor law because there are women paupers; so is religion because there are women devotees; so are agriculture and manufactories because there are women in the factories and fields.[18]

Indeed, she was so successful that male colleagues who might have feared to place a woman on a slate lest she lose the seat, now feared the reverse; that she might cream off so many votes, especially plumpers, that other seats would be at risk. Some of the bickering that Helen Taylor later encountered in Southwark came because she would not renounce plumpers, and was thought to be selfishly endangering the chances of her colleague, the Revd John Sinclair.[19] Elizabeth Garrett had given the women's local government movement a formidable weapon. As party lines emerged on school boards, local Liberal associations who refused to accept a creditable woman on their slate knew that if she ran as an independent, she would either take a seat and be beyond party discipline, or let in a Tory. The LSB had fifty-one members; but outside London, school boards were never larger than thirteen or fifteen members. A solitary independent woman could hold

[17] Minutes of Evidence on School Board elections, 1884–5, vol. xi. Q. 39. As a married woman, it might have been held that she was debarred from public office under principles of coverture.

[18] *Daily Telegraph*, 2 Dec. 1870.

[19] See below, pp. 92–3, Mrs Cowen of Nottingham and Mrs Westlake of London both tried to renounce plumpers in the name of party loyalty.

the balance of power. It was a party manager's nightmare. Bristol Liberals just once refused to include a woman on their slate. An official male candidate was defeated by an independent temperance woman. Thereafter the Women's Liberal Association had a place reserved for its nominee.

Not surprisingly, women, like Catholics before them, and working men, and teachers after them, approved of the cumulative vote. As Helen Taylor delicately put it, '. . . it marks the enthusiasm of the voter'. Said the *Women's Herald*, 'the special position of women renders plumping necessary'.[20] A method of voting designed to protect religious minorities was perhaps to have greater significance as a way of electing social minorities. For women gained twice over. Divisions in London were large and multi-member; outside London, cities went unwarded. Women thus found it easier to be adopted as part of a slate. But as in large divisions the poll was often low (West Lambeth sometimes polled only 20 per cent), and as women ratepayers formed 10 per cent to 20 per cent of the register, woman candidates were nearly always elected and very often at the head of the poll. Then as now, proportional representation irritated majorities and favoured minorities, especially affluent minorities, who could afford to send literature out to 50,000 to 60,000 voters, hire halls, and advertise in the press. Until the Corrupt Practices Act 1883, candidates like Miss Muller could spend several thousand pounds to win a seat. 'A capable, business-like lady, who can spend enough money to bring herself before the electorate will almost certainly be returned', noted the *Women's Signal*.[21]

Party managers, appalled at the anarchy of cumulative voting, much preferred the simple majority vote. They lobbied parliamentary committees to change the law. They tried to discipline candidates not to ask for plumpers. They informally warded their areas, confined candidates to a 'patch' and asked voters to support them exclusively. When women fell into line, they often found themselves with the worst patches, as Mrs Louisa Mallet discovered to her cost. She would have been elected in an open multi-member division of the LSB: she was

[20] Taylor cuttings, 2 May 1885, Box vi. Mill/Taylor papers. *WH* 21 Nov. 1891. The two divisions, Marylebone and Greenwich, returning women in 1870 were also two of the four divisions returning Catholics.   [21] *WS* 23 Sept. 1897.

defeated in a single-handed contest. Annie Besant won her
seat in 1888 by ignoring party managers, saying 'I think it is
simpler running alone'; and collected votes across the
division. When, at the end of the century, education was
transferred to town councils, Chamberlain welcomed the
move partly because it would end cumulative voting and stop
the return of members not 'representative of the general
opinion of the constituency': like women.[22]

Meanwhile, Elizabeth Garrett had written her thank you
notes to party workers, and less decorously scribbled a line to
Anderson, 'I heard yesterday of a plumper given by a man
who had never spoken to me but had once seen me in a new
gown at a party. For the gown's sake only he voted. And then
women are thought to be unfit for the franchise'.[23]

The new board stood high in social and political prestige.
MPs and peers, eminent clergy and large employers, were
among its members. Only the gentle Benjamin Lucraft, a
former Chartist, could be said to represent working people.
The first meeting was at the Guildhall. After the 'intoxi-
cation' of the elections and the warm good wishes from
supporters crowded into the antechamber, the two women
now faced a series of skirmishes. When they assembled they
were asked 'to take two seats apart, this we resisted', and
firmly joined the other members around the large table. They
were ignored. Next came the chairmanship. Elizabeth
Garrett, with the highest number of votes, thought she might
be called to the post. Emily Davies told her she was being
cheeky even to think of it. In the event, Lord Lawrence,
former Viceroy of India, was nominated by pre-arrangement.
Elizabeth Garrett shrewdly slipped in a few words of congratulations to show she was not sore, but the atmosphere was
decidedly hostile and they left at the end by a back door. In
retrospect, she wished she had said more, as the press was
present, but confessed to being 'a little awed by the whole
thing being so extremely like Parliament and by having to
spring up so quickly to get a hearing after someone else had

[22] *WPP* 17 Nov. 1888; *SBC* 25 Apr. 1896.

[23] Anderson, *Elizabeth Garrett Anderson*, p. 156. The draft thank you note, 'I am very
glad, happy, both for the victory itself, and also for it being given to me to have a
share in it. I am sure it will do the women's cause great good.' u.d. draft, Garrett
papers.

finished. The whole difficulty of speaking is concentrated in that moment of swift self-assertion.'[24]

The women joined the LSB well aware that its work would not only shape the lives of London children, but would set paths and precedents, 'be a polestar and guide' for boards throughout the country to follow.[25] The LSB faced formidable problems. Perhaps 40 per cent of London's children irregularly attended the 'efficient' church voluntary schools of the National, and British and Foreign School Society. These had been grant-aided and inspected, their pupil teachers trained. Even so, the better schools were still often only two classrooms, crowded with children sitting on tiers of benches, and crammed by poorly educated pupil teachers in the mechanical rote learning thought to earn payment by results. Children chanted to one visitor that the Pacific Ocean was sixteen million square miles; but no one, including the young teacher, could tell him whether that made it larger than Europe, England, or the adjacent field. The class decided it was about the size of the field.

A further 45 per cent of children put in the occasional appearance at private dame or adventure schools, which received no grant and no inspection. 'None', said the Assistant Newcastle Commissioner for South London in 1861, in a notorious passage,

are too old, too poor, too ignorant, too feeble, too sickly, too unqualified . . . to be unfit for school keeping . . . Domestic servants out of place, discharged barmaids, vendors of toys and lollipops, keepers of small eating houses . . . milliners, consumptive patients in an advanced stage, cripples almost bedridden, persons of at least doubtful temperance, outdoor paupers, men and women of 70 or 80 years of age, persons who spell badly, who can scarcely write, and who cannot cipher at all,

made up London's teachers.[26] They 'minded' children in the

[24] *The Times*, 2 Dec. 1870, cf. *The Graphic*, 9 Dec. 1876. E. Garrett to J. Anderson, 15, 16 Dec. 1870, Garrett papers.

[25] For a useful account of elementary education, see J. S. Hurt, *Elementary Schooling and the Working Classes 1860–1918*, 1979; P. McCann, ed., *Popular Education and Socialization in the Nineteenth Century*, 1977; S. McClure, *A Hundred Years of London Education, 1870–1970*, 1970; D. Reeder, *Urban Education in the Nineteenth Century*, 1977; D. Rubenstein, *School Attendance in London, 1870–1904*, 1969.

[26] Helen Blackburn, *EWR* 15 Dec. 1879; T. Spalding, *The Work of the London School*

dank basements of shops and chapels, in temporary shacks
under railway arches, in crowded living-rooms along with
wash tub and the odd chicken, where children huddled
without slates, sometimes sitting on forms, in the poorer
schools squatting on earth floors. To the dismay of investigators, working-class parents often preferred these small
grubby private schools to the large efficient and cheaper
voluntary schools, just because such schools accommodated
themselves to working-class needs. They offered uncensorious
'minding', practical literacy, a more homely atmosphere, and
did not try to impose formality, inspection, discipline, and
punctuality upon families and their children.[27] They were free
of religious instructions, free of the monitorial method. They
were 'family-friendly.' Professional educationalists despaired.

The residual children, some 15 per cent, lived and slept on
the streets, running wild but occasionally reached by ragged
schools if they were hungry enough and the weather bad
enough. Most poorer working-class children might attend a
school two or three times a week over two or three years,
seldom staying anywhere for more than a few months. Emily
Davies found that of the first eighty children, of all ages, at
Greenwich's new board school, just nine knew the alphabet,
and almost all were verminous, unruly, and underfed.

So the LSB faced a quantitive roofs-over-heads crisis as its
first problem; a complex tangle of attendance, compulsion,
and fees as its second; and a 'quality' problem, a shortage of
trained and competent teachers willing to struggle with large
classes of dirty, restless children, as its third.

The educational census of 1871 showed that London had
some 680,000 school-age children and a shortfall of at least
100,000 school places. As at the same time, the board passed
by-laws requiring compulsory attendance to the age of ten,
and compulsory fees (of 1d. to 4d. a week), the new board had
not only the problem of building sufficient schools, but of

Board, 1900, p. 78; *Newcastle Commission Report*, vol. iii (1861), p. 483. The Commission's 'writing down' (and out) of private adventure schools is challenged by
P. Gardner, *The Lost Elementary Schools of Victorian England*, 1984, pp. 118 ff., who
argues that such schools were neither ephemeral nor ineffective.

[27] See P. Gardner's paper on 'The Family and the School', repr. *Bulletin of the Society
for Study of Labour History*, Spring 1984; and more generally, *The Lost Elementary Schools*,
chs. 3 and 5.

getting children to attend them with pennies in their hands. It succeeded. By the later 1880s, the LSB had won the numbers game. Virtually 95 per cent of all eligible children were enrolled and some 80 per cent of them were regularly attending well-built schools. The quality of teaching within those schools was another matter, and one that greatly concerned later women school board members.

The first school board inevitably saw its work in quantitive terms. Its six standing committees (finance, statistical, works, school management, industrial schools, and by-laws), mapped out their tasks. Emily Davies sat on the statistical committee which drew up the profile of London's school needs. She ruefully told her Greenwich constituents, 'we were charged with putting schools where they were not wanted, interfering with existing schools, miscalculating the numbers and wants of the population [and] exaggerating the increase of attendance which might come from compulsion'. The works committee had acquired 196 schools within three years, and were building a further hundred, designed with interconnecting classrooms. Though neither woman sat on the works committee, Elizabeth Garrett forced them to accept herself and Huxley as ex officio experts, after she protested that they were building schools with defective drains, bad ventilation, and inadequate playgrounds. By the 1880s, most schools were three storey terracotta, their classrooms opening off a central hall. On tight sites, playgrounds were on the roof. Both women served on the school management committee, the most important and demanding of the committees, as it supervised the content of schooling. Under Huxley's chairmanship, they adopted a generous curriculum which included elementary science, history, geography, music, drill and kindergarten for infants. Though it took many years to implement, as the Education Department would not fund all the subjects, it did serve as a marker of best practice for boards across the country. Neither woman could find the time to sit on the industrial schools committee, responsible for children at risk, but Emily Davies sat on the by-laws committee where she drafted the board's scheme to compel attendance. Attendance officers would seek out the children and bring them to school; their work would be supervised by local and divisional committees, chaired by the member. Erring parents could be brought to court. In her own division, a further 5,500 children were now coming to school.[28]

The result was a gruelling work-load. There were weekly board meetings, committees, sub-committees, and divisional committees to chair; managers' meetings to attend, sites and schools to visit, examinations to monitor, attendance visitors to supervise, and a growing correspondence to handle from parents, teachers, job-seekers, and lobbyists.

A later woman board member noted in her diary:

reached Mortimer Street at 9, looked around, then to Hart Street to see about Pupil and Asst teachers, then to Vere Street for the drawing examination—surprised to find that the second step includes geometry; then saw the Visitor and received reports of street cases caught by other visitors, Managers meeting, made Mrs Buxton's acquaintance, then set off for the Rota Works Committee, and Educational Endowments etc—home reading Blackwoods en route . . . very tired.

So Emily Davies urged women to come forward and help in any way possible. They should not hesitate just because there was 'something public about it' and 'not quite in the beaten track'; they should think only of the work that needed to be done, and 'not care much' for praise or blame.[29] Men's initial hostility had quickly disappeared, she said, and the two women members were now overworked. With a busy professional life outside the school board, they had neither the time nor political capacity to make a major contribution to board policy but sensibly limited their efforts to watching over two principles: compulsory attendance, and equal treatment for girls.

Both women were firmly committed to compulsory attendance. It was not negotiable. Both believed that parents should pay full school fees. Elizabeth Garrett's views were characteristically brisk and direct. To remit fees for poor parents would be 'subsidizing improvidence': parents should pay school fees just as they paid the baker's bill (. . . baker's

---

[28] *Kentish Mercury*, 11 Oct. 1873.
[29] Edith Simcox, MS Diary, 'Autobiography of a Shirtmaker', 9 Mar. 1880, Bodley Library.

bill?) They need only spend less at the public house. If they could not, they should seek poor relief. Lucraft was distressed by her intransigent moralism; and even the Revd Picton, much in favour of the short sharp shock for truant children, was moved to complain that 'Mrs Anderson's argument evidently was that the board ought to drive these people into the workhouse'.[30]

For her part, Emily Davies believed that education would only be valued if it had a price. More pragmatically, she feared that if board schools were free, they would draw children away from the church voluntary schools, and would not then have the resources to reach the unenrolled. On the other hand, she came to learn that middle-class models of the family were often wildly inapplicable to working-class lives. Many truanting children, she found, were beyond parental control. Other families were desperately poor, engaged in sweated and casual labour, street trades, or listlessly seeking any work to bring in a few pence. They were in constant motion from one school area to another, as they tried to escape debts and find work, rooms, and wages. Even more settled families could not only not afford the school pence (a shilling or more if there were several children) but could not afford to lose their children's earnings either, if younger children in the family were to get fed. Perhaps half of the older board school children were in work in 1870, and compulsory school attendance merely displaced their work to the beginning and the end of the school day. Boys sold milk and newspapers, ran errands, worked for fried fish, vegetable, and butchers' stalls, crying goods from 4 a.m. in the morning, till late at night, and coming to school 'voiceless and stupid'. Girls were needed for family chores, to stop infants falling out of windows and into fires. In the words of a modern commentator, compulsory, fee-paying school attendance cut off economic strategies for survival while imposing additional burdens on the very poor. For such families, school fees were an impossible imposition, regular attendance a hopeless ambition. 'I cannot sympathize with the view that we are to stand by, complacently admiring independent poverty struggling with starvation', Emily Davies noted down, but she remained convinced that the right

[30] *SBC* 22 Apr. 1871, 3 Aug. 1872, 3 May 1873.

response was to seek to raise wages and encourage trade-unionism, rather than remit trifling school fees.[31]

Fees, certainly, deterred children from attending school. But the disaffection lay deeper than that. It is clear from autobiographies and oral history that many children found board schooling, with its rote learning, monitorial methods, and rigid discipline, so alien and irrelevant that they sensibly kept their minds on other things. Some children fought compulsory attendance in every way they knew how—truanting, 'larking around', assaulting teachers, vandalizing school property—their defiance broken only by the cane and the threat of reformatory school.[32] Women members of subsequent boards tried to mediate children's experience of school. They sought to introduce child-centred teaching methods, curb corporal punishment, and appoint some working-class parents as school managers. Their efforts softened the system, but board schools continued to be a battleground (both teachers and board members employed the military metaphor) of conflicting cultures, middle and working class, rough and respectable, adult and child.

When it came to the nexus of feminist questions, however, the views of Elizabeth Garrett and Emily Davies were far from conventional. With Lucraft's help, they tried to make sure that girls obtained their fair share of educational endowments; they worried about the hidden curriculum of domestic subjects to which girls were exposed at the expense of their academic studies, and they were not reassured to learn that girls had one hour where boys had two and a half hours for the same history examination, because girls were better at it.[33] They fought for equal pay for women school visitors, the appointment of lady inspectors, women-only teachers for infants and girls; and reserved places for women on local school management committees.

Overall, however, their contribution to policy was modest.

[31] Mrs Hogg, 'School Children as Wage Earners', *Nineteenth Century*, Aug. 1897, pp. 235–44; draft MSS notes, u.d., fo. 152, in Davies papers; J. Lewis, 'Parents, Children, School Fees and the London School Board, 1870–90, *History of Education*, 11, 1982, p. 300; and more generally, Rubinstein, op. cit.; J. Lewis, ed., *Labour and Love: Women's Experience of Home and family, 1850–1940*, 1986.
[32] See S. Humphries, *Hooligans or Rebels? An Oral History of Working-Class Childhood and Youth, 1889–1939*, 1981. [33] *WSJ* 1 Jan. 1880.

They said little about the syllabus, nothing at all about religious teaching (still a source of strife between church and chapel as they competed for future congregations); and it was left to later women members to show care and concern for damaged or delinquent children. There is little evidence from their board work that they were committed to elementary education, cared deeply about children, understood the communities they were serving, or recognized that education might be deemed a national rather than a private good. Compared with many of their colleagues on London's first board, they took a narrow view of their task: to impose, without fear or favour, compulsory school attendance on resentful working-class neighbourhoods. Their vocabulary was quasi-imperialist, of 'taking on the burden' of colonizing and elevating rough working-class districts with sturdy gothic board schools. Elizabeth Garrett, in particular, lacked sufficient empathy with the poorer working class to seek their support for compulsory education, and sufficient imagination to make it child-centred. On her engagement she had been teasingly asked whether she preferred to be 'the elect of fifty thousand or of one' and had confessed, revealingly, 'I rather despise the fifty thousand'.[34]

They had stood for the LSB, not as educationalists devoted to children, but as feminists to advance the claims of women to public life. Their achievement was to stand and be elected at all, something that took considerable courage and a refusal to be intimidated by conventional views of women's proper sphere. Elizabeth Garrett's flair and dash, in particular, served her far better as a candidate than as a member. Both women were overworked. Emily Davies was struggling to refloat her college at Girton, acting as its Mistress while travelling up and down to London for meetings. Elizabeth Garrett was now at odds with many of her feminist friends because as a doctor she would not oppose the Contagious Diseases Acts. She had an expanding medical practice, a fledgling Woman's Hospital, and children of her own. Inevitably, they were vulnerable in the very area that later women members were strongest, the unremitting attendance at committees, the assiduous attention to detail, the first-hand

[34] E. Garrett to J. Anderson, 21 Jan. 1871, Garrett papers.

knowledge of sites, school, and staff. Within six months
Elizabeth Garrett came off one committee, and Emily Davies
reduced her commitment on another. In their last year on the
board, only nine members (mainly MPs) had poorer records
of attendance. Lucraft had put in over 200 attendances, the
average was about 90, Emily Davies could only manage 34,
Elizabeth Garrett 28. They had become a token presence.
They had done what they could, and chose to retire at the
triennial elections in 1873.

However, they had one further task to perform—to find
successors. For Marylebone, Miss Buss suggested Miss Jane
Chessar, one of her school staff, and a part-time training
college lecturer. Quiet, devoted, apprehensive about public
life, and seriously troubled by poor health, she would not be
an ideal candidate, however conscientious she would be as a
board member. Fortunately, Mrs Alice Cowell, Elizabeth
Garrett's sister, had just returned from India where she had
done some educational work; and she was pressed into service as well. Elizabeth Garrett promptly sent out a circular
letter to Marylebone voters asking them to support Miss
Chessar and Mrs Cowell, handed them her committee of
supporters, and started to organize their campaigns.

Miss Chessar proved a disappointing speaker (no 'humour
to hand', complained Elizabeth Garrett), and easily tired (she
was 'sleepless and knocked up'). Doubting whether
Marylebone would return two women and overshadowed by
Mrs Cowell's flair, she sought to withdraw. As she wrote to
Miss Davies, 'Canvassers *cannot* be got to work as they did last
time, and everyone says, "I cannot do now what I did three
years ago"'. Elizabeth Garrett would not agree to this—it
would be 'a wasted opportunity'—but allowed her to reduce
her workload. 'It is lucky', Elizabeth Garrett told Emily
Davies with more than a touch of impatience, 'there is so
much her friends can say for her.' She consoled herself that
her sister, Alice Cowell, was proving 'an exceptionally good
candidate' attracting warm support, making splendid
speeches.[35] Elizabeth Garrett's instinct and organization were
correct, and although women candidates elsewhere in London

[35] E. Garrett to E. Davies, 14 Oct. 1873; J. Chessar to E. Davies, 30 Oct. 1873;
E. Garrett to E. Davies, 9 Nov. 1873, Davies papers.

were unsuccessful, the two Marylebone women were safely elected.[36]

They found the next three years quietly productive. Mrs Cowell made a vigorous though insensitive start by persuading the board to demolish working-class housing to provide a girls' school playground; but then contributed less and less. Miss Chessar, however, gained steadily in reputation. As a professional teacher, she spoke authoritatively if not always audibly, on every aspect of girls' education. She introduced suitable cookery lessons for older girls, suitable music teaching for younger children, and when the board proposed to drop its pilot kindergarten scheme, she persuaded them to expand it instead. As befitted a lecturer for the National Health Society, she joined Mrs Cowell in wanting physiology on the school curriculum, gym apparatus in the playground, and girls' swimming lessons in school hours.[37] Above all, as a teacher of teachers, she wanted to see them properly trained and paid. Buildings were important, competent teachers even more so.

The country needed some two to three thousand more teachers a year than the existing pupil teacher and training college system provided. The LSB employed a thousand adult certificated teachers; but it depended for classroom teaching on its 1,300 pupil teachers, some of them only thirteen years old, undereducated, overworked, often unsuitable, and much resented by parents forced to pay fees for their services. When the school management committee tried to attract more and better pupil teachers by raising their minimum age to fourteen years, and by providing some off-the-job training in specialist centres, Miss Chessar had the scheme referred back. She proposed a minimum age of fifteen years, and a proper training programme. The Education Department rejected all change, refused to permit pupil teacher training centres, and insisted that the children should learn the job on the job, under the supervision of the head teacher. Not until 1885 was

[36] Miss Emily Guest failed to hold Miss Davies's seat at Greenwich, Mrs Arnold (wife of the editor of the *Echo*) could not win Chelsea, nor Mrs Charlotte Burbery (secretary of the London Suffrage Society) the City. Miss Louisa Rees withdrew from Hackney. *Brighton Daily News*, 26 Nov. 1873; Alice Westlake to Helen Taylor, 31 Oct. 1873, vol. xv, fo. 179, Mill/Taylor papers.

[37] *SBC* 10 Jan., 16 May, 6 June, 1974.

the LSB allowed to provide its first pupil teacher centre, along the lines envisaged by Miss Chessar. Slowly, day-to-day classroom teaching passed to trained adult teachers, and pupil teachers became ancillary aides. By the 1890s, the Fabians thought even this was not enough. Pupil teachers were 'victims in a narrow groove', often getting their own schooling and training, in the same elementary school they taught in. They needed a broader and enriched education, perhaps at the new secondary schools, before going off to a training college.[38]

Miss Chessar had helped to direct the LSB's attention to the 'quality' questions in education. With some justification, the *Women's Suffrage Journal* could regard Miss Chessar as 'prominent and most influential'. But her health, always fragile, was breaking down. She had to miss six months of board meetings in 1876 and clearly could not stand again. Helen Taylor, planning to stand in Southwark, was taken around by Miss Chessar on school visits and to divisional committees. 'It grieves me very much not to have been able to stand at this election', she wrote to Helen Taylor, 'I hope you will be successful, however, and that you will find the work as interesting as I have done.' Four years later she was dead.[39]

The Marginalizing of Helen Taylor: 1876–1885

The 1876 triennial elections saw the return of four powerful and attractive women to the LSB, Mrs Elizabeth Surr, Miss Florence Fenwick Miller, Mrs Alice Westlake, and Miss Helen Taylor.

The least known of the four was Mrs Surr, doctor's daughter, wife of an eminent City merchant, writer of children's books, and a motherly middle-aged churchwoman, whose affections went out to truanting and neglected children. Her scathing utterances and dogged tenacity in exposing cruelty to children in certain industrial schools were to transform LSB policy while making her deeply unpopular with the majority Liberal group. Her 'able, clever, thoughtful speeches',

---

[38] Annual Report of the LSB, Oct. 1875; *SBC* 4 Apr. 1885; Fabian Society, *The Worker's School Board Programme*, 1894; *The Final Report of the LSB 1870–1904*, pp. 138 ff.

[39] *WSJ* 1 Jan. 1877; J. Chessar to H. Taylor, 13 Nov. 1876, vol. xv, fo. 35, Mill/ Taylor papers. Mrs Cowell also retired.

together with her keen sense of publicity (London shoeblacks carried 'Vote for Mrs Surr' labels in their caps), won her the fourth of Finsbury's five seats.[40]

Florence Fenwick Miller was only twenty-two years old, trained in medicine and an experienced suffrage lecturer. Frederick Rogers, looking back, thought her the most interesting of the London School Board women, 'young, good looking, brilliant . . . very much a demogogue', 'daring enough to talk frankly on public platforms' about human biology to lecture halls packed 'to suffocation'. The Revd Stewart Headlam brought her forward on behalf of the Bethnal Green working men's associations for Hackney, and despite her mother's objections she contributed £50, and yards of lemon and heliotrope ribbon for rosettes, to the campaign. She fought as an official Liberal, and came in fourth with 15,000 votes. She then married and called herself Mrs Fenwick Miller. 'Very high legal opinion was taken on this matter by the chairman of the London School Board before he would call out her name in the division lists as Mrs Fenwick Miller', and a fruitless attempt was made to upset the results of the next election on the technicality of her improper name. Meanwhile she was soon in trouble with her Hackney division for supporting Annie Besant, on trial for publishing *Fruits of Philosophy* with its oblique contraceptive advice. She carefully explained to those shocked and disgusted that she was supporting the freedom of the press and not family planning. She went on to become a professional journalist, writing for the national dailies, and in 1895 editing the *Women's Signal*. Her debating skills were formidable. Said one reporter, 'I have seen men grow visibly paler as she dissected—or rather vivisected—their halting arguments with her pitiless logic; she would leave nothing but shreds behind'.[41]

Alice Westlake, now in her thirties, inherited the Garrett

[40] *WSJ* 1 Jan. 1877; *The Times*, 1 Dec. 1876; *Ladies Pictorial*, 3, 17 Dec. 1881. Her husband, a Conservative, had stood against a popular Liberal earlier in the year.

[41] F. Rogers, *Labour, Life and Literature*, 1913, p. 51; *WPP* 23 Feb. 1899; *WS* 3 Oct. 1895; *The Ventilator*, 8 Dec. 1876; *Hackney Standard*, 14 Apr. 1877; *Eastern Argus*, 28 Apr. 1877; *Ladies Pictorial*, 17 Dec. 1881. I am grateful to Prof. R. Van Arsdel, University of Puget Sound, for additional references. See her article, 'Victorian Periodicals Yield their Secrets: Florence Fenwick Miller's Three Campaigns for the LSB', in *Warwick's Year Studies in English*, 1985.

seat and support at Marylebone. Her credentials were rather more conventional. She was the daughter of Thomas Hare (barrister and president of the London women's Suffrage Society) and wife of a supportive and eminent Liberal QC. They had no children. She ran an immaculate campaign, had Eleanor Marx canvass for her, and came home head of the poll with 20,000 votes. With orthodox Liberal views and strong party loyalties, she, more than any other woman on the LSB, became an 'insider', whose grasp of finance made her an effective administrator and party spokesman in the 1880s. As party lines on the LSB hardened, she was to place party loyalty above support for her women colleagues in their battles on the board, much to their annoyance.[42]

The last of the quartet was Helen Taylor, some forty years old, stepdaughter of John Stuart Mill. She too had been part of the Garrett feminist circle for the past decade, but her work as Mill's political secretary until his death in 1873 had brought her less conventional contacts, both with the local Irish on issues of land reform, and with leading trade-unionists. Southwark had a strong radical presence (the Mills, Fawcetts, and Dilkes had run the trade-unionist Odger as a Labour parliamentary candidate in 1869). Mill himself had declined to stand for the Southwark school board division in 1870, but when radical working men's clubs now approached Helen Taylor, she agreed, and without scruples used Mill's name to build support for herself. Alice Westlake, in her own words 'anxiously looking out still for candidates', was delighted, Chadwick and the Fawcetts offered their help. Her victory, they thought, would be 'a great service both to radicalism and to the woman's cause'.[43] Unfortunately, she shared the Southwark Liberal ticket with the Revd John Sinclair, a cautious Scottish Congregationalist minister, who was forever begging her to tone down her secularist views, give up her handsome donations, and lay aside her claims to plumpers. She would do nothing of the kind, so recriminations mounted, and party workers in the division lined up behind the one or the other.

[42] *EWR* 15 Sept. 1876, 15 Dec. 1876; *SBC* 30 Mar. 1889; *Cambridge Independent Press*, 10 Dec. 1905. Her father, a barrister and charity commissioner, had urged women to join schools boards, *WSJ* 1 Jan. 1871

[43] Henry Fawcett to H. Taylor, 21 Oct. 1876, vol. xv, fo. 60, Mill/Taylor papers.

The Garrett circle, irritated at reports of her populist style and self-regarding behaviour, came in behind Sinclair. Elizabeth Garrett refused to stand on her platform or come to dinner, and wrote her a harsh letter, warning that 'it will do our woman's cause untold harm if this miserable affair becomes public', it would lose Helen Taylor influence on the board, and 'it will prevent anyone to whom our cause is dear ever again wishing to take a part in public work'. Helen Taylor called for a full committee of enquiry at which she was exonerated, and Sinclair hastily switched seats at the earliest opportunity into Lambeth.[44] Helen Taylor won her seat, coming in third. But from then on she was loathed by the official Liberal party caucus and was referred to behind her back as 'the acid maiden'. She in her turn built up an independent power base of her own, drawing on trade unions, radical working men's clubs, and Irish support; and reported to them regularly in crowded open meetings, which she played like the professional actress she had been. At subsequent elections, 'marked by all the excitement, fervour and verve usually associated with parliamentary contests', said one observer, she stood as an independent Radical Democrat, demanding free education, no corporal punishment, and open competition for tenders and jobs. Despite official Liberal opposition, she came in repeatedly at the head of the poll.[45] One of her supporters wrote to congratulate her, saying that the women on the LSB

wanted a leader of decided opinion and very marked intelligence, because even with the very excellent women who have served on it so far, the manpower as you call it has rather complied if not condescended and that partly out of pure politeness to the matters specially urged of girls and their teachers. You will be able to take the necessary stand and by force of pure reason and the public opinion which is behind you get the girls lifted up a step or two at least. I have no doubt however that women's work lies mainly in committees and in

[44] Revd J. Sinclair to H. Taylor, 17 Oct. 1876, H. Fawcett to H. Taylor, 21 Oct. 1876, E. Garrett to H. Taylor, 7 Nov., H. Taylor to Mr Bayley, 9 Nov., E. Garrett to H. Taylor 6 Dec., Revd J. Sinclair and Mr Long to H. Taylor 13 Nov., Mr Long to H. Taylor 27 Nov., 14 Dec., 1876, vol. xv, fos. 75–123, Mill/Taylor papers. Among Helen Taylor's hate mail was one postcard she kept, 'Helen Taylor / Disgusting creature / Man in petticoats / Satan's masterpiece / Her end / Destruction', ibid., fo. 100. Sinclair in the 1890s was to lead the Nonconformist attack on Riley's Anglo-Catholicism. See below, pp. 119–20.

[45] F. W. Soutter, *Recollections of a Labour Pioneer*, 1923, p. 85.

the local [*sic*], but unless they have at least three or four women in the central board the unobtrusive work might be as it usually has been passed by with inadequate recognition.[46]

The four women were placed on the committees of their choice including finance; and it was a mark of their standing that they seconded the chairman and vice-chairman of the new board.

As ever, the women members kept a watching brief on the welfare of girls, pressured as they were by additional domestic work at home, and additional domestic subjects at school. As three afternoons a week were already spent in needlework, the women resisted any further sewing lessons at the expense equally of eyesight and arithmetic, but instead encouraged basic home cooking taught in special cookery centres. Sceptical male colleagues were won over by cakes baked by the girls for board teas. Women members found that teachers tried to compensate for the poorer school attendances and fewer academic lessons of girls by giving them additional homework in order to maintain Department grants. This tired the girls still further and annoyed their parents, who wanted their help and their earnings. As the 'over-pressure' debates of the 1880s on several city school boards reveal, many girls, malnourished and in poor health, were being torn by conflicting school, family, and community pressures which the women members were helpless to resolve, try as they might.[47]

The same loyalty was extended to female staff. Women members persuaded the board that only women should head girls' schools; that married women should not be debarred from posts on the assumption that they were less dedicated, more pregnant, less reliable, more affluent, and less deserving than either single women or married men; and they pressed equal pay amendments only to be defeated by the market principle that better women were had for less than men. They

---

[46] M. Rendle to H. Taylor, 3 Dec. 1876, vol. xv, fo. 126, Mill/Taylor papers.

[47] *EWR* 14 Apr. 1877 complained that the needlework code was designed by a 'fanatical lady specialist' who wanted seven-year-olds to sew five varieties of seam and six sorts of fancy work. Also *SBC* 24 Mar. 1877, 30 July 1881, 26 Jan. 1884. For the PE debate see *SBC* 29 Oct. 1881. Many of the SBC women were themselves very active in climbing, cycling, rowing, and riding. For 'over-pressure debates', see *SBC* 25 July 1885. The concern led to the Cross Commission's investigation into payment by results, which in turn led to the abolition of fees in 1891, as well as debates on school meals and medical services.

fought to open up new careers for women, as visitors, clerks, superintendents, even inspectors, though here male colleagues let their prejudices rip, insisting that 'the intellectual qualifications of ladies might be sufficient but their moral qualities would not be'. Stanley, the tetchy Liberal leader, thought this all a bit of a fad to show 'that the board were in favour of women's rights'. Another member feared that women colleagues would vote *en bloc* to appoint any woman over any man. When the women protested, they were told that their deplorable temper made the men's point. In disgust, Helen Taylor and Florence Fenwick Miller helped found the Metropolitan Mistresses Association to look after women teachers' interests. Such was the women's tenacity that Stanley resisted in 1880 the inclusion of yet another woman on the school management committee which he thought

was somewhat overweighted by the Trades Union spirit of the lady members . . . They were too ready to support large salaries for female teachers. The ladies did very intelligent and useful work in the committees and their influence was most desirable, but it could not be denied that they were a phalanx who were bound together for certain objects.[48]

It was a sign of the women's strength—nine had been returned in 1879—that they were able to carry their proposal. It was a sign of the men's unease that they should resist it.

For its first years, the LSB had had to devote itself largely to the 'numbers' question—of children, buildings and staff. When the chairman gave his Annual Report in 1879, the board's achievements were summarized quantitively—since 1870 school places had increased by 80 per cent, attendances by 100 per cent, examination passes by 80 per cent to 90 per cent. He also added a quotation:

Wherever a board school was put down, a transformation in the habits of the people soon followed; the child's face was polished until

---

[48] *SBC* 9 Feb. 1879. Mrs Westlake, childless, was not 'sound' on the matter, and when the proposal for 'school board nuns' resurfaced in 1881, the other women blamed her for it. *SBC* 26 Nov. 1881. For the debate on women inspectors, see *SBC* 3 Feb., 12 Mar., 23 July, 1881. The Liberal leader was the Hon. Lyulph Stanley, son of Lady Stanley of Alderley, brother of the suffragist Countess of Carlisle. See A. W. Jones, 'The Work for Education of the Hon. E. L. Stanley', Univ. of Durham M.Ed. thesis, 1968–9.

the parents grew ashamed of their own; habits of order and cleanliness were formed; and in many other ways the little ones became a
channel of civilizing influence to hitherto inaccessible courts and
lanes.[49]

This was a vocabulary that Alice Westlake shared with
Elizabeth Garrett before her. Schools were cultural outposts
'put down' to instruct, discipline, and civilize darkest London,
like 'planting a fort in an enemy country'. Whereas Helen
Taylor supported cookery lessons, for example, mainly
because they were popular with parents, Alice Westlake
approved of them because 'the majority of the working classes
did not know how to make the best use of cheap articles and
how to cook their food to the best advantage'. Benjamin
Lucraft was stung into protesting that working-class girls well
knew how to cook a potato. 'Mrs Westlake was not at all
acquainted with the working classes, or else she could not
argue in the way she did'.[50]

The other women, more motherly, more democratic, or less
arrogant than she, understood that board schools had to win
the support of their local community if children were to come
regularly and willingly to school. Schools must be sensitive to
local needs, Helen Taylor insisted, and that was best done by
making them accountable to local people. For the next ten
years, she alone of all board members judged every issue by
that test. In July 1877 she was arguing for greater local
autonomy for division committees and school managers,
because unlike the board they would keep 'in close contact
with the parents. It was needful for the board to be instructed
to the utmost by public opinion.' The board reprimanded her
for her demogogy. This did not stop her using her power as
divisional member to appoint double the average number of
working men (though few women) to schools in her patch,
where they helped select staff, checked the registers, poked
about in school lavatories, examined complaints, and reported
on their school's special needs. Whenever the board erected a
new institution—a cookery centre, an industrial school—
Helen Taylor added local school managers, to the irritation of
not a few board members, like Miss Hill, who thought it

49 *SBC* 4 Oct. 1879.
50 H. B. Philpott, *London at School*, 1904, p. 40; *SBC* 1 June 1878.

reflected on their conscientiousness. They missed the point. Helen Taylor brought numerous delegations of working people from her division to the board to raise their concerns and their consciousness, though they were treated by the board with scant courtesy. She took guidance from her open constituency meetings on such issues as corporal punishment. She was parent- and community-centred, but not, it is worth noting, particularly child-centred. Corporal punishment, for example, was for her an assault on parental rights rather than on children's bodies; she hesitated to protect children from family demands for their labour; and it was left to other women members to care for the crippled or abused child.[51] She, Mrs Surr, and Mrs Fenwick Miller used their votes on the school management committee to turn school buildings as far as possible into community resource centres, with books for lending and rooms for evening meetings. Each summer the three women moved that the three hundred school playgrounds remain open and the children off the streets. Each time business men on the works committee found that this presented insuperable obstacles, until even other men on the board complained that they could hardly put playgrounds into glass cases and present them to the British Museum for the summer. They stayed open. In November, 1880 Helen Taylor startled clerical members with a sudden enthusiasm for Sunday school lettings, but her ardour was dampened when Alice Westlake reminded the board of a recent occasion when certain Sunday school children took time off from singing Bible verses to turn on all the taps and flood the school. The trio argued for more generous space standards (schools were built at nine to ten square feet a child, compared to today's fifty square feet); and WCs, benches, and handrails in new buildings were appraised with an eye for their community use.[52]

[51] By the 1880s, schools were grouped in twos and threes under school managers, a fifth of whom were women, though they represented nearly half in Marylebone, but only 10 per cent in Southwark. By 1893, a third were required to be women. Very few were working men. For a fascinating description, see 'Lady Manager: By One of Them', *WH* 7 Jan. 1893 and more generally, P. Gorden, *The Victorian School Manager 1800–1902*, 1974. See also E. Bayley, *The Work of the School Board for London 1888–1891*, 1891, p. 11.

[52] For playgrounds, see *SBC* 28 July, 10 Nov. 1877, 27 July 1878, 20 Jan. 1887. For school buildings, *SBC* 14 Apr., 19 May 1877, 17 July 1880.

When it came to wider issues of board policy, Helen Taylor encouraged Mrs Surr and Mrs Fenwick Miller to identify with working-class parents on issue after issue—against those (mainly male) assistant teachers who wanted the right to inflict corporal punishment; against the by-laws committee when it hounded defaulting parents into court and distrained their scanty possessions; against the industrial schools committee when it endorsed harsh and brutal staff and harsh and brutal regimes for their children; and against those board members who insisted, despite growing evidence to the contrary from provincial boards, that to remit fees was to pauperize the poor.

The first of these issues was corporal punishment. It was officially limited to head teachers. Most women members doubted its usefulness and deplored its use. Said Miss Hill, 'These children are already too used to blows'. Children were beaten for arriving late after doing errands for their parents, for their stupidity as well as for disobedience. Male teachers smarted when the women commented that the cane was a crutch for bad teaching; and complained that the women had no sense of the discipline problems they faced in large classes of reluctant and insolent boys. Alice Westlake could argue that alternative punishments, such as standing on a bench for hours, were more harmful. But however rough the district or unruly the child, few working-class parents would accept that a teacher still in his teens had the right to cane their child. Many a father turned up to thrash the teacher in turn. Miss Hill tried but failed to get a head teacher sacked who allowed assistants to cane children. Mrs Homans was still fighting off efforts by the Tories in the 1890s to cane girls at industrial schools—'no beating with sticks will reach the soul of the child'. When Helen Taylor brought before the board the case of a child allegedly dying from illegal corporal punishment, deputations of parents demanded its ban. When, in addition and in the name of economy Helen Taylor suggested that classroom numbers should rise from sixty to eighty, as any competent teacher could manage a class that size; and that teachers' salaries should be based not on the size of the school but their results, as reflected in the actual attendances of children at school, she made teachers her enemy for life. At subsequent elections, teachers issued scurrilous leaflets and funded hostile candidates in the vain hope of unseating her and discouraging other women soft on discipline. 'Teachers', noted the press, 'are her aversion'.[53]

Meanwhile, Mrs Surr had taken up the question of corporal punishment within industrial schools. The voluntary societies had never reached the starving and brutalized urchins who infested London's streets, and who slept rough under railway arches or in packing boxes. Shaftesbury and Mary Carpenter had from the late 1840s tried to corral the destitute child into ragged schools and redeem the delinquent child in family-style reformatory schools, but Dr Barnardo estimated that in 1876 there were still some 30,000 children sleeping on the streets. For a while the LSB had sent such children away to country reformatory schools, but as numbers grew, it built industrial and truant schools of its own. Mrs Surr drew the board's attention to conditions at one of them, Upton House, the LSB's new and highly unpopular truant school, notorious for its eighteen-inch-wide iron bedsteads, poor diet, scanty clothing, and severe discipline, which, she said, reduced the boys to 'human machines'. When the board would take no action, she turned to her Hackney constituents, called highly charged meetings, plastered the walls of her division with large posters, and when she could still not move the board, wrote to the Home Secretary. He insisted on an enquiry. This found the governor 'over-zealous' in his punishments: he hosed boys down with cold water in an open courtyard in midwinter, made them wear boots deliberately too tight, and caned them viciously. Alice Westlake accused the other women of coaching parents to complain, and Lucraft turned on her—'Such treatment was only borne by the children because they were children of the poor. Members denied the cruelty because they belonged to another class'. The Liberal group foolishly closed ranks and took no action.[54]

Tempers rose as the board faced the 1879 triennial elections. Florence Fenwick Miller was excluded from the

[53] *SBC* 26 Oct. 1895; LSB cuttings file, GLC Record Office, 1896–7, u.d.; *SBC* 20 Jan., 11 Feb., 12 May 1877; 10 Aug., 19 Oct., 2 Nov. 1878; 8 Feb. 1879; 17 Nov. 1877; *South London Press*, 23 June 1883; *EWR* 15 Dec. 1879.

[54] *SBC* 11 May 1878; 1 Feb., 22 Mar., 5 Apr., 26 July, 11 Oct., 1 Nov. 1879.

Liberal ticket in Hackney; in Southwark the Liberals ran another woman against Helen Taylor, a Miss Richardson (a protégée of the Garrett circle), and manipulated the electoral register to keep off the very poor who might be expected to vote radical.[55] Helen Taylor in turn marshalled her Irish 'heavies' to steward her crowded meetings and to break up those of the opposition. Even the tactful *Englishwoman's Review* was moved to observe:

The Liberal party is endeavouring to thwart the popular choice to the advantage of its own political organizations. This had added greatly to the heat of the contest . . . and made the popular determination to have women the more marked . . . [occasioning] that odium and abuse which men assure women are consecrated to politics and with which women should have nothing to do.

Helen Taylor swept to the head of the poll. In all nine women were elected, five of them topping the list.[56]

Miss Henrietta Muller stood for Lambeth, the largest school board division covering a quarter of London. Affluent (her elections cost her £12,000, she reckoned), educated (she spoke six languages), experienced as a school manager and in political controversy (she had watched her possessions being distrained rather than pay taxes when she lacked the vote), she was helped by the Fawcetts and 'the efforts of ladies who assisted her day after day and week after week' to obtain 19,000 votes and head the poll.[57] In Chelsea, Mrs Augusta Webster (daughter of a Chief Constable, wife of a lawyer, and herself an accomplished Greek scholar was well as popular writer) refused to withdraw in favour of the official Liberal male slate of four. Helped by Helen Taylor and by Fulham working men, she too came in head of the poll. The prickliest of the new women members was Edith Simcox, who came

[55] *SBC* 8 Dec. 1877, 28 June 1879.

[56] *EWR* 15 Nov. 1879 *WSJ* 1 Jan. 1880. Helena Downing, an experienced suffrage lecturer, fought Tower Hamlets with Helen Taylor's help but was defeated, so became a poor law guardian. Miss Richardson was a partner in Eliza Orme's law firm.

[57] *EWR* 15 Nov. 1870, *WH* 28 Nov. 1891, *The Times*, 17 Jan. 1906. Under the pseudonym of Helena Temple she went on to own and edit the *WPP* which in time passed to Florence Fenwick Miller. She was the daughter of an émigré German business man, and her sister Eva was a Lambeth guardian who married into the Bright–McClaren network. See below, p. 217; *Ladies Pictorial*, 31 Dec. 1881.

forward in Westminster at the request of the Democratic Club as the working men's candidate. A committed trade-unionist, she devoted her working life to founding co-operative workshops to end sweating in the clothing trades, and her emotional life to her beloved George Eliot. Within a month of her election her diary noted that she was 'wallowing in School Board Reports—if nothing else is in reach one is called upon to be an industrious and well informed member'; and again, 'I have been to a few schools perhaps the members are getting to hate me less—but is all hard work—I am ready to give up', but added 'one can hardly be very useful on a committee when one's very existence is openly resented'.[58]

The last (and longest serving) of the new women members was the middle-aged Rosamund Davenport Hill, daughter of Matthew, with her sisters Florence and Joanna, friend of Mary Carpenter, and an acknowledged expert on the industrial training of unruly girls. On her father's death, she moved from Bristol to London where Alice Westlake and Helen Taylor urged her to stand for the LSB. She chose the unlikely seat of the City, where, without a chairman and with no women voters to canvass, her campaign soon faltered. Elizabeth Garrett and a quickly assembled list of MPs and QCs came to the rescue, and she was returned a creditable second. Subsequent campaigns ran more smoothly—she canvassed by post, rather than in person, avoiding the public meetings she greatly disliked, and after her early contests which cost her £1,500, Rothschild picked up the bills. In some ways she had an enviable seat. With only a small resident population, few working-class families, and just one board school to supervise, she was blessedly free of the very heavy local work-load that other members faced. She could thus devote herself to general board work. In the next thirteen years she did not miss a single board meeting. She immediately took special responsibility for industrial schools, and with Mrs Surr's help turned them around to be a model in the eyes of the HMIs for the rest of the country. The board was grateful. As she never spoke at

[58] Miss Simcox castigated 'Ladies who work among the poor, [who] think it right to save their money for charity and buy cheap costumes, made far off by the same sisterhood . . . [thus becoming] parties to more oppression than the district visiting of a lifetime can atone'. *The Nineteenth Century*, June 1884; MS Diary entries 19 Oct., 9, 23 Nov., 10 Dec. 1879, 9, 19 Feb. 1880. She wrote 'Helen Taylor lies . . .'.

board meetings but quietly knitted away, they were grateful for that too.[59]

The women mostly got their committee preferences, including finance and works, and all of them came onto the school management committee, to Stanley's open dismay. Miss Hill, Miss Simcox, and Miss Richardson joined Mrs Westlake in sturdy support for official Liberal policies, and moved party motions, amendments, and the previous question with considerable agility. The press noted that they 'take a large share in the Board's business'. As the voting figures show, Mrs Webster and Miss Muller also usually voted with the Liberal majority, while often being critical of it; but the original trio of Mrs Surr, Mrs Fenwick Miller, and Miss Taylor did not. They were a third party—even the Revd J. Diggle, leader of the church–Tory group, managed to support more Liberal motions than Helen Taylor did.[60]

Table 1. *The London School Board: Attendances* 1881–2

Miss Davenport Hill	634	Mr Hawkins	180
Mr Mark Wilks	524	Mrs Surr	177
Mrs Westlake	492	Mr E. N. Buxton	174
Hon. E. Lyulph Stanley	439	Mr White	165
Mr Edward Jones	424	Mr Henry Gover	159
Miss Richardson	404	Revd J. J. Coxhead	158
Miss Simcox	356	Miss Taylor	156
Mr Freeman	316	Captain Berkeley	146
Col. Prendergast	310	Mr Charrington	146
Dr Gladstone	303	Mr Roberts	141
Revd J. R. Diggle	282	Mr Richardson	132
Mr Lucraft	278	Dr Wainwright	99
Revd T. D. C. Morse	271	Mr Saunders	96
Mr Stiff	242	Mr Potter	94
Miss Muller	236	Mr Pearce	81
Mr Heller	226	Revd Brymer Belcher	80
Mrs Webster	212	Mr Corry	78
Revd H. D. Pearson	212	Mr Bonnewell	78
Mr Ross	211	Mr Mills	65
Mr Spicer	203	Mrs Fenwick Miller	57
Mr S. C. Buxton	201	Dr Angus	36
Revd G. Murphy	190	Mr Sutton Gover	35
Mr J. J. Jones	187		

[59] *EWR* 15 Dec. 1879; *Young Woman*, 4, 1895, p. 131; E. Metcalfe, *Memoirs of Rosamund Davenport Hill*, 1904. For Florence and Joanna see below pp. 251–63. See also D. Gorham, 'Victorian Reform as a Family Business: the Hill Family', in A. Wohl ed., *The Victorian Family*, 1978.　　　　　[60] *SBC* 15 July 1882.

Table 2. *Voting Record of LSB Members, For and Against the Recommendations of Standing Committees, 1881–2*

	For	Against		For	Against
Mr Wilks	132	4	Revd H. D. Pearson	25	14
Miss Davenport Hill	127	4	Mr Pearce	19	3
Mr Hawkins	119	8	Dr Angus	15	0
Mr Buxton	116	13	Mr S. Gover	2	1
Miss Simcox	116	6	Mr E. Jones	6	115
Mrs Westlake	115	6	Mr Bonnewell	6	105
Hon. E. Lyulph Stanley	98	9	Mr White	23	96
Miss Richardson	97	9	Mr Ross	4	66
Mr Freeman	96	2	Miss Taylor	20	65
Mr Stiff	93	4	Capt. Berkeley	3	58
Mr Lucraft	85	19	Mr Charrington	9	53
Revd G. Murphy	73	3	Mr Richardson	30	47
Mr Spicer	68	4	Revd J. R. Diggle	20	47
Mr S. C. Buxton	65	11	Mr Roberts	31	46
Revd T. D. C. Morse	65	25	Mr Corry	10	42
Dr Gladstone	64	6	Revd J. J. Coxhead	39	39
Mrs Webster	61	9	Mrs Surr	26	39
Mr Saunders	53	12	Dr Wainwright	20	25
Col. Prendergast	53	17	Mrs Fenwick Miller	22	23
Miss Muller	50	16	Mr J. J. Jones	10	19
Mr Heller	48	10	Revd Brymer Belcher	16	17
Mr Potter	34	1	Mr Arthur Mills	5	6
Mr H. Gover	33	9			

Source: Report of the LSB Election Committee, 1882.

The two tables taken together show how commitment, contribution, and committee work interlocked. Those members with the highest attendances also acquired the greater influence, and were the most likely to vote the straight Liberal line as they were voting for recommendations they had helped to shape. But they could put in those attendances and acquire that influence only if they represented divisions that made relatively light demands on them for local work, either because the division was prosperous and had therefore fewer board schools, or because it was well organized by a smooth party machine. Miss Hill could acquire such expertise and esteem that when the LSB swung to the Tories in 1885 they wished her to hold on to her subcommittee chairmanships. Mrs Westlake had behind her the most efficient machine in London. Miss Simcox was rewarded for her hard work in committee and her well-known dislike of Helen

Taylor, when she was also finally admitted to the fringes of the
ruling group around Stanley, Wilks, and the committee chairmen. Mrs Webster, though a natural Liberal, had defied the
machine and stood in Chelsea as an independent; she could
therefore rely only on herself to hold her seat. Henrietta
Muller struggled with a division so vast (it had *grown* by
250,000 since the founding of the board) that it had eventually
to be divided and acquire further members, if the local work
was to be done. Both put in a respectable number of board
meetings and were usually but not always reliably behind the
Liberal platform.

By contrast, Mrs Surr, Mrs Fenwick Miller, and Miss
Taylor represented three of the board's four poorest divisions,
Finsbury, Hackney, and Southwark. (Annie Besant was soon
to represent the fourth, Tower Hamlets.) As Booth and later
surveys showed, their divisions had high levels of overcrowding, deep poverty, and underfed children. One board
inspector reported that in some Finsbury schools, three-
quarters of the children lived in homes of one room.
Consequently, these women carried a very heavy local load.
They had more board and fewer voluntary schools to supervise, more requests for remission of fees to process, more
defaulting parents to interview, and more truanting children
to place, than in Marylebone or the City. It meant a heavy
call on their private purse for boots, clothes, penny dinners. At
the board itself they tried to offset some of the deprivation
within their divisions by calling for free schools, school meals,
and a less punitive attitude to rough working-class parents,
policies that were not calculated to be popular with representatives of more affluent liberalism. Florence Fenwick Miller
noted 'the affinity between women and the Radical interest',
and added 'the work of the Board has been principally carried
out by women with the aid of Mr Lucraft, and they were the
only thoroughly representative section of the working classes
at the Board . . .'.[61] Not only did these women have a heavy
work-load, not only did they not have a party machine to help

---

[61] *Hackney and Kingsland Gazette*, 8 Oct. 1879; see also Honnor Morten, *Question for
Women*, 'The silent, hard, conscientious committee work done by women as opposed
to the blatant mouthing of the men on Board days is most noticeable on large bodies.
It is only at election time that words tell more than works . . .', p. 62.

them, but they had also to defeat the machinations of both political parties back in their division. Yet the more time they invested in constituency work, the less time and energy they had for central board work. As their attendances at committee dropped, so did their contribution and so did their chances of winning support for their views. When at the main board, they tried to overturn committee decisions from which they had been absent, they naturally got short shrift. Locked into a loop of confrontation, rebuff, and defeat, the three women played their last card, appealing in populist style to local electors and the press, irritating their colleagues yet further.

From the board's point of view, they could not have been easy members. They cut corners. They ignored committees where the hard work was done but from which the press was excluded, thus throwing more work on the other members, but made very long speeches moving hopeless amendments at meetings where the press attended, raising items out of order, challenging the chairman's rulings, holding up the agenda to place on it their own crotchets which they had not sorted out in advance, and making wild allegations without evidence (though evidence had often an annoying habit of turning up). Most of the board could not decide whether the trio was more exasperating when they were right or when they were wrong. And the fact that Helen Taylor could always buy her way out of trouble with top-class lawyers when she was wrong did not endear her to them either. A more shrewd ruling group would not have let them become marginalized; would have given them a job of work to do; and would have shown sensitivity rather than sarcasm to the very real pressures of poverty they represented. Eventually Helen Taylor's talents were recognized when she was belatedly given the chair of the education endowments committee at which she laboured heroically, but by then it was too late.

Mrs Surr, meanwhile, continued to probe conditions at other industrial schools, and she and Helen Taylor were now receiving 'leaks' about one of the worst of them, St Paul's, a private church school which received mainly LSB children. When Mrs Surr first raised the matter, she was met with sceptical laughter. The boys then took matters into their own hands and set fire to the school. Helen Taylor financed their defence counsel. Mrs Surr gave evidence; and the staff were '*most severely* reprimanded by the Magistrate', as she reported back to Helen Taylor. The school was owned by the very chairman of the board's industrial school committee, Mr Scrutton, and the board became polarized, some members regarding the women as 'the champions of the outcasts of the metropolis', while Stanley thought they reminded him of the old fairytale of two girls, one with jewels and pearls in her mouth, one with toads and vipers. 'He would leave the board to decide as to which Miss Taylor resembled. (Laughter)'. Under instruction from the Home Secretary, Sir William Harcourt, the board was yet again pushed into an inquiry but they packed it with Scrutton's friends. So Helen Taylor scribbled Mrs Surr a typical note, while waiting at the finance department to check some vouchers—'The committee is a whitewashing committee, and if you can lay the facts before the public in any less laborious or equally effectual way, you may bring pressure to bear on the board better than through such sham Committee'.[62]

The findings were worse than at Upton House. Boys were in chains, on bread-and-water diets for weeks, flogged if they complained. They were so emaciated they looked like 'a lot of images' (ghosts). Boys stole dog scraps to eat; dying boys were whipped for malingering. Rather than face cross-examination, Scrutton agreed to accept local managers, two to be nominated by the board, three to be nominated by Mrs Surr herself. But the Home Secretary again intervened, closed the school down and wrote a personal letter of thanks to Mrs Surr. Helen Taylor, who had accused Scrutton of manslaughter, settled a libel action out of court for £1,000; the judge, however, praised her motives and her actions. The provincial press carried full reports of the story, claiming that this vindicated the work of women on school boards.[63]

[62] *SBC* 19 Mar. 1881; Mrs Surr to H. Taylor, 20 Sept. 1881, H. Taylor to Mrs Surr, 24 Sept. 1881, vol. xxii, fos. 676, 677, Mill/Taylor papers; *SBC* 15 Oct. 1881. When it was rumoured that Helen Taylor might in later years come back to the LSB, Mrs Surr wrote reminding her, 'It is a wearing, wearing life and there is so much friction in it that I really could wish you more restful employment'—but then went on to reminisce about the 'amusement' they got out of 'the many disagreements we went through together'. Mrs Surr to H. Taylor, 17 Nov. 1888, ibid., fo. 680.

[63] *SBC* 22 Oct, 5 Nov., 19 Nov. 1881; *The Times*, 15 Nov. 1881; *SBC* 1 July 1882; *Nottingham Evening Post*, 16 Dec. 1881.

Any institution, out of sight and out of mind, was at risk. Miss Hill learned that Brentwood, where younger boys were detained, was also highly unsatisfactory. She started spot-visiting, and with Mrs Surr's help, quietly sacked all the senior staff. Within a couple of years the HMIs reported that 'an excellent spirit prevails'. By now Miss Hill knew every child, sent them on holiday, paid their medical bills, took them to the zoo, gave them sausage-and-mash parties at her home and was writing to her boys for years thereafter. She was dearly loved. When the school was relocated and rebuilt, it was renamed after her. Helen Taylor with Lucraft's help began to argue for day, rather than boarding, industrial schools, so that children might remain within their own community. Her motives were suspect of course in the eyes of the board, so London's first day industrial school did not open until 1895, years after the pioneering work of provincial boards.[64]

Harder to solve was the continuing and intractable tangle of compulsion, attendance, and fees. If the board excluded children without pence, it damaged the very child that school visitors struggled to bring to school. But if it did not, parents were unpunished since arrears of fees were hard to recover by fine, distraint, or imprisonment. Guardians refused to pay the fees of the very poor, magistrates refused to punish parents for children's absences. What could the board do? They could remit the fees of the individual parent, but given the views of the COS and penny-pinching vestries, this could only be after laborious and much-resented inquisition, and even then probably reflected the mood of the member as much as the poverty of the parent. The board could perhaps remit the fees of an entire neighbourhood by building a free school, but that would confuse innocent with culpable poverty, segregate the school (because as poorer children came in, the better off and more respectable withdrew), and in any case the Education Department was reluctant to give its consent. Or they could press parliament to abandon fees altogether. But if fees were abandoned only for board schools, that would damage voluntary schools; if abandoned for voluntary schools, who would then need to receive grant-in-aid, they would be receiving

[64] Metcalf, op. cit., ch. 5 *passim*; *SBC* 7 Aug. 1880, 4 Feb. 1882, 2 May 1885.

public money without public accountability. In either case, the rates would go up. Advanced Liberals (including Stanley himself) were coming to favour free education; but the LSB surrounded by parsimonious vestries, unfavourable court cases, worried churches, and Helen Taylor, tried to muddle through on a combination of remissions, arrears, and absences.[65]

Helen Taylor brought scheme after scheme for remitting fees before the board. She tried to apply educational endowments not to clever children but to poor children. The board said no. She next brought forward a scheme whereby fees were automatically related to the family wage. With heavy sarcasm 'she apologized for occupying the time of the board with a question so unimportant as the needs of 20 or 30 thousand poor families in London, but she would never scruple to express her opinion of the cruelty of the board in this matter.' The board rejected both her language and her scheme. The following year she asked the board to petition parliament for free schools. They refused. She tried to establish a free school in Tower Hamlets (whose individual members were notoriously mean about remitting fees) but failed, as this time the Education Department would not agree. Slowly she began to win the argument, if not the votes, with members conceding that fees stood in the way of attendances. One male member 'regretted to find his friends the clergy on the Board going against Miss Taylor's motion because it was moved by Miss Taylor . . . He was not going to be frightened from voting on the right side of this particular question because Miss Taylor or anyone else was in favour of free education . . .'.[66]

By now, the next triennial elections of autumn 1882 were approaching, and board meetings became increasingly unruly. Mrs Webster proposed to retire on grounds of ill health, Miss Simcox on grounds of overwork, but Mrs Surr refused to stand again because she stated that all decisions were being stitched up by an official Ring (comprising leading

[65] Anon. (H. Muller?), 'Some Aspects of the LSB', *Westminster Review*, Jan. 1888, pp. 704–8.

[66] For endowments (which H. Taylor thought could remit a third of all fees) see *SBC* 28 June 1879, 23 June 1883; family remission, *SBC* 19 Mar. 1881, 22 Apr., 13, 22 July, 5 Aug. 1882. London remitted five per cent of its fees, Manchester 15 per cent, Birmingham nearer 30 per cent. For free schools, *SBC* 8 Apr., 15 July 1882.

Liberals, committee chairmen, Misses Hill and Richardson, and Mrs Westlake). The opposition came from some Tories, some radicals, and the three women members. Mrs Fenwick Miller happily joined in. The main decisions on the board, she asserted, were not church against chapel, school board against voluntarism, or education against anti-education, but the Ring against the Rest. Elizabeth Garrett inevitably sprang to the board's defence.

> She said they had heard about what was called 'The Ring' in the board. She never was in the Ring herself, as she had not the time for the hard work in committees, but those who were thus spoken of were those who gave most time and attention to the work in committees and elsewhere. No one could get into the Ring who was not prepared to give much valuable time to hard work and patiently to deal with the matters that were before the committees.[67]

Southwark as usual experienced the rowdiest campaign and highest poll. Anonymous leaflets (probably financed by teachers) urged residents not to vote for Helen Taylor. She retaliated in a rather unpleasant little incident in which she accused Miss Richardson of corruptly bestowing a works contract on a manufacturer who was paying for her campaign. Miss Richardson was able to show that the contract had been awarded before she came on to the committee. Many of the Liberals slipped down the polls, and only Helen Taylor came home head of the poll. Miss Hastings, a charity worker, total abstainer, and Tory daughter of an Army baronet, came in third in Tower Hamlets. The *School Board Chronicle* calculated that the Liberals had been reduced to thirty-two of the fifty-three seats, counting all the women with the majority.

Not surprisingly, the first trial of strength came when Diggle, the Tory leader, moved, and Helen Taylor seconded, that committee chairmen be limited to two years tenure. The board split 25:25, and only the casting vote of the chairman saved the Ring. Most of the women stayed with their former committees. Four women joined a new special committee of the board for higher-grade schools which marked the board's renewed interest in extending elementary education upwards,

[67] *Reports of the London School Board Election Committee*, 1882. (The committee included the Garrett Andersons, Westlakes, Sir John Lubbock, W. E. Forster.) *SBC* 11 Nov. 1882.

to offer, in Huxley's phrase, a ladder of opportunity from gutter to university. Mrs Fenwick Miller protested that the sharp-elbowed middle classes would appropriate any resources for higher elementary education, at the expense of the poorest children in the most deprived areas. In practice, of course, more middle-class users of the board school system would probably generate larger rate resources and political goodwill: and slowly the board developed a more generous and comprehensive educational philosophy. As in the process it undermined existing endowed secondary schools, overlapped with the responsibilities of the newly formed LCC, and failed to take out the insurance of amending legislation, the LSB was in time to find itself in the quagmire of the Cockerton judgement and the ultimate dissolution of school boards. Meanwhile, complaints mounted about the extravagance of the board, the cruelty of its industrial schools, and the activities of the Ring. Support ebbed away, and in the 1885 triennial elections the Liberals lost control of the London School Board.

The Tories and Annie Besant: 1885–1897

The 1885 elections wrought destruction. In Chelsea, Mrs Webster returned to the board; Miss Muller was defeated in Lambeth, the austere Miss Hastings in Tower Hamlets. Mrs Surr and Miss Richardson had stood down, Florence Fenwick Miller with two children, three books, and a financially inept husband in hand, turned full-time to lecturing and journalism. Helen Taylor, who with Mrs Surr had done much to undermine public confidence in LSB Liberals, also refused to stand as she preferred to try the women's cause as a parliamentary candidate in Camberwell. The biggest shock was in Marylebone. Here the school board Four had held together and avoided asking for plumpers as increasingly worried Liberals were doing elsewhere. Alice Westlake scraped home, the others including Stanley, the Liberal leader, were defeated. Only Miss Hill achieved anything like a personal triumph in the City where her work with industrial schools was held to have reduced delinquency and the rates. She came in second. The School Board Election Committee, headed by the Garrett Andersons and aided by Mundella, did their best to stem the rout but with a large number of candidates pursuing a large number of plumpers, only twenty-four of former board members, and three of the eight women standing, were elected. The *School Board Chronicle* wrote in an editorial: 'We have a London School Board which upon the whole represents reaction, sectarian interests, and a spirit of antagonism, to the development of national education. It is a Board unworthy of London and unworthy of the work'.[68]

Alice Westlake and the Revd Mark Wilks became leaders of the Liberal opposition, defending past decisions, seeking to demolish Tory budgets, and striving to protect levels of service. When the new Tory finance chairman produced his first budget, Mrs Westlake left it in shreds. They could not 'bus' children from Tower Hamlets to empty places in Westminster, she said, freeze teachers' salaries, peg the child population, or raise the fees. To cheers from her side she told the new Tory bench that they had a lot to learn, as she demolished their 'businessmen's budgets' in powerful and effective speeches.[69] She had in some ways the hardest job in politics, preventing cuts to services that had seemed safe, at a time when the political vocabulary of the day had swung against her. She faced a demoralizing, depressing, and arduous task. She naturally became more radical herself in the process, coming to favour free schools for example; but the alliance of voluntarism, Anglicanism, and vestry economy rolled on. Temperamentally more suited to administration than opposition, she stood down in 1888.[70]

This time the former School Board Election Committee of the Garrett Andersons widened its base to include the ingredients of a progressive alliance, Radicals, Liberals, trades councils, Nonconformists, temperance advocates, vigilance workers, educational reform leaguers, and a secretary in Sidney Webb. The Committee was bidding for the new LCC as well as the LSB. Although Miss Hill was safely returned, Mrs Webster lost her seat, and the Labour candidate Mrs Amie Hicks and an admirable guardian, Mrs Evans, both

---

[68] *EWR* 14 Nov. 1885; *SBC* 10, 17, 31 Oct. 1885, 23 Jan. 1886.

[69] *SBC* 13 Feb. 1886, 18 Feb. 1888.

[70] *SBC* 13 Mar., 29 May, 16 Oct., 13 Nov. 1886, 30 June 1888. Her division marked her retirement with goblet and cups of repoussé work. She moved to Cambridge where her husband took up a chair. *SBC* 30 Mar. 1889.

failed to make much headway. The Tories held on with a
reduced majority. Three strong women candidates did win
seats. Mrs Maitland, mother of six with considerable experience as a school manager and Liberal party worker, took over
the Garrett/Westlake seat in Marylebone: and devoted herself
to handicapped children. Miss Muller's Lambeth seat was
won by the young widowed Mrs Ashton Dilke, active in local
Liberal, radical, free-thought, and suffrage groups, whose
blithe and breezy style was reminiscent of the young Elizabeth
Garrett.[71] But the star of the election was undoubtedly the
charismatic Annie Besant, standing in Tower Hamlets. She
had been the heroine of free thought and birth control
controversies, was now a Socialist, and during the summer of
1888 had inspired the match girls' strike to a dramatic conclusion. Like Elizabeth Garrett and Helen Taylor, she was
already a national figure looking for an elected public
platform, choosing to stand for the LSB rather than the LCC
with all its legal difficulties, on a platform of free, secular,
popularly controlled schools, and free meals. Mothers' Unions
and clergy picketed her crowded meetings, claiming that any
woman declared unfit to bring up her own child was not fit to
educate others. She responded defiantly, 'I am told that I
should be more likely to win with a programme less advanced,
and that I frighten people by declaring myself to be a socialist.
The timid and frightened people had better not vote for me—
that is all I can say'.[72] But they did—16,000 of them, taking
her to the head of the poll. Her supporters paid her modest
bills of under £100, as well as a salary of £150 a year.

There was now a good handful of members representing the
Labour interest,[73] who were often able to attract advanced
Liberal support; and over the next three years under Annie
Besant's leadership, they scored some notable victories.

[71] *WPP* 23 Aug. 1890, 10 Nov. 1888.

[72] *WPP* 17 Nov. 1888; *Link* 29 Sept. 1888; and more generally, D. Rubinstein, 'Annie
Besant and the LSB Elections of 1888', *East London Papers*, 1970, on which this paragraph draws. She had been denied custody of her children on separating from her
husband, because of her immoral Malthusian views. She thought free school meals
could be financed from City charities. *National Reformer*, 25 Nov. 1888; *East London
Observer*, 24 Nov. 1888.

[73] They included the Revd Stewart Headlam in Hackney, A. G. Cooke of the
London Compositors, Mr Conybeare the Lib.-Lab. Finsbury MP, the ageing
Benjamin Lucraft, Mrs Dilke, and Annie Besant herself.

The first of these was the fair wages clause. It began innocently enough when Mr Cooke asked whether a firm winning a contract paid trade-union rates. Nobody knew. Annie Besant was insistent: 'They wanted now to lay down the principle that the Board should only deal with those houses which dealt fairly with their men. In spending public money they should be reasonably economical, but there was no economy in employing firms who sweated their employees, for then they got bad workmanship'. They carried advanced Liberals and many Tories with them; Stanley and Miss Hill joined the Tory leader, the Revd J. Diggle, in the minority. The *School Board Chronicle* noted in surprised but approving tones, 'The sweating question has become in a manner everyone's business', public bodies should be as high-minded as private individuals. By February 1889, board advertisements carried a fair wages clause, permitting, said Annie Besant, 'decent firms . . . to tender with some chance of success'. The new LCC promptly followed suit, drawing up a list of approved contractors from whom tenders might be sought. When Eyre and Spottiswoode tendered for the printing of scripture cards, Annie Besant and Mrs Dilke, who both knew about the newspaper and printing business, were scathing about the firm's pay scales: 'There is something especially revolting in cheap labour at work on cheap bibles'. Served with a writ in the middle of a board meeting, the women solemnly asked the board to consider whether girls who received low wages from folding Bibles might be forced into prostitution. The firm prudently backed down.[74]

The second issue to concern Labour members was inevitably the matter of fees and attendances. At her divisional meetings Annie Besant was distressed to find that she had to choose between the welfare of the child and the welfare of the family, when asked to determine whether a child might leave school underage. 'A girl, nearly 13, has a father who has earned £3 in the last 18 months; mother averages 2s. a week at washing and charring; a son of 14 earns 5s; the girl is wanted—only too

[74] *SBC* 8 Dec., 15 Dec. 1888; *National Reformer*, 16 Dec. 1888, 19 May, 8 Dec. 1889, 8 Feb. 1890. Following the great building strikes of 1892, the Tories tried to dilute this commitment. Progressives argued that Tories 'professed to educate London's children at the cost of London's parents', G. Wellas, *The Case against Diggleism at the London School Board*, pp. 103–5, 132.

obviously—to "assist and support family"... Another's mother earns 8*s.* a week, father out of work, rent 5/6*d.* . . . :. it makes one's heart sick'. A couple of weeks later she described in her *National Reformer* one of her Notice B meetings in the Isle of Dogs.

Notice B means 'Why does not your child come to School? Come and explain'. And they come—gaunt, hunger-pinched men and women, all, but one, decent folk who did not want to keep their children ignorant, but sometimes there were no boots, sometimes there was a baby, sometimes there was no food, sometimes there was sickness. Ah me! how patiently these people suffer, and why are they so patient? No summonses were ordered at those meetings, and I think, I hope, some went away feeling that the administrators of the law were friends, not foes. But it is heart-breaking work'.

Quoting the Cross Commission, she mobilized party support to stop children being excluded from school while their fees were being pursued.[75] The board agreed they would consider alternative schemes.

As the Tory chairman of the by-laws committee would not produce one, Annie Besant presented a package of her own: fees would be remitted automatically on grounds of poverty, parents need not attend in person to seek their remission, and summons would be issued only by the by-laws committee itself, and not by local divisional committees. She negotiated every item through the board. The result, she boasted in the *National Reformer*, was that fees were now virtually voluntary. Children were not excluded from school if they did not pay, and fees could not be extracted from parents if they would not pay. A motion brought by a Unitarian minister for free education was narrowly defeated in the board. However, late in 1889 Annie Besant brought forward her own motion, that the board should petition parliament for free board schools. She reminded members that it was incumbent on the LSB to give a lead as the matter was now on the parliamentary agenda, and suggested that as most of the cost of schooling was already publicly financed, removing the last pence would not prove 'very fatal to morality' any more than board

[75] *National Reformer*, 30 Dec. 1888; 20 Jan. 1889; *SBC* 15 Dec. 1888; *National Reformer*, 17 Mar. 1889, *SBC* 9 Mar. 1889; *National Reformer*, 21 Apr. 1889.

members were themselves pauperized when they enjoyed free parks and street lights. The only question was whether parents paid for schooling 'privately, or united and paid them as a community'.

Even those who disagreed with her, admired her able and 'temperate' speech. The debate continued into February as amendments, and amendments to amendments were taken in due order. The amended motion was finally carried by 24 votes to 16, to great applause. 'Most of the Progressives set to work wondering how they had done it', Mrs Besant told her readers. 'The wrath of the *Times* and the *Morning Post* is the most eloquent testimony to the value of the victory'. By now, almost a quarter of London's children were having fees remitted; within the year Government offered boards an additional grant in lieu of fees, and all London board schools were free.[76]

Annie Besant had caught the tide on the free schools issue; but more difficult was the problem to which she next turned, the poor health and stunted physique of many London children, the malnourishment of poverty. One strand of the problem was the need for physical exercise and since Elizabeth Garrett's days women members had pressed for sufficient playgrounds, more generous space standards, drill, swimming, and games. Another strand was inadequate clothing, especially boots. In some schools women teachers were making the children's clothes themselves. When Mr Peek offered the Charity Organisation Society £1,000 a year to spend on boots in the mid 1870s, the COS had proved so niggardly, that kindly women members took matters into their own hands, and ran their own boot club.[77] They tried to protect older children, especially girls, from over-pressure. But the most intractable problem was malnutrition. As Sydney Buxton wrote in a letter to *The Times*: over-pressure, was generally 'only another word for underfeeding'. With compulsory attendance, school boards were 'year by year tapping a lower stratum of society, bringing to light the distress, destitution and underfeeding which formerly had escaped their notice'. Most women members and many male

[76] *SBC* 18 May 1889; *National Reformer*, 9 Feb. 1890; *SBC* 30 Nov. 1889, 1 Feb. 1890.
[77] J. Hurt, 'Drill Discipline and the Elementary School Ethos', in P. McCann, ed., *Popular Education and Socialization in the Nineteenth Century*, 1977.

members dug deep into their pockets and their leisure to feed hungry children. As the depression deepened, dozens of feeding charities sprang up to provide breakfasts and dinners. Women noted how quickly children put on weight when regularly fed by industrial schools, and that crippled children given decent meals could be integrated back into ordinary board schools.

It was immediately obvious to Annie Besant, as it had been to Helen Taylor, that you cannot educate a hungry child. In her division there were seven thousand of them, and she immediately wrote to the press seeking funds. Within a fortnight she was feeding three thousand children with soup, bread, and jam. 'This sort of thing does nothing to solve the problem of poverty', she accepted, 'but all these hundreds of children will be warmed and filled and that is always something'. The school management committee reported in August 1889 that a third of the children in Finsbury, Hackney, Southwark, and Tower Hamlets were having their fees remitted on grounds of poverty; and that some 44,000 children (or 13 per cent) were underfed. Less than half of these were being helped by charities. The rest went hungry. The board now co-ordinated existing agencies in the London Schools Dinner Association—'through whose meshes we hope no hungry baby will slip', said Annie Besant—and over the winter provided 250,000 free or cheap meals. Further board surveys in 1894 and 1898 revealed a similar picture, with 31,000 children underfed in 1899; and in 1900 the LSB set up a permanent co-ordinating committee to supplement voluntary efforts.[78]

Food, healthier buildings, and exercise (together with slowly rising real wages) were gradually reflected in the children's energy and intelligence. But most London board school children remained a reservoir of ill health, suffering from constant colds, ringworm, vermin, ophthalmia, sores, and impetigo. Measles, diphtheria, TB, and respiratory diseases (all of them killers when accompanied by malnutrition) were the more common since many board schools would

[78] *The Times*, 16 Sept. 1884; *National Reformer*, 20, 27 Jan., 3 Feb. 1889; *SBC* 26 Jan., 3 Feb., 24 Aug. 1889; *National Reformer*, 8 Dec. 1889, 29 June 1890, *Final Report of the LSB*, p. 323.

not close for fear of losing some grant. Perhaps a third of London's school children had defective hearing, a half had defective vision, and almost all had bad teeth. Girls in the poorest homes had the poorest health. London appointed its first part-time Medical Officer in 1890 but not until the LSB captured the vigorous Dr Kerr from Bradford as a full-time MO in 1902, and not until fears of racial degeneration seemed to be confirmed by the poor health of recruits in the Boer War, did demand build for school medical and meal services.[79]

In a multitude of other ways, Annie Besant pressed forward Helen Taylor's causes. She sought to make parents school managers, open buildings for public use, increase the pay of caretakers, and make the school board accountable to local opinion.[80] Her final successes were perhaps symbolic but none the less impressive. When a member of the public wrote to complain that other than working-class children were attending board schools, and costing the ratepayers money, Annie Besant drafted a reply which the board approved: that all children were equally eligible to attend board schools and that it was desirable that they should. Slowly elementary schools were becoming common schools. Her finest speech was made on behalf of a Mr Moss, a school visitor and free-thought lecturer. The by-laws committee required him not to lecture in his own time against the very religious teaching that he enrolled children to attend. In a passionate plea for religious liberty, Annie Besant persuaded a Tory board, dominated by clergy, to tolerate him.[81]

How did she do it? After all Helen Taylor held many of the same views, and had Mill's name and her own impeccable private life going for her; and yet could record few successes. Indeed, the causes she adopted, such as free education, she thereby defeated; and she stiffened Liberal resistance to even modest schemes of fee remission. Annie Besant joined a Tory board as a secularist, neo-Malthusian, trade-unionist, and with a disastrous marital life behind her, yet managed success after success. Essentially, she built alliances: a Tory–Labour

[79] *Final Report of the LSB*, pp. 325 ff.; see also F. B. Smith, *The People's Health 1830–1910*, 1979, pp. 178 ff. School MOs were only allowed to inspect, not to treat, children.
[80] *National Reformer*, 17 Feb., 7 Apr. 1889.
[81] *National Reformer*, 3, 17 Mar. 1889.

alliance on the issue of sweating to squeeze orthodox political economy Liberals; and a progressive alliance on the issue of free schools. Her efforts were the more impressive because she chalked up many of her victories on the works committee, traditional male territory. It was also a matter of style. Her speeches (and she was, after all, one of the most eloquent orators in London) radiated a sunniness, moral generosity, and sincerity that was clearly infectious, and which brought out the best in otherwise sour and dour colleagues. She had all of Helen Taylor's intellectual independence, but none of her hectoring, waspish, or abrasive way. She was supportive to fellow members, Tories included, quick to give credit and gracefully to accept her allies where she could find and persuade them. She always tried to move things forward. No doctrinaire, she abandoned her own motions to support more modest but more winnable ones, and never sacrificed the good for the best. Stanley, who loathed Helen Taylor's high-minded righteousness, was charmed by her, while many old-fashioned clergy who suspected a she-devil were obviously bowled over. She remarked in her autobiography that she 'largely conquered public prejudice against me by my work on the London School Board' and she was right.[82]

But she was also fortunate in her times. She would have had a tougher task had Stanley and the Liberal Ring been still in power, sensitive to criticism and reluctant to adopt policies not of their own deliberation. Helen Taylor might have been more successful had she come in with a newly elected Liberal administration; if the Liberals had been in opposition; or if she had early been recruited into responsibility. As it was, there was no space for her. By contrast, an incandescent Annie Besant was welcomed as an outrider by the formal Liberal opposition who were themselves becoming increasingly radicalized by the wider political and parliamentary scene, as reflected in the Newcastle programme of 1891, by the depression which broke COS orthodoxies with impunity, and by the cumulative effect of survey, investigation, and inquiry.

The 1891 triennial elections approached, and Annie Besant, increasingly meditative and theosophical, did not stand. But

[82] *Autobiography*, 1908 edn., p. 342. (However, she still found herself in legal actions with clergy outside the board, and never won over the Revd J. Diggle.)

she was involved in one last scrap, pianos on the rates. The pianos were for drill and class singing in school halls (not, as Huxley remarked back in 1871, to accompany the harp, sackbut, or presbytery) but amid charges that labourers' children would be playing the Moonlight Sonata, the Tories managed to blame the Progressives for them and held on to their majority. Mrs Dilke had retired, Mrs Maitland was defeated; but Miss Hill was joined by Miss Eve, a teacher and Liberal party worker in Finsbury, and by Mrs Ruth Homan, an ardent Progressive, who headed the poll in Annie Besant's Tower Hamlets. The widowed daughter of a former Lord Mayor (Sir Sydney Waterlow), Mrs Homan was an experienced school manager dispensing free dinners and eight hundred pairs of free second-hand boots a year; and had taken cookery and nursing lessons to equip her for supervising girls' education. When the piano debate reopened, Mrs Homan was to argue for a more generous curriculum: 'people were only now beginning to understand that education did not only mean the teaching of dry facts, but making the children something better than they were before they entered the schools'. Miss Lidgett, a poor law guardian, made the same point at a women's NUWW conference in 1891: 'We seem to be training clerks and mechanics. We want men and women . . . Some little portion of our teaching ought to be reserved for enlarging the mind, for awakening the imagination, for learning who and what we are . . .'.[83]

Also elected was a young high-church lawyer on the radical right, who was to throw the board into turmoil, Athelstan Riley. Back in 1871 the LSB had hammered out a compromise formula for religious teaching in schools—simple non-sectarian Bible teaching only—which was copied by most provincial boards. Even Helen Taylor and Annie Besant, avowed secularists, had hesitated to reopen the issue. Clerical members on Christmas school visits, however, found that children were being taught that the father of Jesus was Joseph.

[83] For the piano debate, *SBC* 11 Oct. 1890, T. Gautrey, *Lux Mihi Laus*, 1937, pp. 108–9, cf. Huxley, *SBC* 24 June 1871, *WH* 5 Sept., 14 Nov. 1891, *The Young Woman*, 1895, pp. 129 ff. For Miss Eve, *WH* 5 Dec., 1891. *London*, 24 Sept. 1894, 4 Nov. 1897, LSB cuttings file, GLC Record Office. Mrs Homan, *SBC* 5 Mar. 1892; Miss Lidgett, NUWW Conference Report, Liverpool, 1891. Many provincial boards had pianos—Norwich had had a harmonium in one school since 1882.

A nativity without divinity. They demanded that the Trinity be taught as well. Nonconformists, Liberals, and teachers rallied to the 'No-Popery, No Theological Tests' banner; and the matter monopolized board meetings for months. Despite the fact that the Moderates were starving poorer schools of resources, that schools were more and more understaffed and overcrowded, the 1894 election was fought almost entirely on the religious issue.[84] With a record poll of 50 per cent and amid virulent debate, the Progressives achieved virtual parity on the board; the moderates began to fall apart; and the whole controversy began to suggest that cumulative voting and multi-member divisions produced childish sectarian boards. Local authorities might be a more sane and equable body to deliver education.

As ever, the real work of the board was borne by its committees. The four women between them sat on four committees and twenty-seven subcommittees, putting in between two hundred and two hundred and fifty meetings each year (perhaps triple the load of most modern councillors). Miss Hill was in and out of her local schools, introducing them to phonetic reading schemes, Froebel methods, and practical social economy. She went off to Sweden to study the hand-and-eye training, and then paid for the materials herself when there seemed a risk of surcharge. She built up Alice Westlake's cookery centres from twenty to one hundred and sixty, and added fifty laundry centres, all under her personal supervision. The industrial schools committee placed two thousand children a year in six London and sixty-six country schools and she read the case notes of every child, the official diaries of every officer, and spent her holidays visiting every school. In addition she wrote articles and gave evidence to government commissions, starting work on her correspondence at 6 a.m. and finishing her reports at midnight. Mrs Homan, to take

---

[84] The debates are covered in the *SBC* for 1893. For women-led deputations, see *SBC* 21 July 1894 (the women to their fury were kept waiting 4½ hours). More generally, see J. E. B. Munson, 'The LSB Election of 1894', *British Journal of Education Studies*, 1975; Gautrey, op. cit., ch. 7; (G. Wallas), *The Case Against Diggleism*, 1894. Given the fears of anti-suffragists that women were obsessed with religious issues, their reticence is perhaps worth noting. At the elections, seven Progressive and three Socialist women stood; the three existing women members and Mrs Maitland were elected.

another example, was by 1902 vice-chairman of the industrial schools committee, snatching every break in board meetings to make random visits. When Miss Hill retired, Mrs Homan personally supervised all the housewifery teaching in London schools, seeking to make it relevant and attractive to girls and their families. She played a major part in developing schooling for handicapped children; and in addition had two industrial schools and fourteen board schools under her charge, to which she nominated teachers, ancillary staff, and resources. Few men, let alone women, had such responsibility, power, and patronage.[85]

The work-load was awesome. Was it necessary? Not unfriendly critics who sat on the LCC believed that a broken window could not be replaced without LSB members' approval. Policy-making, then as now, shaded into administration. Members, especially the women with leisure, handled functions (remitting fees, checking attendances) that would be later delegated to officers. The result was perhaps more sensitive but also less consistent than would be tolerated today. Members also expected to scrutinize and supervise their officers every step of the way—checking bills and receipts in much the same way as they checked the cook's accounts at home. Miss Hill, and not a senior officer, personally fired bad staff. But this says as much about the officers as it does about members. Local government in general, and school boards (which were not fully within the local government system) in particular, did not possess the battery of internal audit, management services, establishment departments, and performance review, which would in later years monitor and co-ordinate service delivery. Until the very end of school boards, senior staff were often part-time, retaining their private work in law, in medicine, in engineering. As the Metropolitan Board of Works found, corruption could be rife; as the vestries knew, nepotism was prevalent. Staff and teachers embezzled, bullied and drank; and only members' vigilance stood between them and the children. Compared to later local government service, members were of higher quality

---

[85] Metcalfe, *Memoirs of R. D. Hill*; R. D. Hill, 'Technical Education in Board Schools', *Contemporary Review*, May 1888; *Women Members on the LSB*, WLGS pamphlet 1896; *SBC* 4 Oct. 1902.

and officers of lower quality—in social, occupational, and educational terms—than would now be found. Professional administration, in education as in other fields, took time to develop.

Members were not only quasi-professional administrators; they were often more experienced and expert educationalists than the staff. Members, such as Stanley and Wilks and especially the women, would, with the help of an inspector, initiate pilot schemes of kindergarten, Slodj hand-and-eye training, Swedish exercises, cookery, technical drawing, temperance teaching, infant crèches, swimming classes, the tonic sol-fa system, new reading schemes, all drawn from their experience of other boards, other voluntary organizations, other countries—and personally select staff and send them away for special training. These staff would then teach others and the member would withdraw to a broader supervisorial role. As bureaucratic structures developed with self-sustaining routines and as additional inspectors, superintendents, and in time special advisers, were appointed, so the input of members could fall. Given the vastness of London, the size of its divisions, and the fewness of its members, Helen Taylor was right to argue that either London should have eleven boards, or two hundred members on the LSB, or substantial devolution to local committees. In the meanwhile the memories of the industrial schools scandals had burnt deep. Women who were devoted to the welfare and well-being of children willingly took on a work-load which in the phrase of Lord Reay, a board chairman, 'No paid official could be asked to undertake . . . and no Trades Union would allow'. When Lord George Hamilton, the LSB's first external chairman, retired to become secretary of state for India, he commented gracefully,

As this is the first time that I have ever had the pleasure of being associated with ladies in an administrative capacity, I should like to say that there is no part of the work which is more effectively performed, where authority is better maintained, and the amount of work done in the time consumed is greater than in those committees upon which the ladies serve. (Cheers).

Women *London* concluded, were 'the most conscientious,

hardworking members on the board. This school board work is their business'.[86]

Progressive Revival and the End of the School Boards: 1897–1904

In 1897 control finally swung back to the Progressives. Their election manifesto promised free, efficient, and properly resourced day and evening schools, maintenance of the 1871 religious settlement, and a fair labour policy of trade-union rates and a direct labour organization. Miss Eve and Mrs Homan headed their divisions. Elsewhere, women were returned in divisions with experience of women members. Miss Constance Elder, ex-Girton and founder of a Southwark women's settlement, was elected in Westminster. Mrs Mary Bridges Adams, former head teacher and now warden of a women's settlement, was returned as a socialist for Greenwich (and Woolwich) by Will Thorne and the Royal Arsenal Co-operative. Miss Honnor Morten arrived on the LSB as a professional nurse, journalist, and COS worker, though she claimed that her school and settlement experience in Hoxton made her 'socialistic'. She was head of the poll in Hackney; but smoking in Fleet Street for a dare, she was seen by a strait-laced chapel-goer and diplomatically transferred to Southwark instead. Miss McKee took over Miss Hill's City seat. And the first declared Tory woman was elected to the board, Mrs Dibdin, wife of a solicitor, active in parish charities and the Primrose League, and a progressive in everything except official politics. Graham Wallas and the Revd Headlam became committee chairmen, the Revd J. Diggle lost his seat.[87]

The Progressives moved speedily to remedy years of Tory neglect, building new schools, raising salaries, starting new higher grade schools, improving pupil teacher training—and closing one of their two truant schools for lack of customers. By now the LSB had 15,000 teachers; and its forty-one

[86] Lord Reay, *SBC* 4 Oct. 1902; Hamilton, *SBC* 3 Aug. 1895; *London*, 20 Sept. 1894 p. 601.
[87] The Progressive programme was virtually the same as Helen Taylor's, twenty years before. *London*, 14 Nov. 1894, 14 Oct. 1897, *Municipal Journal*, 17 Mar. 1899; *WS* 17 Jan. 1895 (Miss Elder); *London*, 4 Nov. 1897, *Dictionary Labour Biography* (Mrs Adams); *WPP* 29 Nov. 1890, *London*, 11 Nov. 1897, Gautrey, op. cit., p. 79 (Miss Morten).

committees were educating 550,000 children by day and 109,000 at night.

In particular, they expanded two areas of work. The first was evening classes. In the 1870s, evening schools had served as continuation classes for early school leavers, but as day schools became more efficient and children remained longer at school, they had fallen in disuse. In 1885 they were revived by the Recreative Evening Schools Association offering liberal studies and lantern slides to some 10,000 people a year. The Progressives now promoted free secondary school classes of an evening; and about 109,000 (15 per cent) of London's fourteen to twenty-one-year-olds were thought to attend. Women members were supportive but not prominent in this work.[88]

The other major development, however, of special schools and special care for the handicapped, was women-led. From the early 1870s, the LSB had offered remedial and specialist teaching for blind and deaf children. The Royal Commission on the Blind and Deaf (1885–9) paved the way for the Act of 1893 which formally placed all handicapped children under the care of school boards rather than poor law guardians. Mrs Maitland and Mrs Homan found blind children foster homes near their school centres, broadened their curriculum to include music and typing as well as basketwork, and sent the most able on to training colleges.

The blind and deaf had elaborate charitable networks to help them, but mentally and physically handicapped children were neglected. London was thought to have some five thousand mentally defective children—'the last stage in the gradual deterioration of a once vigorous stock', said one commentator—damaged by genetic disorder, parental VD, birth injury, or infant abuse. At the request of the board, a gifted head teacher, Mrs Burgwin, started two special schools with small classes of twenty. By 1903 London had seventy schools for feebleminded children. National efficiency and eugenicist worries had given the matter added salience.[89] Women members helped her obtain resources from the board.

The LSB's Medical Officer suggested that physically

---

[88] Bayley, *The Work of the School Board for London*, p. 13; *Final Report of the LSB*, pp. 278–9; Gautrey, op. cit. p. 165.

[89] Philpott, *London at School*, p. 278.

handicapped children should attend these special schools for the retarded, but women board members would have none of it. Mrs Humphrey Ward, doing a private survey of her own, had found young crippled children locked into rooms for fourteen hours a day: a lad 'whose back had been injured by an accident, alone all day after his discharge from hospital, feebly dragging himself about his room, in cold weather, to find a few sticks for fire, with tears running down his cheeks from pain. His father and sister were at work, and he had no mother'. She offered the board spare rooms for these children in her Passmore Edwards Settlement if the board would provide staff. The women board members and Graham Wallas persuaded the LSB to agree. Miss McKee and Mrs Burgwin designed the equipment—cane wheelchairs, couches, a go-cart for the playground. The Invalid Children's Aid Association brought the children to school; and with good food, rest, and care, 'these maimed and fragile creatures began to expand and unfold like leaves in the sun'. The Hon. Maude Lawrence, daughter of the board's first chairman, who took over Miss Elder's seat, made these children her special care. She taught them, served their meals, scoured London looking for new school sites, and started an after-care service for them when they left school and sought jobs. By now eleven thousand handicapped children were under the board's tutelage, mainly in residential homes. Newer women members, like Honnor Morten and Mrs Adams, urged that children wherever possible should be integrated into ordinary schools, as they needed families in which to flourish as well as special skills with which to survive.[90]

The LSB was one among several educational bodies with concurrent and conflicting responsibilities. While the LSB had been Tory-controlled and lethargic, an uneasy peace prevailed; but as the Progressives rapidly expanded their work in the later 1890s, they dismayed the voluntary schools who were losing children to more attractive and better resourced board schools, they annoyed the private endowed and grammar schools who saw London's forty-four higher grade

[90] C. Trevelyan, *The Life of Mrs. Humphrey Ward*, 1923, pp. 139–40; *SBC* 12 Nov. 1898; *Final Report of the LSB*, pp. 177 ff.

schools for older and more able children as trespassing on their territory, and they undercut the LCC's technical, trade, and evening work with evening continuation classes of their own. Vestry ratepayers complained about soaring school board rates. None of this work had been sanctioned—or anticipated—by the 1870 Elementary Education Act, and the boards had failed to protect themselves by taking out amending legislation. With a supportive government they enjoyed the flexibility to pioneer. With a hostile government they were vulnerable to legal challenge, disallowance, and surcharge, the more so because LSB accounts were audited not by the Education Department but by the Poor Law audit section within the Local Government Board. The auditors cared not whether expenditure was desirable, or even whether it had been approved by the Education Department, but whether it was sanctioned by statute. The notorious Cockerton Judgement of 1899 found that much LSB expenditure was unsanctioned and illegal. School boards should confine themselves to a narrow interpretation of elementary education. They were not entitled to offer a ladder from gutter to university.[91]

The Conservatives had won the general election of 1895 (and were to win the Khaki election of 1900). Faced with an aggressively radical board with which they were in constant friction, the government showed no disposition to throw out a lifeline and bring the law into harmony with school board practice to safeguard their work. The London school board therefore faced deep political hostility from central government. But it also engendered doubt among committed educationalists who wanted a 'co-ordinated and connected' educational system, and were not at all sure that building it upwards from working-class elementary schools was the most effective and generous way forward. As Beatrice Webb (admittedly not a neutral observer) put it, 'We don't believe you can raise the standard of elementary education and save it

---

[91] For a fuller account of the Cockerton judgement and the end of the school boards, see E. Eaglesham, *From School Board to Local Authority*, 1956; T. Taylor, 'The Cockerton case revisited', *British Journal of Education Studies*, Oct. 1982; D. Pugh, 'The Destruction of the English School Boards', *Paedogogica Historica*, 12,1972; P. Gosden, 'The Origins of Cooption to Membership of Local Education Committees', *British Journal of Education Studies*, Oct. 1977. It seems plausible that the LSB was 'set up' by its critics, including the government.

from mere mechanical efficiency unless you have the University in organic connection with it'. The schools faced a further source of challenge from the town and county councils, already responsible for technical and large areas of secondary education. School boards had been devised in 1870 precisely because the countryside possessed no rural local government worth the name. Since 1889 it had possessed county councils. Many local government men like Chamberlain were worried that rural education had remained rudimentary in small single-school boards where the vicar held sway. They were contemptuous of the Athelstan Riley débâcle, and disdainful of the way the cumulative vote returned antivivisectionists, temperance fanatics, and other single-issue 'faddists'. They were anxious that elementary and secondary education should develop under one authority—the borough—so that elected councillors might make proper choices between services and needs, and not merely levy the school board precept. Just as health had passed from boards and commissioners in the 1840s, and utilities from private boards in the 1870s to town halls, so it now seemed politically expedient, educationally desirable, and administratively appropriate for education to come into the mainstream of local government. A Birmingham Quaker banker wrote to a friend,

The Plan, borrowed from America, of electing specialists by popular vote to manage special work, is *always* a bad one, whether in railways, hospitals or schools or any other department of work. The talkers and extremists are sure to be elected that way. The specialists should be chosen by a superior authority representing Common Sense and they should report to and obtain sanction from that central authority.

Gorst, the Tory Education Minister, did not disagree. 'I have never regarded it as possible that School Boards could be a permanent institution—like Boards of Guardians they are a modern anomaly in local government.'[92] He introduced

---

[92] B. Webb, *Our Partnership*, 1948, p. 261; Correspondence of G. Lloyd, 1896, quoted in Gordon, *The Victorian School Manager*, p. 206; Sir J. Gorst, 1900, quoted in P. Gosden, *The Development of Educational Administration in England and Wales* p. 173. See also evidence of the Revd Jos. Wood, Minutes of Evidence on SB Elections 1884–5, vol. xi, Q. 5046. Gorst's first bill, privileging voluntary schools, was so mauled by Dissent that he dropped it.

legislation to abolish school boards and place education under counties and county boroughs. Education committees would be required to co-opt specialist members.[93] Nonconformist hostility was allayed by excluding doctrinal religious teaching from the new LEA schools. Mrs Maitland raised at conference 'one point' which defenders of school boards always overlooked—women would be 'cut off from the possibility of election on these bodies, and education would be carried on entirely by men'. The WLGS, fearing that women's educational work was at risk and that their poor law work might soon follow, lobbied extensively. Balfour, himself a suffrage supporter, offered to allow women to be co-opted as specialists on LEAs, but the WLGS promptly pointed out that of 128 county and county borough technical education committees, only fourteen had so far managed to appoint women. Alice Westlake came out of retirement to argue that co-option was no substitute for direct election. Miss Marion Townsend, both on Bristol's school board and a co-opted member on Bristol city council's technical instruction committee, explained the difference.

On the School Board I am on an exact equality with every other member—man or woman; there is no difference between one member and another, save what they make themselves, whether they are more or *less* hardworking, more or *less* the educationalists. My power of voting money or of opposing such a vote is the same as that of the men members. If I initiate a scheme, I have a fair prospect of carrying it through and seeing the result of my work. There is a sense of security about my position; I know that after each election I have three years of uninterrupted work before me. There is also a sense of responsibility which is most wholesome. The parents and the ratepayers have a right to tell me what they consider best for their children; I have a duty to listen. If I endeavour to carry out their wishes, if they are satisfied with my work, I expect them in return to support me at the next election. Altogether, women thus publicly elected, and thus responsible to the public, have a fair field for permanent, useful, solid work. . . . The position for a co-opted member is *essentially* different. On the Technical Instruction Committee I have not the sense of equal right, the sense of security, or the feeling of responsibility that I have on the School Board . . . If I go against the opinions and prejudices of other members, I know

[93] *SBC* 23 May 1896; WLGS pamphlet, *Women and Technical Education*, 1897.

that at the next yearly appointment such members can refuse to re-co-opt me.[94]

Even Stanley, who had made life so difficult for so many women on the LSB was heard to agree that it would be 'a great loss' if women were excluded from educational policy-making.

The Women's Local Government Society fought long and hard against the bill, only to find that many of its senior women members (Emily Davies, Lady Aberdeen, and Mrs Fawcett among them) approved of the principles of the bill, and were willing to waive women's claims for the greater good. Unable to defeat the bill, the WLGS now tried to ensure that women would be statutorily co-opted to the new LEAs. As Lady Laura Ridding put it to a meeting of the National Union of Women Workers:

The obvious danger of women being excluded from sitting on these local educational committees if their presence there is not made compulsory, is the unfortunate fact that the money vote and money rate will cause all municipal and county rating authorities to wish to monopolize the representatives on these committees. All the political wirepullers, all the electioneering agents, will want to grasp the new power; the party spirit, Conservative and Radical will spin its spider webs round each seat on the board, and unenfranchised women will, by an obvious process of exclusion, be squeezed out.[95]

In response to their pressure, the government agreed to require LEAs to co-opt women. The women's network did their best to ensure that women were allocated a reasonable number of seats. Mrs Elmy commented bitterly that Lancashire proposed to appoint only two women to seventy men, and were it not for statutory co-option, 'scarcely any of the Councils would have recognized the existence of women', and even so, she expected that '*men* will only appoint women who will say "ditto" to themselves'. By April 1903 she was reporting to Mrs M'Ilquham that of the first 109 schemes submitted, two-thirds proposed to co-opt two women, half of the rest only one woman, and the remainder up to five. Very

---

[94] *SBC* 15 Feb., 20 Apr. 1902, WLGS Minutes, 11 Sept. 1902.
[95] *School Government Chronicle*, 24 Jan. 1903; Lady Laura Ridding, NUWW Conference, Brighton, 1900.

many of the newly co-opted women moved straight across
from their local school boards; others came on to represent
denominational and higher education; sometimes they were
deemed to represent professional associations. Within the year
Mrs Elmy was conceding that in many places 'no woman had,
till the passing of the Act, any official power in the direction of
education work, nor would have now but for the compulsory
provision of the Act—so that the way is being prepared for
women's activity whenever they are made eligible for town
and county councils—and thus out of evil good may come . . .
Several times as many women must now be serving under co-
option as were ever returned to the schools boards . . .'.[96] Few
boards appointed more women than they had to. They had
other pressures to appease. In 1902 there were thought to be
370 women elected to school boards, but their incidence was
very various. They formed 12 per cent of the membership of
some boards, and were entirely absent from others. In 1907
there were 615 co-opted women among 8,212 LEA members
(7 per cent of the total). No board now lacked a woman, but
women were seldom as co-opted members able to make a
mark. One exception was Mrs Hume Pinsent in Birmingham,
another was Margaret Ashton, who chaired Withington's
education committee—but she had been elected and not co-
opted to Withington UDC.[97]

London's educational reorganization followed a year later.
The WLGS fought a last-ditch campaign to argue that
London was a special case and should retain a directly elected
educational authority, but the Education Department would
only concede them 'an honoured and large place' in school
work as co-opted members. It was clear that almost all men
thought this was a highly satisfactory outcome.[98] As co-opted

[96] Mrs Elmy to Mrs M'Ilquham, 27 Jan. 1903, Add. Ms 47453, fo. 98, Lady Day
1903, fo. 111, 1 Feb. 1904, fo. 228.

[97] *EWR* 15 July 1907. LEAs co-opted only 69 more women than their statutory
minimum, WLGS Minutes, 26 July 1906. Often women lost seats in rural areas—
Norfolk had six to a dozen women on its small rural boards in the 1890s, but only two
women sat on Norfolk's LEA after 1903. For Mrs Pinsent and Miss Ashton, see
below.

[98] According to the Webbs, the Progressives and Labour wanted to keep the
LSB, the Tories wanted education to pass to the boroughs, the Webbs personally
wanted it to go to the LCC; as that turned out to be everyone's second choice,
that was the outcome. They were forgiven by the school board women; *Our*

members, women would offer service without being able to claim any rights. They would be deferential, not assertive. They would remain dependent on men's goodwill to get anything done. Honnor Morten complained, 'I, for one, don't want to be abolished; I prefer to stand for the poor people of Southwark'. And Mrs Hughes of Oxford noted that if women co-opted members 'did anything not quite pleasant to the men they could be dismissed. What they wanted was women with the popular vote behind them'.[99] That was no longer possible. Co-opted women members continued to offer loyal and conscientious service to children with special needs, to older girls, to women staff. But they no longer were elected by, accountable to, or speaking for working-class communities, could no longer show a sensitivity to neighbourhood needs. To some extent that responsibility had passed to the newly elected Labour members arriving on town councils. Women who had come out in public on school boards were being returned to an essentially philanthropic role within education.

Partnership, pp. 256 ff., 293. 21 LSB members were co-opted to the LEA—Stanley pointedly was not one of them.

[99] H. Morten to Portsmouth branch of the NUWW, *Occasional Papers of the NUWW*, May 1902, p. 19; Mrs Hughes, *Oxford Chronicle*, 25 Oct. 1907.

3

English Provincial School Boards 1870–1903

In May 1881, members of Wolverhampton school board met to fill a vacancy on the death of a member. The chairman suggested that they should co-opt a woman to help with girls' schools and industrial schools. 'The Board needed strengthening', he said, to criticism of his proposal from clerical members. Miss Amy Mandor, Cambridge-educated daughter of a Liberal manufacturer and councillor, joined the board.[1]

The great cities and many of the older boroughs had, like London, quickly formed school boards during the winter of 1870, and their triennial elections occurred within a few weeks of each other to form education's general elections.[2] Nine women were among the first to be elected. At Bath, the Quaker niece of John Bright, Miss Anne Ashworth, active in suffrage circles, and the daughter of a Methodist councillor, Miss Caroline Shum, whose work with ragged schools gave them 'considerable knowledge of the poorer classes', were pressed to stand. Ladies vigorously canvassed for their 'sister candidates' and they were duly elected.[3] Thereafter they worked quietly and without publicity. The fiercely evangelical Miss Catherine Ricketts stood in Brighton; Miss Lydia Becker, parliamentary secretary of the women's suffrage movement, was elected in Manchester; Mrs Huth, founder of a female educational institute, came home unopposed in Huddersfield. Miss Eleanor Smith, sister of an Oxford don, came on to a school board that was determined to have no board schools; and Miss Temple was elected by one vote at Exeter, leaning literally as well as figuratively on the arm of her brother, the bishop. Two women joined small rural boards

---

[1] *SBC* 4 June 1881.

[2] Boards were formed by the town council, on petition of ratepayers, or imposed by the Education Dept. where educational provision was sparse.

[3] *WSJ* 1 Mar. 1871.

in Cornwall and Essex in 1872. 1873 saw two women on Birmingham and Leeds boards.

At each triennial election, their numbers grew; twenty-four by 1876, forty-one by 1879. By 1881, when Miss Mandor joined the Wolverhampton board, women held seats in Bristol, Birmingham, Bridgwater, Grimsby, Hastings, Huddersfield, Ipswich, Leeds, Rochester, Tavistock, and Worcester, as well as on some thirty-eight small rural boards; and they had come on and off boards in Brighton, Bath, Exeter, and Northampton. They were soon to find seats in Sheffield (1882), Nottingham (1883), Bradford (1885), and Norwich (1887). They were well represented in the home counties, in Norfolk, and in Cornwall. During the 1880s nearly ninety women were serving at any one time. Not until the 1890s, however, did Lancashire cotton towns (Salford, Oldham, Bolton) and the great ports and dockyards (Plymouth, Portsmouth, Southampton, Hull, and Newcastle) place women on school boards. By 1903 some 220 women were elected school board members, sitting on almost every large urban board and on dozens of rural ones.[4] Over the years, some five hundred women in all had been elected; single women to begin with, but increasingly joined by married women, around three hundred of them, especially in rural areas. At each election about half of the women (and the men) retired, more so in the towns, though there were others with records like that of Miss Kate Spiller of Bridgwater, who was first elected in 1876, chaired her board in 1895, and was co-opted with an unbroken record of service on to Somerset's LEA in 1903.

Women's experiences of school board work was inevitably very various. At one end were the great fifteen-member[5] city boards of Manchester, Leeds, Birmingham, and Liverpool which had child populations of over 50,000, where boards were highly politicized and the work-load not dissimilar to the LSB, demanding two or three meetings a week as well as considerable local work. Perhaps more typical were the

[4] e.g. Paignton SB in 1895 had three women among its seven members; its chairman and vice-chairman were both men.

[5] The size of school boards was determined by population—five members up to 5,000 pop., seven up to 15,000 pop., nine up to 40,000 pop., 11 up to 70,000 pop., 13 up to 100,000 pop. and 15 for cities over 100,000 pop. Three-quarters of all school boards had only five members.

smaller borough boards, such as York and Yarmouth, Oxford, Southport, and Wolverhampton, with perhaps nine members, who were schooling a few thousand children, who met fortnightly, and where long-standing church–chapel rivalries for many years inhibited the growth of overt party loyalties on their boards. Different again was the experience of women on small five-member rural boards, often based on a single parish with a single church school, chaired by the vicar and meeting only quarterly. Such parishes could sometimes only draw their women from one or two families. Mrs Broadhurst of Foster and Scropton, in Derbyshire, took over her husband's seat (and chairmanship) on his death; when she died, the four other farmers co-opted her brother, and on his resignation, his wife. Sometimes husband and wife or father and daughter served together, an indication, one suspects, not of nepotism but of desperation. Even single women without property could make a contribution disproportionate to their numbers, as they were often rather effective at intimidating parsimonious farmers. In 1883 Miss Aldrich of Diss in Norfolk, for example, stopped the farmers lowering the school leaving age in order to acquire cheap and docile labour. Mrs Christie of Westleigh in Devon became chairman of her board and found herself paying for the school's books and chalks. Other women introduced feeding schemes in their villages that were to be copied by larger boards. For the most part, however, the work of women on rural boards went unreported and is almost impossible to recover.[6]

School board women in the towns were usually the staunchly Liberal relatives of prominent Liberal men, invariably suffragist, mostly Nonconformist, and at least nominally teetotal. They were drawn from the town's social and civic establishment. Mrs Talbot, Citizen Sarah as she was known locally, was a member of Kidderminster's leading carpet family; her husband was a solicitor and councillor. Mrs Manfield of the Northampton shoe family was the daughter of the county's surveyor, and had two councillor sons; with her husband, she built the Unitarian chapel and school.[7] Mrs

---

[6] *SBC* 6 Oct. 1883; R. R. Sellman, *Devon Village Schools in the 19th Century*, 1967.

[7] *Sunday Mercury*, 14 May 1972; *Kidderminster Shuttle*, 21 Jan. 1972; *Northampton Daily Reporter*, 13 July 1899.

Wycliffe Wilson had married into the great foundry family of Sheffield, whose men dominated the city's industrial, municipal, and parliamentary scene; she was to become perhaps the most effective member of Sheffield's board. In Darlington, the women of the Pease and Fry families went on to the board while their men went on to the council. Very often, elected women were the wives, sisters, and daughters of solicitors, doctors, Nonconformist ministers, and newspaper editors. Miss Florence Balgarnie was the sister of a Scarborough Nonconformist minister and on the national lecturing circuit for the suffrage movement. Miss Harriet Grimwade was the daughter of Ipswich's town clerk, and having defied the Liberal town caucus to stand as an independent, she earned their forgiveness by introducing cookery teaching throughout the town. In York, Miss Wilkinson's brother was a solicitor and together they brought the university extension movement to their city. Miss Lucy Hunt of Croydon and Miss Charlotte Avery of Barnstaple were daughters of newspaper editors. Miss Avery's father was also mayor of the town, and a leading Methodist; she was so outraged by the Bulgarian atrocities that she penned some of his most devastating articles, becoming a 'keen politician' in her own right. She later became the town's first woman poor law guardian, and no one was 'better known or better loved'.[8]

Only rarely were school board women in waged work— having run private schools of their own (Miss Elizabeth Connell of Gateshead, Miss Annie Shepherd of Middlesborough, Miss Maria Findlay of Stockton), or written books of their own (like Marianne Hearn of Northampton and Catherine Buckton of Leeds). Their fathers, brothers, and husbands were to be seen and sometimes heard supporting them on the platform. They were inevitably middle- and upper-middle-class women, with a sprinkling of the titled in rural areas. They had confidence, education, leisure. Like LSB women, they were increasingly university-educated, fluent in several languages, and able to finance private study visits abroad.

In the later 1890s, a handful of women came out of the

[8] *Wings*, 31 Nov. 1883; *Yorkshire Herald*, 6, 10 Mar. 1911; *North Devon Journal*, 30 Apr., 7 May 1903.

Lancashire and Yorkshire textile unions and the ILP, such as Sarah Reddish of Bolton. Margaret McMillan, the most outstanding of the ILP's school board members at Bradford, had recently arrived from London as their paid lecturer. Mary Jane Dixon who served on Keighley school board from 1896, was manager of a warehouse, a Methodist lay preacher and ethical socialist, who married an Anglican socialist clergyman. But ILP, trade-union, and co-operative efforts were devoted rather more to poor law than to school board work; and though the token working man was regularly included on Liberal school slates, the working-class woman seldom was.[9]

With a few exceptions, school board women were active Liberals—'Of course I am a Liberal', wrote Marianne Hearn of Northampton in her autobiography while standing as an Independent—and in the mid 1880s they were often founding members of their local women's Liberal association. In return they acquired party workers and a party machinery which allowed them to negotiate with the local male Liberal caucus for a place on the slate. Such Liberalism was a natural extension of their Nonconformity and their evangelical faith. For the Unitarian Guilford sisters of Nottingham and the Quaker Sturge daughters of Bristol, the chapel and the meeting house were at the core of their political and social commitment.[10]

Even where women were not Liberals, it seemed to make little difference to their school board work. Mrs MacCallan of Nottingham, Miss Bignold of Norwich, and Mrs Evans of Rochester were three of the dozen or so women elected on the Tory ticket to school boards, but they were devoted to the well-being of board schools, and refused to support church claims or favour voluntary schools in board debates. Mrs Hancock was one of the Conservative six elected at Sunderland, and 'the best man among them' according to the local press.[11] She supported women's suffrage, and non-denomina-

---

[9] For Sarah Reddish, see *Bolton Journal*, 18 Feb. 1913, 24 Feb. 1928. I owe the information on Mary Dixon to David James, Bradford City Archivist. Liberals and Labour spoke a common language in education; but in poor law work they disagreed on the scale of outdoor relief so Labour tried to strengthen its presence there.

[10] M. Farningham (Hearn), *A Working Woman's Life*, 1907, p. 278. For school board women who founded WLAs, see e.g. Miss Sterling in Falmouth, Mrs Evans in Leicester, Miss Mandor in Wolverhampton, Miss Wilkinson in York, Mrs Spoor in West Hartlepool, Mrs Leach in Yarmouth, Mrs Byles in Bradford, Mrs Connon in Leeds, Miss Ryley in Southport.   [11] *Sunderland Herald*, 7 Jan. 1892.

tional teaching. She opposed false economy. She stood, she said, 'because this was a woman's age'. Why she stood as a Conservative is less clear. Lydia Becker was herself a churchwoman as was Catherine Ricketts of Brighton, but both were marked Liberals. Given that Liberal women candidates acquired votes from Tory women, the Church–Moderate–Conservative party often tried to find a woman candidate in the 1890s to balance its slate, along with the inevitable working man; but such women were reluctant to come forward, often refused to canvass or speak on public platforms in the name of propriety, and once elected could not be trusted to observe party lines. The service ethic of education was often stronger for them than formal party loyalties.

Liberal and Conservative women alike usually had extensive school experience before standing for the board. They had run Sunday schools, served as school managers, sponsored girls' clubs, written school textbooks, lectured to local societies on temperance, sanitation, physiology, and domestic economy, and were the local hon. secs. of Emily Davies's higher education movement, entering girls for public examinations. Many were beginning to organize university extension lectures for young women teachers. Some were raising funds for women's university colleges. Week in, week out, they had walked the streets of their town, and as the benevolent wife of a great employer, as the parish visitor of the local church, as a voluntary worker for a relief society, they had, in phrases that occur time and again, 'a great acquaintance among the poor of the city', an 'intimate knowledge of their needs'. On the board, they could speak of the town's social geography, the streets, the sites, the buildings, the traffic, the families, and the homes in a way few men could rival, and all of them came to respect. Take Miss Anne Davies, known as 'the little lady', an uncompromising Welsh Calvinist co-opted to Liverpool's board in 1879 (and subsequently head of its poll). She worked in Sunday schools, bands of hope, and temperance missions; was a manager of the local training college; and highly active in the Children's Country Holiday Fund, the Ladies Sanitary Association, the Home for Welsh Servants, the RSPCA, and the NSPCC. The *Porcupine* devoted its obituary on her to praising the usefulness of 'old maids'.[12] Many long-serving city board women sustained a similar network of educational, religious, and philanthropic organizations which allowed them to map their city and speak to its needs. Accordingly, women found themselves representing not just the women's cause, not just the Liberal interest, but also the voice of working men's associations, families, and neighbourhoods. As Miss Ricketts said, 'If she represented any constituency, it was the working part of the population'. Mrs Marion Huth was persuaded to stand for Huddersfield in 1871 by her working men's association just as many LSB women were, and they ensured that she was returned unopposed. The lone working man and the lone woman seconded and supported each other's motions; would try together to resite schools to meet neighbourhood need, fight for more generous remission of fees, remind the board of underfed and over-pressured children. For some women, like Mrs Byles of Bradford, it was a painful duty. Others, like Mrs Evans of Leicester, revelled in the cut and thrust of board work and 'loved school board meetings as other women love hunting or bridge. There was an exhilaration about the meetings of those early years, for principles had to be discovered and methods worked out.'[13]

Isolated though some felt at first, women members developed networks through which to pool experience and from which to draw support. They turned to the women's and the educational press, particularly the *School Board Chronicle*, which faithfully reported not just the LSB but provincial boards as well; and which found its way even to small rural parishes. During the 1870s the suffrage movement, in the 1880s the newly formed women's Liberal associations, and in the 1890s the National Union of Women Workers, brought women together in educational debate, ran them as candidates, and supported them in their campaigns. Behind them was the informal network of the women's lecture circuit (suffrage, sanitation, temperance) bringing leaflets, gossip, and messages of support from the interlocked cousinages. Far

---

[12] *Liverpool Daily Post*, 25 July 1898; *Liverpool Mercury*, 25, 26 July 1898; *Porcupine*, 30 July 1898.

[13] *Brighton Guardian*, 29 Mar. 1871; *Records of Nineteenth Century Leicestershire*, 1935, pp. 106 ff.

more than councillors today, school board women travelled: to visit their extended family and friends, to attend christenings, marriages, and funerals, and to take extensive holidays; and invariably they took the occasion to visit local school boards, cookery centres and industrial schools and to study kindergarten methods. Far more than male colleagues on the board, women members brought a breadth of experience and perspective to educational work. They had the time, the money, and the interest, and made good use of all three.

Over the years, women's work on school boards changed. The women who early came on to school boards before party lines hardened were often highly individualistic, unmarried, uncomfortable, educated, and formidable. Not always as talented as Helen Taylor, they often provoked similar reactions among male colleagues. They stood as independents, able in resort and cathedral towns to count on a municipal electorate a quarter of whom were female and loyal to them. Such women soon took their boards in hand, organizing the educational census to estimate the local shortfall of school places, structuring the curriculum, and operating across the whole field of school board work. Miss Ricketts of Brighton, Mrs Buckton of Leeds, Mrs Leach of Yarmouth, and Miss Ryley of Southport found themselves the toughest and most sophisticated members of the board, effectively becoming party leaders, planning election strategy as well as initiating board policy. Lydia Becker of Manchester and Eleanor Smith of Oxford were only inhibited by the fact that they were in permanent opposition which crippled somewhat the contribution they could make. But when these women withdrew from school boards, they were often not followed by other women, unless they had built up a machine and a local association to fill in behind them. They were as loathed as they were loved, and when they departed, board minutes sometimes reveal a collective sigh of relief.

By the time a town went to the polls to elect its second or third school board in the late 1870s, party lines had usually hardened. Instead of a host of individual temperance, liberal, chapel, and working men's candidates, each searching for plumpers at the others' expense, the party caucus now ran a disciplined slate (often running a bare majority of candidates, such as eight for a fifteen-member board), on a common manifesto and with a common machinery. Voters were asked to confine their votes to certain candidates, perhaps casting all their votes for one person, or five for three. The cumulative vote allowed women, and women's Liberal associations, to insist that at least one seat pass down the female line. Sometimes the caucus resisted and was humiliated as at Oxford, Ipswich, Bradford, and Swindon, when women not only displaced the official candidate but even came in head of the poll. Where women were incorporated, they became the most loyal and diligent members of the party portfolio. Miss Emily Sturge of Bristol, Mrs Cowen of Nottingham, Mrs Wilson of Sheffield, were admirable party women whose abilities catapulted them into responsibility. And there were dozens of other quiet, hard-working, well-informed, loyal Liberal women who on committee made a reality of the Liberal party's election promises. But the stronger the caucus and the more prominent the leading Liberal men, the more modest and retiring appear the women. Miss Kenrick and Miss Sturge of Birmingham were highly competent, but rendered virtually voiceless and invisible at board meetings dominated by Chamberlain, George Dixon, and the Revd Dale. In Norwich, too, where the political scene was controlled by George White, the shoe manufacturer, the Liberal women board members were relegated to a minor supportive role, though they performed impressively at women's meetings of their own. If women were critical of the party line, they were marginalized; if they were supportive, they were silent, but allowed to work hard behind the scenes at unreported committees. It was a repressive style of tolerance.

By the early 1890s, board work was becoming increasingly specialized even on the smaller boards, and the style and nature of women's work changed again. Early school boards, like the LSB before them, had seen their problems in essentially quantitive terms, the numbers of children to be taught, schools to be built, teachers to be hired, grants to be earned. Only when theological controversy intruded into debates on the school curriculum, or when scandals were uncovered in industrial schools, did members reflect more widely on their role. But over the years, children were disaggregated, so to speak, and seen to contain special needs and differing groups. During the 1870s, the need of young children for kindergarten methods was identified, then that of older girls for a relevant curriculum. Delinquent and truant children were another distinct group, as were, in the North, the half-timers and early leavers who needed compensatory evening education. During the later 1880s and the depression years, the needs of the underfed, and then of the physically and mentally handicapped child, were recognized. While men managed the accounts, erected the buildings, awarded the contracts, and intermittently argued theology, women members became acknowledged experts on kindergarten and on domestic economy; on industrial schools for damaged children, and special schools for delicate and 'defective' children. They pressed for school feeding, physical exercise, adequate playgrounds, and swimming-baths, all designed to improve child health. In their fields they became policy-makers, holding seminars, attending conferences, writing learned papers, giving evidence to committees of inquiry, and bringing back to their boards best practices from other towns, other countries. They brought all the breadth of their philanthropic training, all the courage of their suffrage work, all the detailed knowledge of their city and its streets, all the hard-won familiarity with codes, circulars and by-laws, and all the warmth they valued in their own family lives, to the educational welfare of the working-class child. Mrs Evans of Leicester persuaded her board to start kindergarten in the 1880s by reminding them that it was only by 'conventional and industrial necessity that children are not making daisy chains'. Children needed sunshine in their schools and their lives.[14]

It is to the contribution of leading women on English provincial boards that this chapter now turns. To Manchester, where Lydia Becker did solid work as a Progressive on a Conservative board, making good the claims for women to a public life; to Brighton, where the campaigning Catherine Ricketts effectively led the Liberal group; to Oxford where Eleanor Smith tried in vain to get her clerical school board to build a board school; and to Birmingham, where Eliza Sturge and Caroline Kenrick were relegated to 'tokenist' roles by the

[14] *Records of Nineteenth Century of Leicestershire*, p. 106.

caucus. In Leeds Mrs Buckton and in Bristol Emily Sturge made policy; in Yarmouth, Mrs Ethel Leach propelled her board through twenty-five years of educational development in five years. In Sheffield, Mrs Wilson was pressed by her family and in Nottingham Mrs Cowen by her chapel into substantial school board work. At Birkdale, near Southport, Miss Kate Ryley was at the centre of acrimonious feuding as she tried to bring wider Liberal and educational principles to bear on a small insular board and a self-important NUT. Meanwhile, in Norwich, the NUT ran Mrs Pillow as a candidate in the teachers' interest. The final study is of Bradford where Edith Lupton, Mrs Byles, and finally Margaret McMillan forced themselves on a hostile Liberal caucus and a hostile school board, and to varying degrees carved themselves a role.[15]

Manchester

Manchester, horrified by a parliamentary survey of its schools in 1869, was the first town council to appoint a school board. As the town's politics had become since the mid 1850s anti-Gladstonian, anti-Irish, and protectionist, its school boards were always to be conservative and sectarian. But Manchester was also heir to an older educational tradition of statistical investigation and inquiry, which stretched back to the days of Cobden and the Anti-Corn Law League. The result was an enlightened Conservative board under an impressive chairman, Birley, who soon acquired a national reputation. Its members were progressive in most educational matters, generous in their sympathies with children, and conservative and sectarian only in their sensitivity to the fortunes of the voluntary schools and the teaching of the Trinity. The board was proud to pioneer voluntary feeding schemes for malnourished children as early as 1873, higher grade schools, day industrial schools, pupil teacher centres, evening schools, and (inspired by Mary Dendy) special schools for the mentally handicapped, in all cases years ahead of London.[16]

---

[15] The studies therefore include seven of the ten largest towns, but not Newcastle and Hull where women's contribution was both late and negligible, nor Liverpool where Miss Davies served as a co-opted member without election until 1897. The other five studies are of women remarkable in their own right.

[16] This account is drawn from Manchester school board reports; and C. Dolton, 'The Manchester School Board', M. Ed. thesis, 1969.

Lydia Becker was in 1870 one of the successful candidates. In her early fifties she had become effectively the national organizer of the women's suffrage movement and her *Women's Suffrage Journal* the most important of the women's papers. Herself an unsectarian Liberal churchwoman, she had the support of the Cobden–Bright–McClaren network. Jacob Bright had been briefed by her to enfranchise women ratepayers in 1869. Like Elizabeth Garrett and Emily Davies she had stood for the school board to stake out women's claims to public and political life. Over the years, she was to acquire an influence second only to Birley himself. By 1877 she was laying foundation stones, by 1879 admirers sent in enough money for her to endow a scholarship bearing her name. To her delight it was won by a girl. Her influence lay in her quiet, unthreatening ladylike style, hard work, attention to detail, and reliability. She refused to shun the unpleasant, but equally refused to score points off male colleagues who over the years recognized that she 'was perhaps better known to the world at large than any other member of the board'.[17]

Like all school boards, Manchester's first task was to conduct its educational census, and by virtue of the strength of the voluntary school system found an immediate shortfall of some eight thousand places, rather less than in most great cities. However, children attended only irregularly and the child population was outpacing provision. By the 1890s Manchester had 90,000 children to be schooled. Miss Becker was one of the subcommittee which planned school provision, organization, and curriculum. She tried but failed to have older girls taught alongside boys, to the moral gain of the boys and the educational gain of the girls; and voted with the minority to exclude all religious teaching from schools as she wanted no part in 'religion making'.[18] These issues aside, the board agreed harmoniously to receive existing schools, build new ones, and aid denominational ones, and to adopt a liberal curriculum.

Like the LSB the Manchester school board had next to turn to the tangle of attendance, compulsion, and fees. Manchester's voluntary education societies had long paid the fees of

[17] Lydia Becker papers, Fawcett Collection; *WSJ* 1 Dec. 1879;
[18] *SBC* 9, 30 Dec. 1871.

poorer children, and this the new board continued. Lydia Becker devised their scheme of remission, based on the concept of a family wage. Manchester was soon spending more on remitting fees than the rest of the country put together. Chamberlain at Birmingham denounced them, Helen Taylor of London applauded them. It was not clear which was the more embarrassing.[19] Harder to solve was the matter of attendance, since Manchester had many children in half-time work in the mills, many other Irish children hawking in the streets, and yet had passed by-laws requiring thirteen-year-olds to attend school. (London initially allowed children to leave school at eleven.) Miss Becker noted sadly that 'the duty of parents to provide for their children of tender years is not generally recognized by the very poor; and the contrary practice of making them work for their bread is lamentably frequent'. But she also appreciated that the welfare of the child was inextricably linked with the welfare of the family, who might need its domestic labour or modest earnings if they were to survive. At her attendance committees, widowed fathers, Irish hawkers, poor German tailors, blind mothers, and tubercular parents pleaded with members to let their children leave school to earn a few shillings. She struggled to find helpful suggestions—morning-only lessons, evening classes, different domestic arrangements—and tried to stop poor parents being hounded by the courts, but often confessed herself baffled as to what should be done for the best.[20]

Like women members everywhere, she took the slum child to heart. She petitioned the town council not to salt the roads, as it hurt barefoot children. She checked and tasted school meals. She worried about industrial schools, and when Mrs Surr's revelation of vicious brutality in certain London reformatories hit the Manchester press, she redoubled her visits, seeking to send Irish children who were hawking and begging to ragged schools instead.[21] Above all, she fought for the right of girls to education. They needed geometry and science, just as she would have the boys taught cooking and sewing. At

[19] *SBC* 2 Nov. 1872. From 1876 guardians were supposed to be responsible for the remission of fees.

[20] *WSJ* 1 Aug. 1873.

[21] *EWR* 15 Oct. 1890; *Fifth Triennial School Board Report*, 1885.

Mechanics Institutes, Co-operative meetings, and speech days she begged parents not to short-change their daughters.[22]

She died in 1890, one of the last two members of the original board; and *in memoriam*, a Conservative sectarian board co-opted a succession of Liberal women to take her place, a tribute that Miss Becker would have wanted. Rachel Scott, one of the first Girtonians and wife of the editor of the *Manchester Guardian*, held the seat until 1896, when it was passed to Mary Dendy, sister of Helen Bosanquet and a powerful voice in the national eugenics movement.

Mary Dendy had seen simple-minded children hanging back in the playground on her school visits. She started to make their welfare her special care, personally inspecting forty thousand of Manchester's children and finding five hundred mentally defective among them. She persuaded the board to follow the good practice of Bristol and Nottingham by building special schools. She realized that the bigger problem came when mentally defective children left school, to return to the streets, procreation, and prostitution. She started a resident colony, Sandlebridge, where, strictly segregated, they learnt to read (though not to write, in case boys and girls communicated with each other), count, and practise simple crafts, 'children all their lives, but happy harmless children, instead of degraded and dangerous ones'. She went on to found the Lancashire and Cheshire Society for the Permanent Care of the Feeble-minded, for 'the weakest link in our social life was the mass of mentally feeble persons, unguarded and unguided, who perpetually propagate their species'. Her public meetings, pulpit lectures, eugenic congresses, learned papers, and extensive correspondence bore fruit in the Royal Commission on the Mentally Defective in 1904, to which Mrs Hume Pinsent of Birmingham was appointed, and following which Mary Dendy became one of the first commissioners in lunacy.[23] In 1903 she, along with a more recent school board woman, Mrs Pankhurst, were co-opted on to Lancashire's new LEA. By now, HMIs found that Manchester's education was lagging behind: only in its special schools under Mary Dendy was it leading the way.

---

[22] *Manchester Faces and Places*, 1889–90, vol. i. p. 169–70.
[23] *WS* 21 July 1898; H. McClachan, *Records of a Family 1800–1933*, 1935, pp. 135–84.

Brighton

Brighton in 1870 was a town of some 90,000 people, of whom 11,000 were children. When it became known that a woman was standing for the school board, the *Brighton Gazette* proclaimed confidently that she would not be elected, but that if she were, 'her solitary position on the board would procure for her unenviable prominency such as few ladies care to achieve'. The lady in question promptly replied that she wanted only to serve her country, its women, and its children. 'She believed a true patriotism would exist when women were allowed to be useful in public matters . . . there would be less following of what is expedient instead of what is right' With such sentiments she was naturally returned head of the poll, and the only member of the board with a clear view of what to do and where to go.[24]

Her first success was to persuade the bemused board to subscribe to the *School Board Chronicle*. Whatever the LSB did one month Miss Ricketts did her best to implement in Brighton the next. She turned to the matter of the educational census. She hired seven assistants (six of them women, of course), cut up a map of Brighton for their use, supervised her team, and in three weeks completed her report to the board. The board took over temporary buildings and some local schools; Miss Ricketts had them altered and furnished, and added infant classes; and she pressed her colleagues to start building new schools to meet the shortfall of 3,350 places. Most children coming to school for the first time, she said, would be 'untidy, disorderly and miserable; and it would produce a bad effect upon them to put them into ugly workshops'. The poorer the child, the better must be the school.[25]

She next turned to school management matters, and persuaded the board to adopt the London syllabus, the London religious compromise, the London rules restricting corporal punishment, and the London staff structures. When it came to the question of attendance, she realized, as Lydia

---

[24] *Brighton Gazette*, 9 Dec. 1870; *Brighton Guardian*, 14 Dec. 1870. Able Liberal men such as Alderman Ireland gave priority to town council work. Miss Ricketts was a member of the National Education League.

[25] *Brighton Guardian*, 11 Jan., 15, 22 Feb., 8 Feb., 8, 22 Mar., 31 May 1871.

Becker did, but Elizabeth Garrett did not, that the board must win the assent of families to compulsory schooling. So in an imaginative step, she hired motherly women visitors to talk to mothers in their homes. She brought parents on to school management committees where they received a large measure of devolved responsibility, so that their school might reflect their neighbourhood's needs, and not be too 'mechanical' in form. Attendances rose satisfactorily. She had masterminded the first three years of the board's work.[26]

Her role was recognized as the triennial elections approached. Street literature, cartoons, squibs, posters, and letters to the press, asked voters to choose between 'Miss Ricketts and Co.' and the Tory high-church vicar of Brighton, who had tried to defend denominational schools on the board. As 'Queen' of the board, she made a series of major authoritative speeches on education policies at large public meetings where she was greeted with 'hearty, loud and prolonged cheering'. Her male colleagues were not entirely pleased to be left fumbling for things to say. She told her audiences that Tories and clerics wished to relegate board schools to the marginal task of reaching and teaching those down-and-out children for whom they had no wish and no place in their voluntary schools; whereas she believed (as Annie Besant was to argue many years later) that board schools should offer teaching to all children of all classes, and choice to all parents.

A keen politician as well as a keen evangelical, she decided to raise the political temperature by playing the 'No Popery' card. She accused her clerical opponents on the board of nasty Puseyite tendencies and of seeking to manufacture churchmen rather than educate children in their voluntary schools. In her spare time she gatecrashed Tory election meetings, and spoke from the floor, to the consternation of Tory clerical opponents who thought they had been safely abusing her in her absence. She invariably came out the better in such confrontations, and the hissing with which she was greeted when she stood up had turned to applause before she left.[27] At least as importantly,

[26] *Brighton Guardian*, 29 Mar. 1871.

[27] *Brighton Guardian*, 26 Nov. 1873, *Brighton Gazette*, 3 July 1873, *Brighton Examiner*, 2 Dec. 1873, *Brighton Gazette*, 27 Nov. 1873. Four thousand people at a time were

she ensured that Liberal polemical work was underpinned by a systematic canvass of the entire borough, by an orchestrated letter campaign in the press, by a planned sequence of public meetings, and by public endorsement from chapels, temperance groups, women's organizations, and teachers. Local papers acknowledged 'the businesslike as well as ladylike qualifications' of Miss Ricketts, as on a high poll she swept the Liberals back to a majority which they were to hold for another twenty years.[28] Deservedly and significantly, she nominated both the chairman and the vice-chairman of the new board. The phrases and policies she had devised for the 1873 campaign were still being quoted and employed by Brighton Progressives as late as 1898.

Over the next five years, she continued to inspire board policy, building more and better schools, caring for truant and neglected children. Sensibly, the church party did not challenge Liberal hegemony in the 1876 elections and a contest was avoided. But in the summer of 1878 Miss Ricketts resigned, to take her energy and convictions as a missionary to the Chinese. With much pleasure the board gave her a gift of money to speed her parting. Not for another decade were women to reappear on Brighton's school board, and then only as carefully chosen, deferential, and loyal members of the Liberal slate.[29]

Oxford

Miss Eleanor Smith was the sister of one Oxford professor, and the surrogate sister of another, A. V. Dicey. She was a highly educated, humane, gently liberal, and morally tough lady, dedicated to public service. For many years she had laboured as a district visitor, financed district nursing, and sustained a provident dispensary in the town, bringing medical care within the reach of Oxford's poor. She was also an ardent advocate of girls' higher education, and with the support of the Green–Toynbee circle helped to found Somerville College. As Dicey himself said 'She laboured more

meeting in the Dome to protest against ritualism, confessionals, and high-church practices.

   [28] *Brighton Herald*, 29 Nov. 1873, *Brighton Guardian*, 20 Dec. 1873.
   [29] *Brighton Gazette*, 6 Mar. 1878.

for the town than any other woman' and could count more friends among rich and poor alike, University and city, than anyone he knew.[30]

When late in 1870 it was clear that both sectarian and non-sectarian parties were refusing to include a woman on their list, she decided to stand as an independent. She canvassed hard for the working-class vote in the poor district of Jericho; she asked women ratepayers to plump for her. Oxford's new board of nine had three members nominated by the University, three clergy, and three unsectarian members of whom one, Miss Smith, displaced an official candidate.[31]

She quickly found that the school board was to be merely a school attendance committee. Its educational census showed that Oxford had an adequate supply of voluntary school places for its 5,000 children, and University and clerical members in unholy alliance on the board wanted only to use school board powers to compel children to attend their voluntary schools. Eleanor Smith took every opportunity to argue that the board should build a non-denominational board school. It would improve educational quality in the city and offer parental choice. Too many church schools, she said, were in run-down buildings, with one teacher supervising five classes and dependent on child assistants. Parents were refusing to pay school fees to have their child taught by another child and were keeping them at home. She also pointed out that church schools were seldom willing or able to reach the slum child. It was left to her, with her experience of Oxford's poorer streets, to call school managers together in the summer of 1873 to encourage them to train children in lavatory habits, to inspect them daily for infestation, and to send in weekly attendance reports.[32]

She had one success in 1877. Denominational schools were refusing to admit difficult children; and she persuaded the board that rather than send them away to desolate, and expensive, residential industrial schools, they should open a day industrial school of their own. It was one of the first in the

---

[30] R. S. Rait, *Memorials of Albert Venn Dicey*, 1925, p. 26; *Oxford Journal*, 19 Sept. 1896.

[31] *WSJ* 1 Mar. 1871.

[32] *Oxford Chronicle*, 16 Mar. 1872; *SBC* 5 Apr. 1873, 27 June 1874, 2 Dec. 1875, 6 Dec. 1873.

country. Not everyone approved of keeping such children at home with their families (many poor law women quite ruthlessly sought to break the child's contact with families they thought damaging), but the saving both of rates and of family life carried conviction with a clerical board. She found it a never-ending struggle. As she confided to a friend in 1881, 'Education is a great burden still . . . and worst of all is the misery of elementary education'. Only the day industrial school 'for waifs and strays' gave her any pleasure, turning out children 'more handy and educated than all but the children from the very best homes'.[33] Church school teaching remained bad and bigoted, and she could do little to lift it.

She came off the board in 1883 and died in 1896, without ever seeing the non-denominational board school she had worked for. But her example brought other women into poor law and town council work, where the influence of the University and the clergy was less crippling.

Birmingham

Birmingham had in 1868 some 65,000 school-age children, of whom a third were at efficient schools, a third at bad or very bad schools, and a third at no school at all. It was this finding by its local Educational Society which had generated Birmingham's Education League and much of the energy behind Forster's 1870 Elementary Education Act. By 1870 Birmingham also had developed its caucus. Liberals went into the school board elections with fifteen candidates expecting to sweep the board. The denominationalists ran only nine men, concentrated their voting, and for the first and last time ever returned a majority in the capital city of Liberalism. Stunned but undaunted, the Chamberlain Six spent the next three years out-talking and outmanœuvring the majority. At the 1873 triennial elections, the Liberal Unsectarian Eight (including Miss Eliza Sturge) came home to power.[34]

Miss Eliza Sturge, thirty-nine-year old niece of Joseph Sturge, daughter of a former mayor, and an active suffrage

[33] *SBC* 12 May 1877; E. Smith to Mrs Pearson, 8 Oct. 1881, fo. 191. Pearson papers.
[34] A. E. Taylor, 'The History of the Birmingham School Board 1870–1903' Univ. of Birmingham MA 1955.

lecturer, had been brought forward by a newly formed women's Liberal committee. She was accepted by the caucus, spoke on platforms with Chamberlain, and came third in the 1873 poll. Like women members everywhere, she helped to appoint women staff and started domestic economy teaching. Less usual was the articulate support she gave to Chamberlain's demands for free and universal elementary education, comprehensive in its reach and not confined to the poor.[35]

When she stepped down in 1876 to spend more time on suffrage work, Miss Caroline Kenrick (sister of the city's subsequent MP and sister-in-law of Chamberlain) stood ready to take her place. The men could find no room on the slate, but if she agreed not to run as an independent, would offer her the next vacancy. She agreed, and they did. Impressively competent, Miss Kenrick held the customary watching brief for infants, girls, and women teachers; and used her experience as a health lecturer further to develop domestic studies, chairing national conferences at which she kept T. H. Huxley and Edwin Chadwick in order and to time, with grace and a magnificent watch. Above all, she knew her schools, spending part of every day on her visits, taking

the general inspection of the girls' schools as her special duty. Her judgement in the election of school mistresses is relied upon by her colleagues; and it may be said that the present satisfactory condition of the girls' schools is mainly attributable to the sound principles which she has inculcated. Miss Kenrick is fluent as a speaker, and is an admirable woman of business.[36]

When the sectarians offered not to contest the 1879 elections if the Liberals restored Bible teaching to schools, Miss Kenrick was bitterly distressed but loyally abided by her party's agreement. After fourteen years of quiet, undemonstrative, hard work on committees, her own health broke down. She resigned, and died a few years later. The press noted that she

[35] *SBC* 11 Oct. 1873, 14 Nov. 1874, 25 June 1875.

[36] *Edgebastonia*, 1 July 1881; *WSJ* 1 Jan. 1877; *SBC* 15 Jan. 1877, 15 Mar. 1879. See also B. Webb, *Diary*, vol. i. p. 109, 16 Mar. 1884, for comments on the Kenrick–Chamberlain families—'the aristocracy and plutocracy of Birmingham. They stand far above the town society in social position, wealth and culture, and yet they spend their lives as great citizens, taking an active and leading part in the municipal, political and educational life of their town'.

'cared for the welfare of the children with all her heart' and gave her life to them.[37]

Her seat was inherited by Miss Dale, daughter of Dr R. W. Dale, an equally faithful party worker. During the 1890s, the Tories ran two women of their own. But by now, Birmingham's education had stagnated. It had still not built enough schools by 1903, its attendances were below the national average. It had refused to devolve power down to school managers (thus placing very heavy burdens on conscientious board members to carry all the local work), and kept even the tiniest decisions firmly within the party machine. Once the charismatic energy of Chamberlain and Collings was diverted to the national scene, the Birmingham board seemed increasingly old-fashioned, bureaucratic, stilted, and centralized. It contributed little to national education policy-making after the first decade. As for the women, co-opted into minor responsibility, with little disposition and no power base to challenge the machine; with Birmingham politics bitterly torn, even within families like the Kenricks, between Gladstonian liberalism and Chamberlain's liberal unionism; they kept their heads down, worked hard, and acquiesced.

Leeds

Leeds was a staunch Liberal and Nonconformist town, with a weak Church presence and few denominational schools. So with only 25,000 voluntary school places for the town's 58,000 children, Leeds faced a worse shortfall than any city outside London, except Liverpool. The town council promptly set up a school board in the winter of 1870; and among the candidates were two women, both of them active in the girls' higher education movement, Mrs Catherine Buckton, the Unitarian wife of a cloth merchant, and Miss Lucy Wilson, who was working closely with Josephine Butler in the Contagious Diseases campaigns. Amid a flurry of candidates without party organization, neither was elected; but in 1873, with a firmly established Liberal caucus, Mrs Buckton was included on their slate.

She soon became one of the most effective women on any

---

[37] *SBC* 10 Mar., 16 Dec. 1879, 12 Apr. 1884; *Birmingham Daily Post*, 12 Dec. 1894.

school board. She was already an established lecturer and writer on health and cookery; and as a founder-member of the Leeds Ladies Educational Association, was running classes for young women teachers which foreshadowed the university extension movement. The board by now had twenty-eight new schools in the pipeline, had adopted the London syllabus and its religious compromise. Mrs Buckton inevitably started domestic economy teaching for girls, and sent the ablest of them to the Yorkshire school of cookery which she had founded. She also 'topped up' the school syllabus by offering voluntary classes of her own on health and sanitation. As they took in colliery explosions, the fermentation of beer, and human physiology as well, they were packed out. She reran them and reprinted them; as *Health in the Home*, they became a best-seller adopted by other school boards.[38]

Her credentials established in womanly work, she was soon contributing to the full field of educational policy. Facing the same tangle of compulsion, attendance, and fees as other boards, she led the argument in Leeds for free schools, insisting that the cheaper the education, the better the value, as children would attend more regularly and earn the board more grant. She wanted, she said, 'a truly national system of education and free schools . . . She felt that it was impossible for widows, deserted wives, fathers disabled by accident or disease, to pay the school fees when the children were starving or half clad.' She later presented a petition to the board to reduce fees to 1*d*. for younger children, 2*d*. for older ones. When told this would damage the parents' self-respect, she asked whether the parents of children who attended free grammar schools or enjoyed scholarships and endowments, lost theirs. 'She maintained that the working classes paid much more than any other class, in proportion to their means, for the education of their children.' Her proposals were overwhelmingly carried; but the Education Department, who were not exposed to her eloquence, overturned them (as they did in Norwich and London), insisting on standard 3*d*. fees and individual remission for the poorest families.[39]

Meanwhile Mrs Buckton was encouraging the board to speed up its building programme—she for one would not prosecute any parent who kept their child from home when the only place for it was a denominational school place. Like Eleanor Smith of Oxford she persuaded the board to open a day industrial school for difficult children, where regular feeding and proper discipline would rebuild the child's physical, moral, and family life.[40] Prompted, like Miss Becker of Manchester, to redouble her visits to existing industrial schools in the light of the London scandals, she found scandals of her own. She suspended the superintendent for starving and birching feeble-minded children; and vetoed the appointment of a former prison warder as his replacement, since it would give 'a wrong colouring to the school . . . and inspire the children with fear and distrust instead of confidence and affection. The man who filled the position should be one who had the greatest experience of child nature, and who understood the infirmities which arose from starvation and exposure.'[41]

Having made study visits to London and the Continent, she brought kindergarten teaching into Leeds schools, and held seminars and public meetings at which she explained to parents and staff Froebel methods: how to teach geometry with bricks 'in a kindly way'. She introduced phonic reading schemes. The *Leeds Mercury* reported that four-year-olds could read words like 'punishment', 'exquisite', and 'perpendicular' (a curious run of examples). She inspired classrooms of children with her own love of flowers and nature study, by giving them pots, cuttings, and seeds. In 1878 and 1879, the children displayed over a thousand window-boxes, board members gave prizes, and the children went home clutching their geraniums. Her colleagues paid her the unique tribute of devoting sections of their triennial reports exclusively to her work; 'every lady teacher welcomed her to the school'. She consistently headed the poll.[42]

She was not only a professional educationalist, she was also a professional politician. In 1876 she wrote the Liberal

---

[40] *SBC* 6 Nov. 1880; *SBC* 1 July 1876; Mrs C. Buckton, *Address*, p. 81.
[41] *SBC* 28 Jan., 18 Feb. 1882.
[42] *SBC* 10 Apr. 1875, 28 June 1879.

election address, which was then adopted by the Leeds caucus, and by the Free Church candidates. On the platform, the keynote speeches fell to her, and by 1879 she was so overworked that she was pleading for a woman colleague to stand with her. As one friend reported, 'Mrs Buckton has worked in season and out of season, jaded and harrassed by some of the present members of the Board' who were anxious only to protect church schools. Mrs Alice Scatcherd, organizing the women's Liberal vote, was willing to run, but space could not be found and she was too loyal a party worker to stand as an independent.[43]

However, at the 1879 elections the Liberals lost control, and Mrs Buckton found herself in opposition; seeking (like Alice Westlake later in London) to protect board schools and thwart sectarian religious teaching.[44] She had built well, and the Tories were unable to demolish most of her work.

For several years she had been running homes in both London and Leeds, and now chose to settle in London. In her nine years on the board, she had done more than any other member to maintain the impetus of the building programme. When she retired, Leeds had forty-nine board schools to Liverpool's sixteen, Birmingham's twenty-one, and Manchester's twenty-two. The board had won high attendances and high grant, and had fewer prosecutions of defaulting parents, than any comparable board. The problem she had left unresolved, as she was the first to admit, was the quality of teaching staff. Too few of them were trained, too many of them were young pupil teachers; and to her dismay, she saw the Tory majority worsen staff ratios and increase overcrowding, the better to promote their voluntary schools. Her parting gift to her colleagues was a powerful fifty-one page *Address* on education policy in Leeds, with which they fought the 1882 election and with which they returned to power. Without her.[45]

---

[43] *Leeds Mercury*, 20 Nov. 1876, 19, 22 Nov. 1879; *WSJ* 1 Nov. 1879. For Mrs Scatcherd, see above pp. 35–6.

[44] *SBC* 1, 29 May 1880.

[45] No women immediately followed. Mrs Connon, who founded the local WLA in 1889, was co-opted in 1893. In 1897 she was joined by Mrs Robert Hudson, a churchwoman, who devoted herself to physically and mentally handicapped children. *Yorkshire Observer*, 16 Sept. 1922.

Bristol

Bristol in 1871 had sufficient voluntary school places for most of its 27,000 children, but only half the children were regularly attending school. Its unsectarian (Liberal) male board enforced attendance, introduced the London religious compromise, special cookery classes, temperance teaching; and limited corporal punishment. Its difficult children were placed in Mary Carpenter's schools nearby at Kingswood. Much of the work customarily undertaken by women board members was in Bristol achieved by a male board, in part because the town's network of women's philanthropic organizations had put early and effective pressure on the board. For Bristol possessed one of the most impressive women's movements in the country. They came from Quaker and Unitarian philanthropic families—Mary Carpenter, the Sturge cousinage, the Hill sisters, Mary Clifford, the Priestman and Winkworth sisters; and were among the first in the country to fight the Contagious Diseases Acts, to campaign for the suffrage, organize their municipal vote, and form women's party political associations.

At the January 1877 triennial elections, two board members retired. To the dismay of existing members who wished to avoid a contest and its costs, two working men and Miss Helena Richardson insisted on standing for the board. Miss Richardson was an experienced youth worker, an obsessive temperance advocate, and a fierce anti-ritualist evangelical, much in the mould of Miss Ricketts of Brighton. The church party discreetly distanced itself from her. The Liberals were clearly cross that they had not thought to include a woman of their own. Her election meetings were sparsely attended, but she claimed 'a knowledge of the wants of the working classes' and believed 'there was room and work for a lady on the board'. The gentlemen spared no punches. Even the unexceptionable nature of her election address was held against her. She favoured temperance, pure air, and economy, but then who did not? asked the Liberals rhetorically. They circulated rancorous leaflets; infiltrated and heckled her meetings, which then refused to endorse her as a candidate; and both Conservative and Liberal agents struck women voters off the burgess list as fast as they dared, in case they voted for her. Women themselves were divided. One suffragist wrote to the press advising women not to lose sight of educational principles by voting for Miss Richardson out of misguided sentiment. Miss Sturge more generously hoped women voters would return a woman to the board. After a rough campaign in which she reckoned she acquired a sympathy vote, Miss Richardson was elected, having displaced one of the unsectarian slate. She held the balance of power on the board.[46]

Over the next three years she voted more and more with the unsectarians, using her casting vote to stop children without fees being excluded from schools; trying, like women members everywhere, to reduce the time spent on needlework and required, like women members everywhere, to supervise the cookery teaching. She worked hard. When she stood for re-election she claimed to be a firm supporter of board schools. Apart from references to temperance, her election address was indistinguishable from that of the Liberals.[47]

But the Liberals had learnt their lesson and never again went to the polls without a woman candidate of their own. They selected Miss Emily Sturge. The thirty-year-old daughter of a Quaker surveyor, she was the hon. sec. of the West of England suffrage society, a poised, confident, and fluent Liberal lady. One of her sisters, Elizabeth, had trained with Octavia Hill, another became a doctor, and a third was active in philanthropic work and was herself to stand for the town council (unsuccessfully) in 1907. Elizabeth accompanied her sister as she campaigned, noting that Emily 'made an excellent impression on the various audiences which we found assembled in dingy club rooms in the upper storeys of public houses, where many of the meetings were held. She threw herself into her new work with ardour . . .'.[48]

Miss Richardson realized that Miss Sturge would take her votes and struggled to hold on to her plumpers. Miss Sturge in her turn commented that 'she was there as a Liberal candidate, and did not wish to interfere with Miss Richardson who

[46] *WDP* 16, 17, 20, 22 Jan. 1877.

[47] *WDP* 24 Mar., 28 Apr. 1877.

[48] E. Sturge, *Reminiscences of My Life*, 1928, p. 82; S. J. Tanner, *The Suffrage Movement in Bristol*, 1918, p. 14; Mrs A. Beddoe, *The Early Years of the Movement*, 1911. The renegade was brother Clement, a Tory member of the LSB and LCC.

was the temperance candidate', but when asked her own views on temperance could not resist adding, silkily, that 'she could hardly see what that question had to do with education. (hear, hear)'. Inevitably, Emily Sturge was returned, Helena Richardson was not.[49] Three years later, the Liberals added to their slate a working man, and conscious that Miss Sturge lacked a certain something, the suffragist Mrs Alice Grenfell, a mother who would 'understand the motherly relationship to children'. In desperation, the Tories found a working man and a lady. Their chairman explained to the inaugural campaign meeting:

> It has been felt for some time that the Conservative party had laboured under a disadvantage because they had not a lady candidate . . . Miss Douglas had great experience of parish work, the only condition that the lady had imposed being that she should not be required to appear on any platform lest she be asked to make a speech.[!]

Miss Sturge characteristically rejoiced that there were three women candidates, and that all three represented political associations 'which certainly', she said, 'was a vindication of their position'. All three were safely elected; and over the years their seats were passed on to a sequence of able women.[50]

As in Birmingham, therefore, the Bristol women, Miss Richardson apart, were loyal party ladies. Unlike Birmingham, they were not dwarfed by the national stature of their male colleagues, and, again unlike Birmingham, they had a powerful independent base of their own. Indeed Bristol women Liberals were among the first to refuse to work for male colleagues who were anti-suffrage, to the dismay of the Liberal caucus. Inevitably, women members taught cooking and supervised sewing, but they also insisted on working over 'a large surface', as Miss Sturge's obituary noted.

Though regarded by her family as a cool and somewhat

---

[49] *WDP* 14, 19 Jan. 1880.

[50] *WDP* 13, 15 Jan. 1883. Miss Douglas's seat passed to Miss Georgina Taylor, a parish worker, who would not speak on mixed platforms, though an effective speaker within the NUWW; and then on to Miss Mabel Wait, a prominent churchwoman. *WDP* 20 Jan. 1892. Mrs Grenfell's seat went to Mrs Swann; and Emily Sturge's to Miss Marion Townsend.

reserved lady, Miss Sturge took the children to heart. She found them additional playgrounds, got the board to build two swimming-pools for them, promoted gym, and sent them off to country camp. Helped by her sisters, she organized school dinners every winter. Anxious that children should climb Huxley's ladder from gutter to university, she ensured that board schools were available for evening classes; and brought teachers, board members, and educationalists together in the Bristol Educational Council in 1889, where she hoped to develop links between elementary, secondary, and higher education. She made teacher training her responsibility, and long before London had pupil teachers attending special training centres, and women staff attending a day training college she had helped to found.

The women members also made handicapped children their especial care and started special teaching for them. Miss Sturge was right to claim in her 1892 election campaign, that the Bristol school board, while not a 'showy board', was one of the best in the country, free of scandal, and with a solid achievement behind it, no doubt, she added, because all members agreed that 'education was not a thing to be economical over . . .'.[51] The Liberals remained in power. The Tories, somewhat maliciously, moved that Miss Sturge be appointed vice-chairman of the board. The Liberals, who had chosen someone else, were embarrassed; Miss Sturge abstained, the other Liberal women members voted with the majority, and it fell. She died in a tragic riding accident shortly after, and her seat passed to Miss Marion Townsend, who became nationally known for her care of physically and mentally handicapped children. By 1900 members of other boards were visiting Bristol to study her methods.[52] The tribute paid to Miss Sturge at her memorial service was equally a testimony to a succession of able and dedicated Bristol ladies: 'she rendered us as a city—and the children of our city pre-eminently—a service that may deliberately be called a splendid service . . .'.[53]

[51] *WDP* 13 Jan., 4 June 1892.
[52] *SBC* 16 Jan. 1900, 4 Jan. 1902. Miss Townsend taught engraving, ticket writing, shorthand, and typing, as well as the customary craft work.
[53] *WH* 9 July 1892.

Great Yarmouth

Ethel Leach was one of the most remarkable of all the ladies elected. She was married to a local radical ironmonger of Irish roots, who encouraged her in 1881, when she was aged thirty, to join the Yarmouth school board, then one of the most backward in the country. In the 1890s she became a guardian and despite vicious hostility from the board's clerk, transformed its nursing and child care. She failed in 1908 to be elected to the town council, but after the War became a JP, mayor, alderman, and councillor in quick succession. *En passant* she was Helen Taylor's election agent for the Camberwell parliamentary seat in 1885, lectured for the Irish Land League, and welcomed to her house scores of eminent Fabians and Irish who came to talk politics and enjoy Yarmouth's Siberian breezes. Yarmouth's Sixth Form Centre now bears her name. Given her doubts about higher grade schools, this would have appealed to her sardonic sense of humour.[54]

So dismal was Yarmouth's school system, that the Education Department had imposed a school board on the town in 1875. By 1881 its members (six sectarian to five unsectarian) with 3,000 children to school had still only places for 1,780 of them and attendances from only 1,200 of them. The syllabus was spartan, the teaching depressing, the grant earned negligible. Children came to school as young as possible, as seldom as possible, and left as soon as possible.[55]

Mrs Leach's campaign was stormy, but she was elected as an independent on the women's vote. The press noted that 'she quickly made herself heard'. Within six months, she asked the board to introduce kindergarten methods, by now widely established elsewhere. Half of Yarmouth's school children were under seven and she would have them learn form, number, colour, and sound through handling bricks, sticks, and soaked peas. She passed round model houses,

[54] Ethel Leach to Helen Taylor, letters 1887–90, vol. xviii, Mill/Taylor papers; *Yarmouth Mercury*, 18 Apr. 36. Information from her grandson. Her husband was working long hours, her son was at a Quaker boarding-school, she had two well-trained servants, and was deeply depressed and bored. See also her charming *Notes of a Tour*, 1883.

[55] *The Board's Triennial Report* for 1884 noted that 'the principal causes of irregular attendance were stated to be due to the want of means, clothing and boots, and in many cases want of food' (p. 10). Board schools merely 'minded' small children.

churches, and towers at the board, built by Norwich children, and commented how sorry she felt for Yarmouth's infants who looked 'so weary and miserable' and who often were eager to escape from school. The board could not, and agreed. She added cots for the two-year-olds, toys for the three-year-olds, and books for the older children.[56]

An avid reader of the *School Board Chronicle*, she redoubled her visits to Yarmouth's industrial school at Cobholm and forced a public inquiry when she found one lad had been hit with a trowel. The board's chairman carefully explained that no other irregularities were to be found, and the *Yarmouth Times* sourly commented, 'Like Mrs. Surr, some provincial representatives of school boards are seeking to gain popularity by a little raid on industrial schools generally'. Mrs Leach retaliated with the Home Secretary's comment that such schools required eternal vigilance. She asked for and obtained a motherly woman as matron, reduced the level of corporal punishment, and drew up a new dietary for the school. She did too well: eighteen months later male members of the board thought the school had become soft, and placed a male superintendent in charge. She was always uneasy about industrial schools, believing they 'treated the boys like pariahs, and making a separate class of them, only tended to cause them to lose their self-respect'. She asked the members to board out difficult boys with respectable families instead.[57]

Attuned to the national debates on over-pressure, she toured Yarmouth schools with Dr Collier to find teachers keeping children in after hours to cram them for grant-earning exams. She persuaded the board to ban home lessons and after-hour teaching. Like women everywhere, she was asked to start cookery teaching and took on the task gracefully, if somewhat cynically, given her own dislike of housework. Finding many Yarmouth children half-starved, she started penny dinners—they 'entail heavy work upon me' . . . she confided to Helen Taylor—and a year later persuaded the board to sponsor penny breakfasts funded by voluntary

[56] *Yarmouth Times*, 5 Feb. 1881; *Yarmouth Journal*, 18 Apr. 36, *Yarmouth Times*, 19 Mar., 23 July, 1881; *SBC* 20 Aug. 1881.

[57] *Yarmouth Times*, 10 Dec. 1881, 7 Jan. 1882, 15 Oct. 1887, 24 Mar. 1888; *SBC* 23 June, 7 July 1883; *Yarmouth Times*, 24 Dec. 1892. Fostering difficult children became widespread only after 1945.

subscription as well. She had understood instinctively that schools must be attractive if they were to be effective; and that meant attending to the quality of teachers and the matter of fees.[58]

Most small school boards found it difficult to train or attract good teachers. Yarmouth's Triennial Reports regularly lamented the poor quality of their pupil teachers. Mrs Leach's personal experience prejudiced her, she said,

> more and more against their system of teaching pupil teachers, and the placing of more boys and girls as teachers in their schools after they had merely received elementary instruction itself. The educational authorities were crying out against the mechanical instruction prevailing in their elementary schools, and she thought the pupil teacher system had largely to do with it.

A month later, her proposals for a pupil teacher centre, 'like Liverpool, Birmingham and London', she encouraged the board, were accepted.[59]

On only one issue did she not have things her own way, and that was the matter of fees. Yarmouth faced the same problem as everywhere else of attendance, compulsion, and fees. Yarmouth's solution was to refuse to remit any fees whatsoever, to refuse to admit any child who failed to bring fees, but not to strive too energetically to bring that child to school. Within weeks of joining the board Mrs Leach asked members to remit the fees of two truanting boys, but was overruled. She then raised the matter as a substantive motion, saying that though the town was notoriously short of school places, the ones it had were half empty. Parents could not afford the fees. The guardians, who were empowered to help the most destitute, refused to do so. The church members on the board, including a private-school proprietor, accused her of seeking remission 'not for the sake of the poor but to ruin denominational schools in order to set up a godless education', a response that dismayed even the Tory local press.[60]

However, by October 1886, the number of barefoot children in the street was increasing, and the town's economic distress

---

[58] *Yarmouth Times*, 9 Mar., 6 Apr. 1889, 22 Mar. 1890. Ethel Leach to H. Taylor, 10 Jan. 1889, vol. xviii, fo. 44, Mill/Taylor papers; *Yarmouth Times*, 15 Oct. 1887.

[59] *Yarmouth Times*, 21 Dec. 1889, 25 Jan. 1890.

[60] *Yarmouth Times*, 24 June, 22 July 1882; *SBC* 29 July 1882.

forced the board to think again. The chairman now reported that voluntary schools in Leeds and Ipswich were happy to see the fees of poor children attending board schools remitted, as this cleansed their own schools of the most destitute and irregular children. The board thought this a good idea, and Mrs Leach wryly seconded the chairman's shiny new scheme. Attendances rapidly increased, from less than 72 per cent to around 77 per cent of eligible children—though this was still a very low figure.[61] Mrs Leach tried to reduce the level of fees to 2*d.* a week rather than pursue individual remissions. Given the poverty of the town and the high cost of high fees in poor attendances, lost grant, truant officers, court summonses, and special schools, she suggested that it would not be 'so very ruinous to Yarmouth if they had free education'.[62]

One board member tentatively suggested that it was time Yarmouth started a higher grade school. Mrs Leach lashed out. Of 163 boroughs, only 23 were content to let children leave school once they had reached the fourth standard, and that included Yarmouth. She 'had always thought it was a disgrace to the town and the community that Yarmouth should have such a low standard—the same as that which prevailed sixteen years ago'. Most children were leaving school by eleven, many now reached the fourth standard at nine. They then got 'small jobs as errand boys but, generally speaking, being free of the discipline of the school, they loitered about the streets and lost all the good obtained at the schools, and became a nuisance and danger to the community'. Juvenile crime was down 50 per cent in the country as a whole, up by 50 per cent in Yarmouth. Under pressure from her, the board agreed to raise the school leaving age to the fifth standard in 1890.[63] Only when Mrs Leach was satisfied with the level of elementary education would she contemplate higher grade schools. She wrote disingenuously to Helen

[61] *Yarmouth Times*, 9 Oct. 1886. Attendances rose again to 84 per cent when fees were abolished. Mrs Leach tried, like Miss Becker and Miss Taylor, to switch endowments to the poor. *Yarmouth Times*, 21 Dec. 1889; speech at the Liberal Club, 25 Mar. 1893.

[62] Ethel Leach to H. Taylor, 7 Aug. 1888, vol. xviii, fo. 42, Mill/Taylor papers; *Yarmouth Times*, 25 Jan., 5 July 1890.

[63] *Yarmouth Times*, 20 Dec. 1890. Of 163 boroughs, 122 boards including Norwich and King's Lynn kept children till the 5th standard, 18 city boards to the 6th standard, only 23 released children at the 4th standard.

Taylor in 1889, 'I have lately become alive to the fact that I am beginning to be an influence on the board but I have only won it by hard and determined work'.[64] Given Helen Taylor's experience of the LSB, that must have provoked a smile. When Ethel Leach won a place on the Executive Committee of the National Education Association by a fine speech in praise of Stanley, Liberal leader of the LSB, one suspects that Helen Taylor's private comments might have been less than ladylike.

Meanwhile, with each triennial election, Ethel Leach increased her votes. Helen Taylor came down to speak at every campaign. Mrs Leach had one nervous moment in 1887:

I am getting mortally afraid of another woman coming forward. My candidature has become so absurdly popular—on every side I hear the cry Mrs Leach top of the poll, how I wish nomination day would arrive. I think I shall go in on a great wave of enthusiasm. I am lost in astonishment at my popularity. It is fortunate that I have so small an opinion of myself—or rather that I know myself so well, or my head would get turned. I think much of my support will come from the church and Conservatives which is very remarkable. If they know their own interest they will put up a woman, the demand for another seems so great and new candidates are coming into the field every day. I shall be ever so pleased if things turn out as well as they promise for I know that it will please you and it is such happiness to please you. I wish I could feel I had done anything to merit it, beyond plain steady honest work—I mean that I wish I could feel that I had any great qualities like your own to merit it. I think I must try to keep things quiet for the present, as some woman will be tempted to rush in to reap the benefit, I know that there are one or two jealous and ambitious souls, and it is very tempting altho' you and I know that it may all end in smoke.
P.S. I know that another woman has been asked to stand with me, on the ground that she was sure to get in with me this looks bad.

In the event she was able to report back delightedly that no other women stood, but several 'nice' women workers emerged 'round me, of whom I hope to see more in the future'. With one, 'a real live radical', she 'fell in love with straight away'.[65]

[64] Ethel Leach to H. Taylor, 10 Jan. 1889, vol. xviii, fo. 44, Mill/Taylor papers.
[65] Ethel Leach to H. Taylor, 17 Jan., 16 Feb. 1887, vol. xviii, fos. 35, 36, Mill/Taylor papers.

By her 1890 election she was engaging in aggressive publicity. She posted bills around the town giving her attendance figures at meetings (far higher than any man) and youngsters carried sandwich boards with the jingle:

> I mean to vote for Mrs Leach,
> Mrs Leach, Mrs Leach,
> Whose goodness no one can impeach,
> Mrs Leach, Mrs Leach,
> The poor child's friend is Mrs Leach
> And that is why I do beseech
> You'll give your votes to Mrs Leach.[66]

She was now second in the poll, surpassed only by the Roman Catholic, aroused immense enthusiasm at her meetings, where she emphasized not just her womanly work (1,276 boys and ten men, 3,000 girls and infants and only one of her) but her wider concerns. Although she was the outstanding woman Liberal in the town, representing the party at national conferences, she was oddly reluctant to share meetings, votes, or literature with male Liberals, and apparently unconcerned about winning a majority on the board. This was partly because in a small town she did not need party machinery to be effective; as a radical she was impatient with the lethargy of local Liberalism; but also because as a suffragist she was coming to 'distrust political associations . . . both parties are ready to avail themselves of the services of women for their own aggrandisement but neither are willing to give them the power of independence . . . women have to take the utmost care not to become the tools of party to be dragged through all the tortuous paths of compromise called politics.'[67] In seven or nine years, with the loyal help of Dr Collier, an elected member who was also the workhouse doctor, and her own national network, she brought an isolated, reactionary, and sectarian board more or less into line with Progressive boards elsewhere; and managed to retain the affection and esteem of her colleagues while doing it.

In 1895, this tough and determined lady was greatly moved when unanimously she was made vice-chairman of the

---

[66] *Yarmouth Times*, 8 Feb. 1890.
[67] Ethel Leach to H. Taylor, 10 Jan. 1889, vol. xviii, fo. 44, Mill/Taylor papers.

board. She accepted the position with 'a light heart and easy conscience' recognizing it for the tribute it was.[68]

Sheffield

Women came surprisingly late to Sheffield school board, given its radical reputation, partly because it contained very able men and because those same able men had already organized their women into supportive ladies committees. However, in 1882, Mrs Sarah Wilson reluctantly was pressed into service by her family, standing 'not from any wish for personal honour . . . but [from] a conviction of duty' as girls needed ladies on the board. And in a heartfelt addition, 'She wished some ladies other than herself would take an active part in the work'.[69]

The Wilsons were at the epicentre of Sheffield politics. Their money came from a large metalworks. Her husband, Wycliffe Wilson, was a poor law guardian, and later Lord Mayor; her brother-in-law, H. J. Wilson, had been chairman of the board and was later the town's MP. Wilson women had worked in anti-slavery, teetotalism, higher education, chapel, and Josephine Butler's anti-Contagious Diseases campaigns. So Mrs Wilson stood not as an independent, but as part of the Liberal slate, proudly belonging both 'to a sect and a party', supporting free and unsectarian education, and opposed to grant aid to voluntary schools.[70]

For the first three years she worked quietly away on traditional womanly work, bringing women teachers to her house for tea and sympathy, but at the 1885 triennial elections the Liberal John Leader suffered a bereavement and in his absence she was pushed to the fore, making keynote speeches in favour of free schools, unsectarian Bible teaching, and the need for continuing education.[71] The Liberals lost control;

---

[68] *Yarmouth Times*, 20 Apr. 1895.

[69] *Sheffield Daily Telegraph*, 14 Nov. 1882. It included John Leader, H. J. Wilson, and in the wings, A. J. Mundella.

[70] W. S. Fowler, *A Study in Radicalism and Dissent: The Life and Times of H. J. Wilson 1833–1914*, 1961; R. Wilson, *The History of the Sheffield Smelting Company Ltd., 1760–1960*, 1960; J. B. Bingham, *The Period of the Sheffield School Board 1870–1903*, 1949, p. 147; *Sheffield Daily Telegraph*, 21 Nov. 1882. Family goods were distrained in 1904 rather than pay rates to voluntary schools.

[71] *Sheffield Independent*, 2 Nov. 1885, 3 Nov. 1888,

but three years later her husband took their electoral machinery in hand, and warded the city. Mrs Wilson, joined now by Mrs Mary Anne Ripper, a former pupil teacher, swept back to power on a high (52 per cent) poll, insisting that board schools and parental choice were safe only in unsectarian hands.[72]

Grateful colleagues would have made Sarah Wilson chairman of the school management committee, but linguistically troubled by the term 'chairman' she became its president instead. Given the prestige of the post, and the weight of its work (the other three committees of buildings, by-laws, and finance were relatively light), she was largely determining school board policy. At a typical school management committee in February 1889 her members discussed inspectors' school reports by individual school; reviewed school grants, examinations, and senior staff appointments; and appraised school premises, their lettings, furnishings, and ventilation. Much was delegated to her personally. Absent teachers, nomadic children, parents' complaints, evening classes, pupil teacher progress, infectious diseases, the scrutiny of punishment books, the auditing of school penny banks, the appointment and discipline of junior staff, temperance teaching, cookery classes, and school prizes were all part of the minutiae of her day-to-day work. Given that Sheffield was schooling 61,000 children, two-thirds of them in board schools, she was carrying individually a work-load and responsibility that today would be filtered through three tiers of officers and four subcommittees. She handled it all serenely, without fuss, hassle, or controversy. She started kindergarten teaching, and with Professor Ripper's help manual skill classes and elementary science lessons. (She sent him off to London to report on best practices.) And with Mrs Ripper she continued to organize eight hundred cheap school dinners every day; and music, sewing, and reading, in recreational classes for school leavers every evening. She made major speech after major speech on the board, not only carrying through her own proposals but, like Mrs Buckton of Leeds, leading the debates

[72] *Sheffield Independent*, 3, 14, 15 Nov. 1888. Mrs Rippon was a Progressive and former headmistress; her husband, a board school teacher, became professor of Engineering and vice-chancellor of Sheffield University.

for free, comprehensive, and continuous education in Sheffield, from gutter to university.[73]

She resigned in 1892 to pay an extended family visit to America, and Mrs Ripper took on much of her work. Like Mrs Wilson, she visited each school a couple of times a year, the only board member to do so; and like Mrs Wilson she insisted that children from the poorest homes needed the sunniest schools and the smallest classes.[74] As in Bristol, the two seats passed down the female line to a series of able women. Meanwhile, Mrs Wilson, on coming back from the States, returned to the philanthropic world of training homes, temperance work, and propagating the Gospel to the Jews, without any obvious regret.[75]

Nottingham

The first years of the Nottingham school board were tranquil, as this lace and hosiery town was well supplied for school places, but in 1877 the borough extended its boundaries, virtually doubled its population, and presented its school board with a shortfall of nine thousand places, overcrowding, and absenteeism.

It was a staunchly Liberal board, with many members active in Birmingham's Education League; and it had already begun to liberalize its school curriculum and to introduce kindergarten teaching, higher grade schools, and improved pupil teacher training.[76] Mrs Cowen was included in their 1883 slate as one of their 'winning cards'. Like Miss Sturge of Bristol and Mrs Wilson of Sheffield, she came from a family devoted to social work, chapel life, and public service. Anne Cowen was the eldest of three Guilford sisters, all of whom held public elected office (Hannah also joined the school board and Sarah became a guardian). All of them were

---

[73] Sheffield school management committee Minutes, 28 Feb. 1889; *SBC*, 4 Apr. 1891; *Sheffield Independent*, 21 Mar. 1890; Sheffield SB Minutes for Jan. 1890.

[74] *SBC* 4 Apr., 2 May 1891; *Sheffield Independent*, 15 Nov. 1894.

[75] Mrs Wilson's seat went to Mrs Hoskins, the nominee of the WLA (400 strong) who was active in the Peace Society, temperance work, and mothers' meetings. *Sheffield Independent*, 11 Nov. 1891; and then to Miss Maud Maxwell, Liberal daughter of a town councillor, who worked especially with delinquent and 'defective' children, *SBC* 10 Sept. 1892.

[76] D. Wardle, *Education and Society in 19th Century Nottingham*, 1971, p. 84.

staunch suffragists—Mrs Cowen formed the local women's suffrage society and continued to organize all its political work; keen Liberals—inevitably, they had formed the local women's Liberal association; prominent in the High Pavement Unitarian Chapel, and in philanthropic work, serving on the Executive of the highly influential Town and Country Social Guild, which interlocked Nottingham's civic élite.[77] Mrs Cowen's husband was an ironmaster and himself a town councillor.

She came on to the school board to care for infants, girls, teachers, and mothers—a 'lady, having greater familiarity with the wants of the poor' would be better able to improve school attendances as she could visit mothers in their homes. She also came on to the school board as a Liberal like 'her grandfather, father and husband . . . Besides she had come out with those with whom she was desirous of acting'. She had no wish to do the ladylike thing and stand as an independent seeking plumpers. Her adoption by the Liberal Association, she said, 'is a much greater triumph for the friends of women than if she had stood alone'. She proved so popular that she had hastily to draft leaflets asking women voters not to plump for her lest they endanger the Liberal majority. 'It would be no advantage to Mrs Cowen', she wrote, 'to be returned by a great majority if one of the Liberal Eight were defeated. She can only be fully effective for good on the board if she has the assistance of the whole of the gentlemen with whom she is associated, all of whom are good friends of our cause.' On polling day as the votes came in, she was to be seen calmly on the platform assisting the town clerk (surely illegally?) to check the voting sheets. She came in second.[78]

Though reticent at board meetings, she worked hard in committee, seeking compensatory education for the most deprived children and deprived areas. She extended kindergarten work, so that children might be 'brought up quietly like

[77] *Nottingham Evening Post*, 16 Dec. 1881; *High Pavement & Christ Church Chronicle*, 23, 269, Dec. 1916; *Biographical Catalogue of High Pavement Chapel*, u.d.; *WH* 14 Mar. 1891, *Nottingham Evening Post*, 13 Jan. 1904. The Town and Country Social Guild had 200–300 members aiding 2,000–3,000 of the city's needy; to it were affiliated housing, blanket, sick poor, and boarding-out societies. Mrs Cowen specialized in public health lectures. Annual Report, 1884, Town & Country Social Guild, p. 26.

[78] *Nottingham Daily Express*, 19, 21, 23, 27 Nov. 1883; *EWR* 13 Dec. 1883.

plants in the garden, and educated without knowing they were being instructed'. She supervised domestic economy teaching, and did her best to help poorer women unable to pay school fees, by seeking their remission, by providing school dinners, and by finding decent cast-off clothes for their children. At the 1886 election campaign, her male colleagues spoke about numbers, of buildings, grants, examination passes, attendances, and fees. She spoke about the developments of the school day and the well-being of the child. 'The girlhood of Nottingham has no truer friend', wrote a local paper, 'nor one so discriminating and sound in judgement'.[79]

But as the church schools found themselves losing ground, so party controversy intensified. Mrs Cowen became a tough and astringent party speaker, carrying much of the platform work for the Liberals. She was joined by another Liberal woman, Mrs Catherine Bayley, who proposed as the mother of 'children in a sunny and happy home', to bring similar sweetness and light to the children of the poor. This was too much for the town's working men, who, irritated by the claims of those with 'silver spoons' to speak for them ran two Socialist candidates of their own. The church party also ran a woman, Mrs Macallan, who claimed she was above party politics and well soaked in the steam of school soup-kitchens, than which, she argued with some justification, there was no greater proof of devotion to board schools. All three women were re-elected; but the Liberals lost control, in part because the church party had the more efficient machine but also because the working men's candidates had split the progressive vote.[80]

Nottingham Tories made modest progress, under pressure from Mrs Cowen who led for the Liberals on the school management committee. When Mrs Bayley retired, Mrs Cowen was joined by her sister Hannah. But the Liberals again went into the 1892 triennial elections with faulty party machinery, leaving Anne Cowen to fight Mrs Macallan in a potentially Tory ward. Hannah Guilford was opposed by a man whose manifesto insisted that the board had too many

---

[79] *Nottingham Daily Express*, 15, 18, 20 Nov. 1886.
[80] *Nottingham Daily Express*, 1, 12, 13, 14, 19 Nov. 1889. Mrs Bayley was also prominent in the Guild and had founded the local orphanage and day nursery.

women. She gallantly called for 'the best and widest and fullest education that could be given', so that children were educated for life and not merely trained for work. Miss Guilford only managed to hold on to her seat when workers in adjoining wards switched to help her late on polling day. Mrs Cowen, in the shock result of the election, was defeated.[81] Two years later she was dead from bronchitis. Her sister was to continue her work, receiving the MBE for services to education in 1918.

Birkdale (Southport)

Birkdale was an upper-class suburb of Southport, near Liverpool. As its voluntary schools only had five hundred places, less than half those needed, the Education Department imposed a school board in 1883. The gentlemen living around the park nominated a sectarian board without the bother of an election, built their one and only board school, and in a rash moment co-opted Miss Kate Ryley, elder daughter of Birkdale's largest house, as one of their seven members in 1889.[82]

Miss Ryley came from an advanced Liberal family, connected to the Liverpool Croppers, committed to anti-slavery, home rule, Josephine Butler, and higher education for girls. In 1885 she had turned the women's Liberal association from a genteel social club to a crusading suffrage and feminist organization. Impatient of opposition, in her abilities and animosities she resembled Helen Taylor.[83] The board she joined had only one school to manage (not enough to keep its members occupied); and most of the developments of the 1890s—day industrial schools, care of handicapped children, pupil teacher training, liberalizing of the curriculum, the opening of higher grade schools—not unnaturally passed it

[81] *Nottingham Daily Express*, 15, 17, 19, 23, 24 Nov. 1892; *Final Report of the Nottingham School Board*, 1902.

[82] W. E. Marsden, 'The Development of Educational Facilities in Southport, 1825–1944, Sheffield MA thesis, 1959; H. J. Foster, 'The Influence[s] . . . on the Development of Schooling in a 19th Century Lancashire Residential Town', Edge Hill College M.Ed. thesis, 1976.

[83] *Southport Guardian*, 17 Oct. 1934; *Southport Visitor*, 20 Sept. 1902. Miss Ryley had a degree from St Andrews, a bicycle on which she pedalled to Wales on holiday, and a fierce feminist Liberalism which she shared with Lady Carlisle and Eva McClaren, in refusing to work for anti-suffrage MPs.

by, though Miss Ryley did manage to start technical evening
classes. Two issues, however, of national dimension were felt
even in Birkdale: fees, and school discipline.

Birkdale had deliberately set its board school fees higher
than those of the voluntary schools. When in 1891 the government offered grant to boards abolishing fees, Birkdale was one
of the very few to refuse it, fearing this would empty the voluntary schools. Having lost the battle at the board, Miss Ryley
published confidential correspondence between the board and
the Department, 'because she wanted the people to know
what was going on, the schools were theirs'. She was criticized
at the next board, 'with due deference to her own sex'. She
interrupted briskly with 'Please don't apologize. Treat me
exactly as another member of the Board.' With the help of the
local Liberal agent, she organized petitions from parents
asking for free places. The Tory board used the Conservative
agent, who happened also to be the attendance officer, to
check the validity of the signatures. Miss Ryley stalked him
around town and found that parents who did not have a
picture of Gladstone on their walls were pressed to withdraw
their signatures. To help things along, she now offered to pay
all the fees herself until the next election, an offer the board
felt able to refuse. So she indulged in some decidedly
unladylike language, and faced a libel action until local aldermen had it squashed lest it bring Birkdale into disrepute.
Finally, the board was forced by the Department to provide
sufficient free places to meet parental demand. At the
elections she had her reward and came in head of the poll.[84]
The sectarian majority did not forgive her.

Her second battle was even more bad-tempered, and as she
herself admitted in retrospect, could have been better
handled. The board, influenced by denominational salary
scales, underpaid its staff, and had therefore to rely too
heavily on untrained and pupil teachers. Discipline was
poor, exam results were worse, punishment was heavy, and
the school over the years was forthrightly criticized by the
HMI's—'A large number of the children are listless . . .
The Head Master does not appear to take sufficient part in the

[84] *SBC* 31 Oct., 19 Dec. 1891; *SBC* 3 Sept., 22 Oct., 3 Dec., 17 Dec., 1892; *Southport
Guardian* 10 Sept. 1898, 7 Sept. 1892.

work . . .'.[85] So when the headmaster, a churchman, rashly applied for a salary increase, Miss Ryley opposed it. She reported that as the only board member ever to visit the board school, she had found a pupil teacher caning a child, made him stop it, and introduced a punishment book into the school (fairly common elsewhere but which the HMI thought the whim of 'a somewhat eccentric lady member').[86] The incident might have blown over—the boy teacher presented Miss Ryley with a chrysanthemum on her next visit—but her two other Progressive colleagues with good cause complained that the school was unsatisfactory. The teachers rebelled, objected to interference, sought to ban board members from the school, ran their own election candidates against the Progressives, and wrote highly charged letters to the press. The head, with the collusion of sectarian board members, referred to Miss Ryley as a 'brawling woman'. Another teacher, who chose to remain anonymous, wrote to the press, hoping Miss Ryley would suffer defeat in the elections.

Those of us who only know her through the reports of the meetings of the Birkdale school board have often wondered at the patience of the men. Mainly through the action of Miss Ryley and one or two who always support her, the meetings of the board have been long, acrimonious, and irritating, most unpleasant to read of, and enough to make sensible men resolve to have nothing to do with the Birkdale school board. A far more efficient, harmonious school board may be secured if it were known that Miss Ryley were not to be one of its members. Zeal and ability are very excellent things; and so also are prudence and consideration for others. Miss Ryley asks to be elected for one reason only . . . She wants to meddle with the teachers; they must be made to keep a record book of all the punishments they administer in the schools. It surely is enough trouble and worry to have to punish idle and bad children, without being in turn punished by being compelled to write down long accounts of all corrections. How can the discipline of the school be maintained when those who have to conduct it are harnessed and perpetually interfered with? . . .

Her supporters rallied, informing Birkdale that Miss Ryley

[85] *The Story of an Educational Boycott*, 1900.
[86] Departmental Memorandum, quoted Foster, op. cit., p. 214; *Southport Guardian*, 3 Mar. 1894.

has 'a reputation wider than this county; she is invited to speak on education, here, there and everywhere'. She again came home second in the polls.[87]

The HMI's were still warning that if the school did not improve, grant would be withdrawn. The church party, unwilling to force the issue, did little to stop a Progressive victory in the 1898 elections. The head was dismissed, the NUT blacked the post, and when a new and effective head was appointed, the staff resigned. By now there was chaos; the dismissed headmaster had died of TB, the public mood had swung, and the board was being freely accused of his murder. Miss Ryley faced catcalls, jeering, and charges of being a dictator when she appeared in public.[88]

Meanwhile, back on the board the Progressives had moved Miss Ryley to the chair as she had 'the largest number of votes . . . wide and intimate knowledge of educational matters. She has ample time, and taste and talent for public work. She is in every way qualified for the chairmanship of the Board.' But, went on her proposer, 'I was told that no woman should have that honour'. The church party shrewdly nobbled the least threatening of the Progressives, made him the chairman, and in return he agreed that a casual vacancy be filled by a sectarian, allowing them to regain control. Miss Ryley was furious, the teachers were jubilant, and the new head was dismissed, to be replaced by a man acceptable to the NUT. Miss Ryley protested vigorously at the board 'being ruled by an outside body . . . It was a grave constitutional matter'. She took legal advice and for a while withdrew from board work (and official Liberalism) in disgust,[89] working instead for women's suffrage, the peace movement, and justice for Ireland.

But there was another twist to the story yet. The ex-Progressive chairman of the board himself died a few months later and Miss Ryley as the vice-chairman was automatically

---

[87] *Southport Visitor*, 2 Sept. 1893, *SBC* 26 Jan. 1895, *Southport Visitor*, 12, 17 Jan. 1895; letter of Miss Ryley to *Southport Guardian*, 3 Sept. 1898; *Southport Guardian*, 24 Aug., 11 Sept. 1895.

[88] *Southport Guardian*, 7, 10 Sept. 1898; *SBC* 30 June, 21 July 1900.

[89] John Taylor, *WS* 13 Oct. 1898; *SBC* 4 Aug., 1 Sept., 15 Dec. 1900; *The Story of an Educational Boycott, passim*; WLA Minute-books, 15 May 1899, letter of Miss Ryley to E. McClaren, 27 Oct. 1899, WLA Minute-books.

called to the post. The church majority could not unseat her, but they could muzzle her; and they promptly excluded the press from all meetings she chaired. The secretary of the Northern Counties Education League wrote to the press:

Such a course is an insult to ratepayers, whose money these gentlemen are spending in the dark. But it is more than that. It is intended to wound and to lower in public estimation . . . Miss Ryley . . . by banishing the reporters the moment this lady opens her lips in the position to which she has been legally elected, and which she will retain by law until the board ceases to exist'.[90]

Time for the Birkdale board was in any case running out. Miss Ryley used her elevated position to campaign with the Women's Local Government Society against the Education bills which would exclude women from elected office. In 1903, however, she was co-opted with Margaret Ashton to Lancashire's new LEA where she remained until 1925, busy and happy on a committee free of the petty jealousies of small communities, working on the larger canvas she needed.[91]

Norwich

Norwich in the 1870s had a well-merited reputation for radicalism. Its Liberal leader, George White, a prominent shoe manufacturer, had long battled with the Education Department to establish free schools for needy children. The 1881 special elections,[92] however, brought in a sectarian board, as well as two women, Miss Charlotte Lucy Bignold and Miss Mary Anne Birkbeck.

Miss Bignold was a churchwoman and daughter of the town's Tory MP, maintaining 'the best traditions of the Victorian era—gracious manners, widespread philanthropy and a keen personal interest in the lives of the poor and unfortunate, and a strong supporter of all educational movements'. She was a faithful school manager, funding out of her own purse several years of school dinners, and providing meat

---

[90] *SBC* 16 Mar. 1901.

[91] WLGS Minutes, 21 May 1901, 11 Sept., 22 Sept. 1902. Miss Ryley had started a local branch of the WLGS in Southport which had placed several women on poor law boards. *Southport Visitor*, 20 Sept. 1902.

[92] The board had dissolved itself because it refused to work with its vice-chairman, Samuel Daynes, when his corrupt property transactions were exposed.

rather than soup for the children. Until party lines hardened, she voted freely for more schools, sites, and resources. She organized the lectures which started the town's university extension movement for the literate; and a probation service for the delinquent. Unlike many Tory women, she enjoyed public performance, and chaired meetings of the Primrose League that were nine hundred strong. She decorated the city walls with fetching lilac posters at election time.[93] The Liberal ladies had rather less staying power, though they supervised cookery and the welfare of pupil teachers, as women did everywhere. Lacking any independent power base of their own, and overshadowed by the men, they supported the male Liberal leadership with silence and hard work.[94]

One woman however, broke ranks: Mrs Margaret Pillow, who ran for the Norwich board in 1893 as the first woman NUT representative, against a background of NUT battles with the Norwich board. Norwich had in 1887 appointed its first school inspector. Though the town had enough places for its 11,000 children, modern buildings, and generous staff ratios, he reported that its exam results and grants were poor because its teachers were poor. The board increased its supervision. The teachers grew more restive. Mr Bidgood, former headmaster of the Duke Street Higher Grade School and now working in Gateshead, stirred matters up when he told Norwich teachers that their board was making them 'instruments for grinding money out of the children'. It was not professional to inspect teachers, he said, make them keep punishment books, deny them automatic salary increases, or expect them to resign if they did not earn enough grant. By contrast, 'the teachers of the North had a power of which his

[93] *EDP* 12 Dec. 1924; *WH* 25 Apr. 1891; election ephemera, Colman and Rye library.

[94] Miss Birkbeck was an unsectarian who lived outside Norwich; she resigned in 1886 and was replaced by Mrs Buxton, wife of the board's treasurer; on her defeat, her place was taken by Mrs Anne Burgess who ran a temperance hotel and whose husband was a former board member; on her resignation due to ill health, Mrs Harriet Green was nominated by the WLA in 1890. She resigned because of poor health in 1896, and Mrs Birkbeck, widow of a former board member, was co-opted in her place, but was later defeated. Norwich SB Minutes 1881–9: *EDP* 8 Jan. 1917, *Norwich Mercury*, 26 Oct. 1881, *EDP* 24 Oct. 1884, 21 Oct. 1887, 25 Oct. 1890, 11 Mar. 1891, 12 Nov. 1896. Between a third and half of school managers in Norwich were women, much higher than in most towns.

hearers had no conception'.[95] He thought they should put a teacher on the board.

In many boroughs, teachers had by now organized the cumulative vote to place their spokesman on the board and to bring 'realism' into debates on discipline.[96] Mrs Pillow was formidably well qualified. A Cambridge graduate, she had taught in all fields of education, ending as headmistress; she was the first woman to qualify as a sanitary inspector, lectured and wrote on health and domestic economy, and was now married to the county inspector of technical education. She stood as an independent Progressive.[97]

Her candidature naturally upset the official Progressive, Mrs Green, who had 'put in three years of hard plodding work' visiting schools, only to be told by Mrs Pillow that she had lost the confidence of the staff 'as she always did what the inspector told her'. The squabbles intensified, the women teachers supporting Mrs Green, the male teachers supporting Mrs Pillow. Both women demanded apologies from the other. George White shrewdly welcomed Mrs Pillow in general, while claiming to have some doubts in particular, fearing he said that her educational expertise would bring conflict with the inspector and disharmony on to the board. The women's Liberal association (of Colmans, Jarrolds, Jewsons, Mottrams, Copemans, and Ryumps) backed Mrs Green, ruefully noting 'the chronic state of open warfare'.[98] In the event, both women and a Liberal majority were returned. Over the next three years Mrs Pillow worked hard to improve teachers' salaries, argued for their right to use corporal punishment, had some sharp things to say about new school buildings, and tried to reduce over-pressure on teachers and children alike. She was, of course, marginalized, and had few results to show for her work. As the next elections were fought

---

[95] Mr. Phillips, *Second Annual Report*, 1889; Mr. Bidgood's speech, *EDP* 16 Oct. 1893. It was said that he was 'petted and pampered' at Duke Street and had only to ask for oil paintings to get them in his school. He just happened to be the disappointed candidate for the inspectorship. *EDP* 25 Oct. 1893.

[96] The *Southport Guardian* reported that teachers were on SBs in Birmingham, Bradford, Brighton, Bristol, Leeds, London, Plymouth, Sheffield, Sunderland, and Tyneside, 31 Aug. 1895.

[97] *EDP* 6 Mar. 1929.

[98] *EDP* 19, 20, 21, 24 Oct. 1883.

not over the rights of teachers, but over the wrongs of Nonconformists under the proposed 1896 education bill, Mrs Pillow found herself deserted by the staff, denied their plumpers, and voted off the board.[99]

Bradford

Bradford in 1870 was a town of some 145,000 and noted for its worsteds, smoke, filth, and high infant mortality. It had recently seen the great Nonconformist leader, the Revd Edward Miall, unseat a corrupt publican as the town's MP with the help of highly motivated women canvassers. Now in late 1870, the town's civic élite proposed fifteen names to the mayor as a school board. To their annoyance, the working men came up with names of their own, one of whom, James Hanson, was incorporated into the slate. The women, radicalized by their work for Miall, also proposed two names, and they were not.

The new board, like the town council, was staunchly Liberal and Nonconformist, and faced relatively few problems. It had a deficit of only two thousand places. By the late 1870s its board had sent members to London to bring back kindergarten, cookery, and music teaching; had a progressive PE syllabus which included swimming; and was the first board to open in 1876 higher grade schools, charging 9*d.* a week for languages and higher maths. Its hardest problem was the 10,000 children labouring half-time in the mills. Girls who worked the morning shift (6 a.m. to 12.30 p.m.) came to school tired, to face afternoon lessons dominated by needlework. Their formal schooling was scanty.

Meanwhile, the Liberal caucus saw no reason to include women on its slate. They were doing nicely. The women's suffrage committee, seeking a seat, wrote in January 1882 to both Tories and Liberals with a memorial signed by a thousand women voters. The Tories refused to consider it; the Liberals said the time was not ripe. The women's Suffrage Society would have let the matter drop, but some of its less docile members called a public meeting to put forward two nominees of their own. The meeting was packed. One woman Liberal moved that the matter be postponed—for one woman

[99] *SBC* 10 Feb., 18 Aug., 29 Dec. 1894; *Norwich Mercury*, 21 Nov. 1896.

on the board would be too few, and two would be too many. An official Liberal from the caucus argued, 'He admired the representation of women on the school boards; he admired still more the principle of a Liberal majority on the school board (applause)'. A woman candidate would split Liberal ranks and let in the Tories. At this point the audience took a hand. One woman said, 'if the contest was put off for three years the Liberal Executive would want to put it off for another three years at the end of that time'. Some of the men urged the women to act independently of party. Some of the women wanted to act independently of the men and sought to exclude them from the hall. Confusion reigned as the meeting agreed both that the time was not ripe to have women candidates, and that they would adopt two women candidates immediately. The husband of one of them withdrew his wife's name, as he believed her health was not too good. The other, Miss Edith Lupton, an artist from London staying with a Bradford brother, remained firm.[1]

Miss Lupton soon found she was rather good at making speeches, managing three a night in school rooms, temperance missions, and co-operative halls, advancing both progressive views on education, and feminist views on her right to stand. For someone not already a member of the board with easy access to current issues, her speeches were robust and well informed. Lydia Becker came over from Manchester to help, and was scathing both at the supineness of the official women's suffrage society and the obduracy of the men. 'Seeing that 6,000 women electors did not in any way form part of the constituency of the Liberal Four Hundred, they were right in bringing forward a candidate of their own . . . The verdict of a constituency could only be obtained by means of a contested election, and the women of Bradford were prepared to stand and fall by the result (applause).' The Liberal Eight must do likewise. One woman rose from the hall to say that as she paid rates, 'it was only right that she have her voice in the election of a lady member of the school board'. It was clear that Miss Lupton's candidature tapped much latent feminism in Bradford. To this was added an unholy glee, shared by many men, that a woman 'should put her foot on Bradford's Mighty

[1] *Bradford Observer*, 8 Nov. 1882.

Caucus'. The mayor tried to persuade Miss Lupton to stand down. When she refused, angry letters appeared in the press at her effrontery in imposing an unwanted and costly election on the town when 'she was not brought forward by anybody of position; she was not even proposed by work people; she was almost unknown here'. Election day itself was quiet—all parties had forsworn the hiring of cabs—though the press noted that some Irishwomen came to the poll obliged to plump for their priest but wanting to vote for Miss Lupton. They decided they were illiterate, and went home without voting, honour and religion both satisfied.[2] Miss Lupton took a Tory seat and the Liberal Eight were elected.

This did not mollify the majority; and as everything she said was taken as criticizing their previous work, relations did not improve. Within a few months, she had identified the pressure put on half-timers to do extra school work, and the caning they got if they failed. She helped make over-pressure a national issue. Parents wanting their children's earnings or domestic help, doctors worried about the children's health and strength, teachers concerned to earn payments by result, all joined in. The board was overwhelmed by letters, petitions, learned medical articles, parents' deputations, and resolutions coming from meetings three thousand strong. One mother wrote saying her son was not to be given home lessons. He was, and was then caned for not doing them. She took out a summons for assault and fought her case successfully through to the Queen's Bench.[3] Home lessons were stopped.

Miss Lupton tried now to stop corporal punishment, but without success. Revealingly, she did not try to limit the half-time system itself—'dull children' she said were 'the most fit for work' and they should be released to do it from the age of ten. This was not because she thought they were economic adults, or because they were from destitute families (indeed, many mill families were thought to be comfortably off); but because she was contemptuous of what the board offered in the name of compulsory education. It was 'futile'. Children were mechanically crammed, their talents ignored, their

    [2] *Bradford Observer*, 16, 21, 25, 27 Nov. 1882.
    [3] *SBC* 17 Mar., 7 Apr; 1883; *Bradford Observer*, 13 Sept. 1883; *SBC* 7 June 1884, 2 May 1885.

moral and aesthetic nature left untouched. 'The children in fact, have been entirely neglected for the sake of so-called religious, political or sectarian interest; and the policy of the board . . . [is merely] a Bricks and Mortar policy.' The board had cut salaries, cut liberal studies, and put its money into Gothic, as monuments to male ego. Every new school, she said, merely added to 'the many architectural triumphs and school failures existing in the town'.[4]

It was a pity that she was locked into the over-pressure debate. Had her flair for publicity and her energy been directed to rethinking pupil teacher training, or to the content of the curriculum, she might well have successfully challenged the narrow, repetitive, mechanical, rote learning that passed for teaching in so many board schools; and which was experienced by so many children and families as harsh, mind-numbing, and irrelevant. As the *Bradford Observer* said, 'she did much work and said many things which it was highly desirable that somebody should do and say'. She stood down at the 1885 elections, sorry to go. Her sentiments were not reciprocated.[5] She was loathed. When the Women's Liberal Association approached the caucus three years later for a woman to replace her, they were again rebuffed. This time, an Establishment woman decided to take them on: Mrs W. P. Byles, leading light of the Women's Liberal Association, whose father-in-law owned the *Bradford Observer*, and whose husband was later the town's MP. Possessing 'the platform manner of a Cabinet minister . . . she moved audiences to enthusiasm by the strength and unmistakable sincerity of her convictions'.[6]

The caucus now called in the women's committee for

[4] *SBC* 5 Sept. 1895; *Bradford Observer*, 15 Dec. 1884, 4 June 1885; *School Board Echo* (ed. E. Lupton), 27 Apr. 1888.

[5] *Bradford Observer*, 26 Nov. 1885, 19 Nov. 1888. See Minutes of Evidence on School Board Elections, 1884–5 (vol. xi), evidence of Mr Myers, Chairman, Bradford SB.

Q. 2658: Can you tell us anything about the lady who got in standing independently?—I can tell you too much about her.

Q. 2660: She had the next largest number of votes when she did come in of those elected?—It was a sort of fluke.

[6] *Yorkshire Observer*, 20 July 1931. Mrs Chanter had come on to the board without a contest in 1885 and left in 1888 without making any public impact. At the 1888 elections, Miss Freeman also stood as an unsectarian teacher, but failed. Miss Lupton offered to stand but then withdrew. See also D. James, 'William Byles and the Bradford Observer', in D. G. Wright and J. Jowitt, eds., *Victorian Bradford*, 1981.

negotiations, and agreed to put Mrs Byles on the slate. Their backwoodsmen rebelled, and the offer was withdrawn. The women were urged to accept a man to represent them instead. They refused. The caucus now suggested that a woman should run quasi-independently of the Liberal Eight but should have exclusive right to the women's vote. The women reluctantly agreed, but found that the men promptly started to poach. Mrs Byles and the WLA got more and more angry, declaring 'the Liberalism of Bradford was a weak and backboneless sort of thing'. Mrs Illingworth, as chairman, went even further. The women

were beginning to realize that the men have really set to work to see if they could put an end to their aspirations and make them retire . . . As far as the school board was concerned, the men now had to deal with the women as well, and they would never again have the school board all their own way, as they had hitherto had it'.[7]

The elderly James Hanson, the solitary working man on the board, was the only official Liberal to support Mrs Byles. He stood with her on the platform, ashamed of his erstwhile colleagues, although even he could not resist adding that 'their previous experience of lady members had not been such as to impress them very favourably'. Bradford Liberals received two polling cards, one from the Liberal Eight, and a pretty pink one from Mrs Byles. On election day she drove around in Mrs Illingworth's carriage, decorated with ivy, yellow chrysanthemums, and yellow ribbons, to find that in some wards women Liberals were working harmoniously with men, in others they were in bitter conflict, poaching each others' pledges. She was elected; and took her revenge at the first meeting of the board, joining with the Church party to elect James Hanson as board chairman over the head of the official Liberal nominee.[8] It must have been a deeply satisfying moment.

On the board she and Hanson worked together to form a radical wing, arguing for trade-union rates and an eight-hour day for staff; double-checking all industrial schools and

---

[7] *Bradford Observer*, 21 Nov. 1888. Arthur Illingworth was the town's Liberal MP.

[8] *Bradford Observer*, 20, 22 Nov. 1888; *SBC* 15 Dec. 1888.

developing special teaching for deaf and dumb children. With Hanson's help she also managed, before she retired, to appoint the country's first full-time school medical officer, Dr Kerr, in June 1893. Mrs Byles's seat passed to another radical woman, Miss Gregory,[9] but the new charismatic figure on the board was to come not from women's liberalism but from the ILP: Margaret McMillan.

Bradford's ILP had its roots in the bitter, unsuccessful Mannington Mill strikes of 1890 and 1891, and more generally, in the failure of local liberalism to accommodate progressive working men. ILP political work was sustained by local ethical societies, labour churches, and socialist Sunday schools. In 1891 they had placed the organizer of the Mannington Mills strike on the school board; Jowett had fought his way on to the town council; Ben Tillett had been narrowly defeated for the parliamentary seat. Now in 1894 the ILP ran three candidates on a programme not essentially different from that of the London progressives—healthy buildings, free and extensive day, evening, and higher grade schools, and sturdy labour policies, including a direct labour organization. To ILP dismay, only one of their slate was elected; as comrades noted 'and her only a woman'.[10]

She had come to Bradford in 1893 as an ILP lecturer, having been sacked from her post as ladies' companion in London, for speaking in Hyde Park on socialism. For her first three years on the board she made little impact, as the Tories were hostile to her views, and the Liberals unforgiving of an ILP intervention which had split the vote and given the Tories control. As she put it, 'My fellow members on the Board forget my existence for the most part and my remarks, when I make any, are listened to, stared at, and allowed to fly into space in the most disconcerting manner'.[11]

However, she attracted their attention when she took up the

[9] For Mrs Byles's work on the board, see *SBC* to Aug. 1889, 2 Aug. 1890, 9 Nov. 1889, 4 Oct. 1890. Miss Mary Gregory was a friend within the WLA, daughter of a Congregational minister, and a strong progressive. With Mrs Byles and Margaret Bondfield she tried to form a branch of the Shop Assistants Union in the town. *Bradford Weekly Telegraph*, 6 Oct. 1911.

[10] K. Laybourn, 'The Defence of the Bottom Dog: The ILP in Local Politics', in Wright & Jowitt, op. cit.

[11] *Bradford Observer*, 2 Nov. 1895; more generally, A. Mansbridge, *Margaret McMillan, Prophet and Pioneer*, 1932; M. McMillan, *The Life of Rachel McMillan*, 1927.

cause of the half-timers—'if parents were so blind as to be ready to mutilate the minds and bodies of their children, the Board ought to prevent it'. No support here for child labour, and no hesitation about child as opposed to parental rights. On the strength of a local Bradford conference on half-timers, she persuaded the board to join a deputation to the Liberal government asking that no child should leave school for half-time work until they were twelve. The minimum age was indeed raised, though the half-time system itself was not abolished until 1918.[12]

By 1896 she had found the issue which was subsequently to obsess her: the bodily welfare of the slum child. Just as Annie Besant had impressed on the LSB that you cannot educate a hungry child, so Margaret McMillan was to teach a nation that you cannot educate an unhealthy child; and like Annie Besant her emotional generosity as well as her oratory won over Bradford's board. The town's infant mortality rates and morbidity were among the highest in the country. She urged that school buildings be properly ventilated and of adequate size. When the Corporation stopped the children swimming in the public baths, she persuaded the board to attach a bath to a local school, as a way of cleaning dirty children and exercising stunted ones. She found that some children had been stitched into their clothes six months before. When first undressed they had to stand on large sheets of paper to catch the vermin. She joined Dr Kerr's medical school inspections, encouraged the board to introduce school nurses and school clinics; and with Dr Kerr wrote and sent out a pamphlet to all parents giving simple guidance on child health. Like most Progressives she also fought for school feeding, special classes for retarded children, and open-air schools for delicate children. As she said during her 1897 election campaign, she would allow others to argue over religion; personally she believed that 'every ill ventilated school should be looked upon as a blasphemy (applause) ... electors ought to be asking themselves why it was that any School was understaffed, any child underwashed, or any teacher died of consumption', from working in stinking, overcrowded classrooms.[13]

By 1902 her own health had broken down and she returned

[12] *SBC* 24 Jan. 1895.   [13] *Bradford Observer*, 3, 6 Aug. 1897, 8 Nov. 1898.

to London to pioneer nursery education in Deptford. Bradford's women board members had shown a courage and exhibited a commitment that was to shame its board.

Women's work in elementary education was as varied as the boards on which they sat—for great manufacturing cities, county market towns, or resort spas. Their work was shaped by the political complexion of the board, its civic pride, its educational ethos, and the sophistication of its members. It reflected the extent to which women had an independent power base of their own, a well-organized female electorate, an extensive philanthropic network, an articulate women's suffrage or Liberal association; and therefore the degree to which women were welcomed, merely tolerated, or actively thwarted by the official party caucus. Those coming on to boards in the 1870s faced a numbers problem that dominated all other issues and which seemed more naturally male territory. By the 1890s women had brought quality questions to the fore, and their work was seen as more valuable and less threatening. Women's contribution also depended in no small measure on the personal skills they could deploy, their willingness to commit income, leisure, and devotion to the service of the working-class child, the support they had from family and friends, their willingness to defy convention, brave heckling and ridicule, and risk their good name. Many women worked themselves into exhaustion—time and again, women retired, their health broken, to die a year or two later. It was said of Florence Melly that no paid official ever worked harder or longer hours than she did. Few would have known, as she bicycled to every elementary school in Liverpool in the worst of weathers, that she was a diabetic struggling always to keep her condition stable.

The political and personal style of school board women varied from town to town. Yet almost without exception, they shared a perspective and commitment in common that set them apart from their male colleagues.

The first of these feminist strands was their clear view, best put by Lydia Becker, that by standing for elected office they were staking out the claims of women to a wider public and political life. Every success was swiftly quoted by suffrage workers and friendly MPs to show that women were fit for the parliamentary vote. During the 1870s women mostly stood as independents, in part to distance themselves from church–chapel controversy, but also to emphasize that they were standing primarily as women. By the 1880s they considered it the greater triumph to be part of an official Liberal slate, able to count on voting majorities for their work on the board. By the 1890s they were usually members of the Progressive alliance. They received loyal support from women voters, women's organizations, and usually the local press; and they took to the platform and public controversy with surprising skill and courage, regarding it as a duty somewhat analogous to religious testimony or the taking of the pledge. It wasn't easy. As Emily Sturge of Bristol said,

the prospect of a contested election is very alarming to many ladies, and it never quite loses its terror for the most hardened. On the first occasion you know so little of the work that you are dismally conscious that you have almost nothing to talk about at the inevitable meetings, and afterwards you feel rather tongue tied, because you dislike to say anything which could hurt the feelings of colleagues with whom you have worked pleasantly for three years, whom you respect, and yet you may be, educationally speaking, in the opposite Camp. And sometimes there is some amount of heckling on the part of your own party to be gone through which is anything but pleasant. But all things have an end, the speeches are done, the votes are counted, and you find yourself one of 15 people of all sorts of different creeds and professions, who have to shake down so as to be able to pull together in harness for the next three years.[14]

Some women found it very hard to expose themselves to the personal comments and familiar remarks of electioneering. The quest for votes, the promises made, the selling of self, the easy friendliness with all and sundry, seemed a form of prostitution from which ladies instinctively recoiled. But they steeled themselves with what was, quite simply, an article of faith: children needed women on the board; the community needed women in public service.

Conscious that they were highly visible, they took

[14] Miss Sturge's speech to the NUWW Conference, Liverpool, 1891, p. 18. Florence Melly, *Liverpool Mercury*, 20 Sept. 1928.

themselves and each other very seriously. Miss Alice Busk advised women:

Be not too ready to speak at the Board and Managers' meetings; loyally abide by the vote of the majority; never be the partizan of a particular teacher; and never discuss members decisions outside the Council Chamber. Should you see something that needs to be done, resolve to accomplish it in the face of any opposition. Marshall your facts carefully; be able to say, I have seen, I know, not I hear, I am told.[15]

Many women like Elizabeth Garrett or Catherine Buckton turned out to be superb candidates; others like Ethel Leach were effective and eloquent polemicists at the board. Most proved to be skilful committee women, well-informed, tenacious, and good-tempered. And all were devoted educational social workers within their patch, speaking to the needs of damaged children and deprived areas with impassioned sincerity.

Secondly, these women stood for public office not just to advance the rights of women, but to promote the well-being of infants, girls, women teachers, and staff. Without hesitation, women played the separate spheres card, insisting that as most children were infants and girls, and most staff were girls and women, they needed the tender loving care of women board members. Few men were willing, at least in public, to deny that education was suitable work for a woman.

So women members brought kindergarten methods, cots, toys, flowers, bright colours, and movement into infant teaching. They taught men that it was unnatural for four-year-olds to sit long hours patiently and silently on hard wooden benches; that children learnt through play, through exploration, through doing. They offered seminars on Froebel methods themselves, and persuaded boards to offer in-service training to staff.

They immediately took responsibility for the curriculum of older girls. Almost to a woman they deplored the three afternoons a week given to fine needlework as befitting only future ladies' maids. Instead they introduced cookery, home nursing, and laundry work to equip girls with the relevant skills to

---

[15] NUWW speech, Brighton, 1900, p. 66.

make their own lives and their own homes more comfortable. Many women members were in fact semi-professional lecturers in domestic economy, and even those whose personal housewifery skills were in somewhat short supply could not avoid the job. A number of them also tried to get boys taught such skills, so that as single men in the Services they could be self-reliant, and as family men would be better fathers. And they conceded not an inch in their claim that girls, along with boys, should have equal formal education, equal access to endowments, equal opportunities to proceed up the educational ladder. They were well aware that on older girls were focused all the conflicting pressures of school, family, and work. They tried to prevent the educational needs of girls being consistently sacrificed to the larger claims of their families. Many girls dropped out of school early. Women members started Kyrle Societies and Snowdrop bands to offer them recreational evenings, which would keep girls off the street and compensate for the sacrifices forced upon them by working-class culture, working-class chauvinism, and working-class poverty.

Brighter girls were recruited as pupil teachers, and women members struggled to prevent them being undereducated and overworked. They brought newly appointed women staff home to tea and cakes, watched over their health and welfare, and sent them off on country holidays or helped them move to easier schools when their health broke down. Women members themselves, with their links to Emily Davies's higher education movement, often organized lectures for pupil teachers and young women staff of an evening, which became the basis in many towns of the university extension movement.

Again, to a woman, they tried to get equal pay and wider opportunities for female staff. Mrs Spoor made some progress in West Hartlepool, but as Mrs Gardner of Hull and Mrs Pankhurst of Manchester pointed out, 'more work and less pay is evidently the fate of female teachers',[16] even though single women faced higher expenses (in clothes, travelling, decent accommodation) than men. However, women were as divided as men on the proper behaviour of married women teachers. Mrs Tubbs of Hastings would have women staff

[16] *SBC* 24 Sept. 1892, 9 Dec. 1899.

resign on marriage; others would have them resign on pregnancy; still others thought that if mothers were asked to resign, so equally should fathers. Miss Shilston of Seacome tried to get the word 'male' deleted from all job advertisements; a colleague said he would agree if she would raise a battalion to fight the Boers.[17] Whenever they could, women recruited other women as school visitors, attendance officers, clerks, and head teachers, though their success here was as patchy as on the LSB.

Thirdly, women board members were strikingly child-centred. They opposed payment by results, not for the sake of the staff, but for the sake of the children, who were, said Miss Hearn of Northampton, 'certainly being taught but were not being educated'. Apart from kindergarten, they introduced manual training, and hand-and-eye work, for as Miss Clephan of Leicester said, 'the awakening of intelligence and the training of the faculties of sight and touch' were more lasting than 'mere book learning'.[18] They persuaded teachers to experiment with phonic reading schemes; and working closely with the inspectorate, they did more than most to make a reality of Huxley's curriculum. It was the women who moved the syllabus away from the 3 Rs, needlework for the girls and drill for the boys. They introduced social economy, physiology, nature study, books, and music into schools. They also tried to introduce temperance teaching, temperance textbooks written by their relatives, temperance prizes which they donated, and temperance Bands of Hope which they had founded, to the mild irritation of male colleagues.

They were aware too that much of what passed as slowness and stupidity among poor children was due to defective hearing, poor eyesight, and the listlessness of malnutrition. Men might pronounce in favour of cheap school dinners; but it was the women members who organized them during the 1880s and 1890s. In their turn, women persuaded male members to abandon military drill and replace it with gym, swimming, and organized games, which were less regimented and offered more exercise, and thus improved the children's moral and physical health. Many a poor child managed to attend school in the decent second-hand boots found for it by a woman member and the clothes cut out or acquired for it by

[17] *SBC* 9 Dec. 1899.   [18] *SBC* 27 Oct. 1900.

women teachers. School medical inspection and school baths were both pioneered by women: Miss Dixon in Keighley and Lady O'Hagan in Burnley persuaded their boards to do what Bradford's Margaret McMillan had done. Those board members opposed to corporal punishment were invariably women (occasionally supported by the board's token working man) and accounted for much of their unpopularity with male assistant teachers who had daily to face the large classes of disruptive and undisciplined boys. Mrs Fielden wanted heavy-handed teachers fined. Miss Soames of Brighton wanted them sacked.[19]

Their hearts went out in particular to the neglected, the difficult, and the handicapped child. They were scrupulously vigilant in their inspection of industrial schools, and almost every case of institutional cruelty reported in the *School Board Chronicle* was uncovered by a woman. Indeed, they were often the only board members to visit the children at all in the more remote country industrial schools, giving up their holidays to do so, and keeping in contact with them for years after, writing letters to them as they served in the trenches on the Somme. Time and again, only a kindly conscientious woman board member stood between a child and its abuse. During the 1890s as the plight of defective children became publicized, it was women members who organized the special classes, paid for the ambulances to bring children to school, and nursed delicate children back into health with home-grown vegetables and tempting meals. They started after-care support schemes for the more fragile and feeble. In return, they were rewarded with the love and affection of children and female staff. When a woman member like Mrs Wilson of Sheffield proposed to leave the board because she was overworked, she relented when begged to stay on by women teachers who saw her as their only friend in court. For unlike most board members, women had and took the time to visit schools, to know every teacher, to recognize many children. The obituary on Miss Wilkinson of York may stand for countless more: she 'regarded the children as her own and expressed great pleasure at the constant recognition she received in the streets from them'.[20]

---

[19] *SBC* 13 June 1885, 17 Aug. 1889.   [20] *York Herald*, 6, 10 Mar. 1911.

Finally, most women members were particularly sensitive to the relation of school and community. They found many mothers so hostile to compulsory attendance that in Bolton in 1879 they boycotted school board elections. As the *School Board Chronicle* noted, 'the very poor of whom we speak never asked to have education provided for their children, never wanted it, have practically nothing to gain by it and much to lose'. Statutory schooling was 'not for their good or their pleasure' but for 'the safety and progress of society'. It was a public good paid for by the poorest.[21] So women members deliberately sought working-class votes, canvassed in working-class areas, and tried to represent the working poor. As they sat on attendance committees—and having the time they sat on them longer than men—they tried to soften the impact of compulsion by seeking parental consent and by remitting parental fees. Miss Avery of Barnstaple sat quietly through committee meetings taking careful notes, and would then spend the rest of the week visiting defaulting parents, gently remonstrating with tired mothers in their homes. Women everywhere walked and cycled their streets, scrutinizing sites, inspecting buildings, assessing the traffic, checking the distances, making sure that new board schools would as far as possible serve their local community, by evening as well as by day. They argued that parents would more willingly send children to school if they had a choice of schools, unsectarian board as well as sectarian voluntary schools; and if the children were taught not by other children but by trained adults. Miss Wilkinson failed to win this argument at York's board meeting, brought in the Education Department, and the town was required to build an additional thousand school places.[22] For the same reason, women usually tried to devolve power down to local school managers, where the voice of parents could more easily be heard. It was noted too that women were among the board members most willing to support good labour practices, trade-union rates, and a direct labour organization, which brought them warm support from working men's associations. The somewhat stilted praise of Mrs Evans from Leicester's senior inspector said, 'She brought to a successful initiation a variety

[21] *SBC* 18 Dec. 1884. The information on Bolton I owe to Dr John Garrard.
[22] *The North Devon Journal*, 7 May 1903; *SBC* 22 Oct. 1892.

of reforms, that none but a woman, and a very highly gifted one, morally and intellectually, would have been bold and skilful enough to do'.[23]

Did it, in the end, make much difference? There were some boards of large towns, such as Hull, Newcastle, and Plymouth, where women came late or not at all. But there was no board regarded as progressive by the press or by the Education Department which did not, for much of its life, have women members on it. Boards without women or whose women were elected late on, did introduce cookery, did sometimes develop kindergarten, did agree to school feeding, did run special classes for handicapped children; but such developments usually took place years after they were standard elsewhere and often then only because local women's philanthropic groups or the local HMIs put great pressure on them to come into line. If boards lacked women members, they had usually either to co-opt docile female relatives to form a ladies committee or employ professional female staff, to organize womanly work. By the mid 1880s, certainly, boards who liked to think they were progressive were on the defensive if they did not have a lady member. Liverpool, Wolverhampton, and York co-opted women to fill casual vacancies; they then held their seats at subsequent elections. Mrs Sarah Wilson came on to Sheffield's board for the same reason. Once there, it was to women that education pressure groups, HMIs, and working men's organizations looked, as the voice of enlightened policy.

Outside London for most men other than the chairmen of the great city boards, school board work represented their modest contribution to civic life, and took place at the interstices of their professional jobs and their political activity. It was very part-time work. For women, their school board work was their duty, their faith, and their delight. Lady O'Hagan became chairman of Burnley's board in 1901 on the hesitant but casting vote of the board's only Socialist. She wanted, she said, for Burnley 'an education as complete for the poorest in the town as it was for the richest'. She, like very many women, gave her life to that service.[24]

[23] *Leicester Daily Post*, 24 May 1894.   [24] *SBC* 19 Jan. 1901.

THE POOR LAW

4

The Election of Ladies 1875–1914

In Preparation

MALE clerics and male caucuses might be hostile to making
space for women on school boards, but the 1870 Education
Act had at least made it plain that women could stand for
election. Whether women could stand for poor law boards was
simply unclear. When Bristol women gathered in Lewis Fry's
drawing-room in February 1882 to see if they could elect
women on to their local poor law board, the Dean of Bristol,
Dr Beddoe, startled his listeners when he told them that as
a young clergyman in a country parish he had helped to elect
a woman guardian back in the late 1830s. He recalled that the
rest of the board, farmers to a man, had promptly left her to
get on with it.[1] The women's movement could also cite one or
two other examples of women who had been elected or co-
opted to parish and poor law office without legal challenge in
years gone by. But Baroness Burdett-Coutts, the wealthiest
woman in the land, had failed to stand in Marylebone in the
1860s when her own sense of propriety had been reinforced by
the legalistic doubts of vestry officials. She had withdrawn her
name. Now that women were serving with aplomb on school
boards, the next step was to test the field of poor law work,
where, women argued, they were equally needed, equally
suitable.

The work of guardians was conducted within the framework
of the Poor Law Amendment Act of 1834. From the late
eighteenth century, a rapidly growing population drifted off
the land into burgeoning 'frontier' towns in the industrial
Midlands and North. Parish poor relief systems, designed by
the Tudors for small, stable rural communities, broke down.
Demands for outdoor relief multiplied, poor rates soared.

[1] *EWR* 15 Feb. 1882.

Pauperism was thought to be eating up property. The Swing agricultural riots of 1830 suggested that social discipline was in disarray. The 1834 New Poor Law was therefore designed not only to cap poor law expenditure, not only to restructure the administrative machinery of poor relief by joining small parishes into larger and financially more viable unions, but also to discipline the roaming, the restless, and the unemployed by permitting them poor relief only within a penal workhouse. For the poor law assumed that pauperism was a moral disease, unemployment a refusal to seek work. If the conditions of poor relief were harsh enough, workhouse life bleak and severe enough, then the merely malingering and the morally supine would seek work with new-found enthusiasm. Only those utterly degraded and destitute would still seek to enter the workhouse. Said Mrs Shaen, a London guardian, it would be 'a defence between a human being, however bad, and starvation'. Nothing more.[2] It was an analysis obsessed with able-bodied men who should, but apparently would not, work. Scant attention was paid to the much larger groups of destitute elderly, abandoned children, and helpless women who also had nowhere else to go. Chadwick had envisaged separate workhouses for them; they ended up in the general workhouse.

The 1834 Poor Law was loathed and feared by the poor. Claims that if the rich were to enjoy their rents then the poor had better receive the rates echoed through the class bitterness of Chartism, and beyond. Working men were usually as harsh as anyone else on so-called scroungers; but they also knew that most poverty and destitution arose not from idleness but from low wages, erratic employment, large families, sickness, widowhood, and old age. When they joined poor law boards after 1894 they were to demand appropriate responses; generous outdoor relief, public works for the unemployed, and pensions for the elderly. As George Lansbury said, he could not quite see why it should demoralize the poor but not the rich to receive a pension. As for the cost, 'Hang the rates', he exclaimed. The rates were a community chest to be spent on community goods; you paid in when in work and took out when in need.

---

[2] Mrs Shaen, NUWW Conference, Nottingham, 1895, p. 155.

The facts of unemployment were understood perfectly well at local level. In rural areas of arable farming where winter work was short, and in textile towns where work depended on the order book and the trade cycle, outdoor relief continued to top up the family income. In one guise or another, most able-bodied paupers managed to remain outside the workhouse and collect some outdoor relief, thus undermining the principles of 1834 at their very core, despite intermittent attempts by the Central Poor Law Board to tighten up the system. At the same time, the orphaned child, the deserted wife, the chronic sick, the unmarried mother, the simple-minded, and the infirm elderly, however deserving, often had nowhere else to turn. Much of the history of poor law work, and especially of women's work, over the next half century, was the attempt to modify and reform a punitive institution that was misconceived and misbegotten: designed for idle able-bodied men, but occupied by the very young, the very old, and women. Few women guardians were prepared to argue that the principles of 1834 were fundamentally flawed, that it made villains of victims. Instead they tried to impose the principles of 1834 with renewed urgency, believing (falsely as it happened) that pauperism was on the increase.[3] They sought to reduce and if possible abolish outdoor relief to the able-bodied, thus forcing them into the workhouse, while at the same time removing the 'innocent' inmates of workhouses into more suitable and specialized care. The measure of their success was reflected in the Majority and Minority Reports of the 1909 Royal Commission on the Poor Laws, both of which called for the end of the workhouse. The scale of their failure is to be seen

	Paupers			
	Indoor	Outdoor	Total	% of population
1860	113,507	731,126	844,633	4.3
1870	156,800	876,000	1,033,800	4.6
1880	180,800	627,800	808,000	3.2
1890	188,000	587,000	775,000	2.7
1900	215,400	577,000	792,500	2.6
1910	275,000	539,600	916,400	2.6

Source: 31st Annual Report, LGB 1901–2. Appendix E is the main source.

[3] The 1870 figures reflect the Lancashire cotton famine. Rowntree's study of York in 1901 calculated that half of primary poverty came from low wages, about a quarter from the death of the main wage-earner. The 1910 sub-totals exclude casuals and insane paupers.

today when elderly people shun hospitals that were former workhouse infirmaries, avoid old people's homes but queue patiently for sheltered housing. The workhouse continues to cast its shadow over social policy, exactly as the Victorians intended it should.

From the 1820s the Quaker, Elizabeth Fry, and Sarah Martin had been visiting prisons. Now, in the early 1850s, as Chartism faded and Anglo-Catholicism flowered, sisterhoods of women (often modelled on the Kaiserworth deaconesses in Germany) sought to build bridges between rich and poor by offering their lives to service. Other women were inspired by Mrs Anna Jameson's call to a *Communion of Labour*, that the world needed the services of women. Some nursed the sick, others found their way into the workhouse. Frances Power Cobbe, staunch Bristol feminist, and Louisa Twining, daughter of a wealthy tea merchant, went to visit former servants in the House, and were appalled by what they found—inedible food, the sick and dying un-nursed and their sheets unchanged for months on end, the elderly sitting out their lives on hard benches in bleak, empty, echoing rooms, stunted children listlessly mingling with the depraved and the deranged. They were out of sight, out of mind.

Louisa Twining could not walk away. She asked the local guardians if ladies could visit officially with flowers, books, and sympathy. She was refused. The guardians feared interference, the staff feared scrutiny, the clergy feared religious enthusiasm. She asked the Poor Law Board for permission. She was refused. So she began to make private visits, take notes on what she saw, write to the press, and build up a body of concern. In 1857 she found her public platform. The Social Science Association had been formed to promote good practice and benign law at the interface of public and voluntary work. From the first, women were full members of the SSA and following an impassioned address by Miss Twining, the Association set up a subcommittee to promote workhouse visiting. It became the Workhouse Visiting Society (WVS).[4]

---

[4] L. Twining, *Workhouses and Pauperism*, 1898, p. 42. The objects of the WVS are described in the *Englishwomen's Journal*, July 1858:

> . . . to promote the moral and spiritual improvement of workhouse inmates . . . [by] the introduction of a voluntary system of visiting, especially by ladies, under the

As Mrs Jameson said, 'Why should not our public workhouse be so many training schools where women might learn how to treat the sick and the poor and learn by experience something of the best means of administration and management?' The visitors, as much as the visited, would benefit and women would be working for women.[5] From the WVS flowed the demand for women guardians.

Within a few years the WVS had some 340 members visiting a dozen London workhouses and many in the provinces. They tried to win the support of doctors and clergy, and to avoid alienating the paid staff. They introduced lockers and shawls in one workhouse, better lavatory facilities in another, toys for the children and cushions for the elderly in a third; and everywhere pictures, books, and flowers. Miss Cobbe delighted her old ladies by giving them a canary in a cage for the window. Lady visitors tried also to befriend children and raise the standard of nursing. The WVS with Baroness Burdett-Coutts's money opened a training home for workhouse girls entering domestic service. Meanwhile, Louisa Twining toured the country, lobbying guardians to open their workhouses to lady visitors, seeking always a larger role for women within poor law work. Even 'as mere visitors' said John Stuart Mill to the first AGM of the women's Suffrage Society in 1869, women had uncovered 'enormities . . . which had escaped the watchfulness of men specially selected as fit to

sanction of guardians and chaplains, for the following purposes:

1. For befriending the destitute and orphan children while in the schools, and after they are placed in situations.

2. For the instruction and comfort of the sick and afflicted.

3. For the benefit of the ignorant and depraved, by assisting the officers of the establishment in forming classes for instruction; in the encouragement of useful occupation during the hours of industry; or in any other work that may seem to the guardians to be useful and beneficial.

For the SSA, see E. Yeo, 'Social Science and Social Change: A Social History of some Aspects of Social Science Investigation in Britain 1830–1890', Sussex Ph.D., 1972. For L. Twining, apart from her own writings see K. McCrone, 'Feminism and Philanthropy in Victorian England: The Case of Louisa Twining', Canadian Historical Association, Historical Papers, 1976. And more generally, E. Ross, *Women in Poor Law Administration*, London MA, 1955.

[5] A. Jameson, *The Communion of Labour*, 1856, p. 102. Louisa Twining said of her semi-mystical writing, with its notions of sacrament, sacrifice, and service, it 'leavened the whole course of public opinion upon women's work to a degree little thought of or acknowledged by those who have inherited unconsciously her ideas and entered into her labours'. L. Twining, *Recollections of her Life and Work*, 1893, p. 155.

be inspectors of poor houses'. And Miss Cobbe noted in her *Autobiography* that the WVS 'brought by degrees, good sense and good feeling quickly and unostentatiously to bear on the boards of guardians and their officials all over the country, and one abuse after another was disclosed, discussed, condemned and finally in most cases, abolished'.[6]

By 1865 the WVS had disbanded, successful in its immediate aims. Ladies were now visiting scores of workhouses, and they all knew what needed to be done. But while they were unofficial, they could only soften the system at the edges. 'The utmost caution was necessary in conducting the proceedings of this unauthorized and unofficial visitation', wrote Louisa Twining in 1880, 'as the visitors were entirely in the power of the guardians', and their rights could be withdrawn at any time. 'Never be a lady visitor', said another woman visitor to a lady seeking to contribute to poor law work, 'because you have no power and no position'.[7] For power and position, women had to become guardians, as Louisa Twining informed a startled Select Committee on the Poor Laws in 1861. Ladies in fact continued to act as voluntary visitors until the end of the century, gaining experience, mitigating the lot of inmates, boarding out children, introducing Brabazon occupational therapy schemes into the workhouse and after-care schemes (such as the Metropolitan Association for Befriending Young Servants and the Girls Friendly Society) for those leaving it. Not until 1893 was their position regularized by the Local Government Board, and by then ladies committees were rather popular with male guardians who hoped that way to avoid having elected women guardians forced on them. When a newly elected woman guardian innocently sought to start a ladies visiting committee, she was somewhat taken aback to find her proposals resisted by other women guardians who saw it as the soft option.[8]

Apart from the campaigning of Miss Twining and the WVS, three other developments were to bring women into

[6] J. S. Mill, Annual Report, Women's Suffrage Society, 1869, p. 10; F. P. Cobbe, *Life of Frances Power Cobbe as told by Herself*, 1904, p. 318.

[7] L. Twining, *Workhouse Visiting and Management*, pp. 60, 69; Report of Poor Law Conferences, SW District, 1901–2, p. 409.

[8] *Parish Councillor*, 13 Sept. 1895.

official poor law work. The first of these was the professionalizing of nursing; the second, the professionalizing of philanthropy; and the third was the professional inspection of poor law children by Mrs Nassau Senior.

Frances Power Cobbe had found that some of the worst wards in the Bristol workhouse were its sick wards. Low as pre-Nightingale nursing standards were in voluntary private hospitals, those in workhouses were unbelievably worse. By 1869 the sick formed over a third of the workhouse inmates, 'nursed' by their fellow paupers, who were sometimes too illiterate to read medicine labels, often too drunk or infected to follow hygiene instructions, or too elderly and frail to lift, wash, or change their patients. In Yarmouth, where the head nurse was over eighty, Mrs Leach found the almost helpless were nursing the utterly helpless. The sick rotted to their death in verminous beds in foetid wards. The WVS had long argued that sick wards needed proper nurses (Dr Rogers, the gallant Medical Officer of the Strand workhouse, was one of the few to introduce paid nursing), as well as purpose-built infirmaries, and adequate salaries for medical staff.[9] In the mid 1860s William Rathbone, Liverpool's merchant prince, turned to Florence Nightingale for help with his workhouse infirmary of 1,300 patients, one of the roughest and most notorious in the country. Florence Nightingale had used the Crimea memorial fund to start the first nurses' training school. From there in 1865 Agnes Jones went north to Liverpool with a team of trained dedicated nurses. She was desperately overworked, yet 'a picture of happiness' according to her family. She worked wonders of order and cleanliness, before tragically contracting typhoid in 1868 at the age of thirty-five from another nurse.[10] The publicity surrounding her death, the campaigning by the *Lancet*, and the political clout that Miss Nightingale now brought to workhouse nursing, ensured that the need for trained nurses was more widely recognized. Louisa Twining, practical as ever, in 1879

[9] F. P. Cobbe, *The Workhouse as a Hospital*, 1861. For Yarmouth, see below. Dr J. Rogers, *Reminiscences of a Workhouse Medical Officer*, 1889, for his description of the foul wards in the Strand and the cruelty and corruption in Westminster Infirmary. He paid tribute to the lady visitors as among his few allies in providing medical care, pp. 22–3. Mrs Beeton was one of his first paid nurses.

[10] *Memorials of Agnes Elizabeth Jones by her Sister*, 1871.

founded the Workhouse Nursing Association to increase their number, and some guardians such as Eva Muller took training themselves. As standards rose in voluntary hospitals, women guardians struggled to bring pauper nursing to an end, and to set similar standards of care in workhouse wards, but many male guardians remained unconvinced that trained nurses were needed by the elderly infirm or deserved by women off the streets.

1869 had also seen the founding of the Charity Organization Society to repress mendicancy, rationalize and professionalize philanthropy, and hold firm to the principles of 1834. Over the years, the COS encouraged its members to become guardians, and co-opted guardians on to its committees. COS-trained women became rent collectors, district visitors, sanitary inspectors, and guardians. Those women guardians who came from other roots, such as parish work, moral purity movements, rescue work, or temperance organizations, almost always spoke in COS tones, tough, stringent, judgemental; they shared conferences and committees with COS workers; and they never doubted 'that their insight of character could discriminate the really deserving'.[11] At its best, the COS instilled in its women workers both business methods (it was guardians like Miss Brodie Hall of Eastbourne who introduced the case-paper system) and an enlarged sense of social responsibility. As Mrs Armytage said to the NUWW in 1893, 'We ought not to think that it is sufficient to be a Woman, with a capital W, to play the angel on the Board of Guardians'. Public and philanthropic work must lose its image of being glamorous, fashionable, desultory, self-indulgent. It required staying power, hard work, firmness, training, and selfless personal service. 'It is *labour*,' wrote Miss Gertrude Wilson, a Leeds guardian, 'let no one be deceived into imagining it is a dilettante occupation or a harmless hobby for a few hours of leisure; let any one who will do it, do it well, and let her understand that means doing it by hard, persistent, thorough work . . . but work which I, at least, have found practicable, honourable and profitable.' Mrs Phillp of Birmingham reported that on the election of two women, an official told her, 'The first thing one of them wanted to do was look at my books', something no male guardian had ever asked. That was the authentic COS note. When Sophia Lonsdale did precisely that at Lichfield, she found the master and matron were fiddling the accounts, and sacked them both.[12]

However, the COS also generated a cast of mind that many found repellent. Its women often became case-hardened, inquisitorial, authoritarian. Miss Wilson of Leeds noted that women guardians were often harsher than men, especially when COS training was overlaid with moralistic and simplistic temperance views. Miss Augusta Brown, a Camberwell guardian, was asked in 1895 whether women were not too emotional for poor law work. She replied that on the contrary they were 'not sympathetic enough; they are too hard and severe, especially to their own sex.'[13] Miss Brodie Hall, an Eastbourne guardian, was splendidly tenacious in her efforts to board out children. She was also the guardian who persuaded her board not to give a small pension to a worn-out fever nurse of seventy, because 'the women ought to have saved money'. She wanted the names of those receiving poor relief to be circulated in the town to shame some and deter others. Her head for business was widely admired, but the male chairman rebuked her saying, 'It was hard enough to be poor, and it was not difficult to imagine the misery which might, in many cases, by the adoption of this plan, be produced upon people who through no fault of their own had been reduced to penurious circumstances'.[14]

COS women were convinced that pauperism was a state of mind, a habit of dependence, not a cluster of 'penurious circumstances' such as unemployment or ill health. They came on to poor law boards in the 1880s horrified by the dirt

[12] NUWW Conference, Leeds, 1893, p. 231; NUWW Conference, Glasgow, 1894, (Mrs Phillp) p. 80; V. Martineau, *Recollections of Miss Sophia Lonsdale*, 1936.

[13] *London*, 22 Aug. 1895 cf. Revd Johnson of Northampton in 1906, women guardians were 'very hard on their own sex—harder he thought as a rule than the men were'. Poor Law Conference, West Midlands, 1906–7, p. 19. But see below, p. 287.

[14] *Eastbourne Chronicle*, 25 Apr. 1885, 14 Apr. 1883. See also *Eastbourne Herald*, 5 Aug. 1939, 'Whilst no trouble was too great to ensure that the children in the care of the guardians had the best possible start in life, she was no weak sentimentalist in the administration of the poor law'.

and disorder they saw around them, appalled at the casual generosity and equally casual neglect from old-fashioned guardians. They brought high hopes that the pauperized need not always be with us. But they were often in their turn to be cast as reactionaries, to find themselves isolated and often voted off the board, when the post-1894 generation of guardians insisted that the problem was poverty not pauperism, and that unemployment was a failure of society not a lack of moral fibre in the individual. COS women were beached amid the wreckage of an irrelevant ideology.

The campaigning of Louisa Twining and the WVS, the spotlight on workhouse nursing, the arrival of the COS, as well as the granting of the ratepayer franchise to women, and women's work on school boards, all helped to place women on poor law boards. One event in particular aroused women's conscience, and propelled many of them into doing something about it: Mrs Nassau Senior's report on workhouse girls.

Workhouse visitors had for many years reported that children were damaged by workhouse life. They became institutionalized, apathetic, and sullen, unfitted for any life outside the monotonous regularity of the workhouse. They were like watches, 'whereof the main spring is broken. The hands may be directed whither we will, but they have no motion of their own,' said Miss Florence Hill. Girls suffered especially, wrote Miss Cobbe.

Girls want affection, want personal care, want household duties, want everything which can train them to honour their bodies, and keep pure the souls which God has given them. To effect this, we mass them by hundreds where they have no affection, no personal care, nay, hardly a personal existence at all, save as units in a herd, no household duties; and as much degradation as hideous uniform, and cropped hair, and shoes which change the natural lightness of youth to the shuffle of age, can possibly achieve.[15]

Already Joanna and Florence Davenport Hill had begun their long campaign to board orphan children out into foster homes rather than warehouse them in workhouses or send them off to large district boarding-schools. James Stansfeld himself, the

---

[15] F. P. Cobbe, *The Philosophy of the Poor Laws*, 1865. F. D. Hill, *The Children of the State*, 1868, p. 13.

newly appointed Local Government Board Minister, and a noted feminist, was perturbed at reports of ophthalmia and ill health in poor law schools. He appointed Jane Nassau Senior, sister of Thomas Hughes, MP, and daughter-in-law of the noted economist, to inspect them. She visited schools, workhouses, and orphanages, and traced seven hundred ex-workhouse girls, reporting in 1873 that two-thirds of them did badly, failing to hold down a job, drifting into prostitution before returning back to the workhouse to pass their taint on to their children in turn. The reason? They had been deprived 'of the cherishing care and individual attention that is of far more importance to the girl's character than anything else in the world . . . What is wanted in the education of girls, is more *mothering* . . . '.[16] Ill health forced her soon to resign and she died of cancer in 1877 universally mourned—'so bright and beautiful,' grieved Mrs Elmy—but not before she had seen the first women elected as guardians, inspired by her Report to mother the children of the state.

Obstacles to Election

Though women could have tested the law at any time after 1834, it took the work of Louisa Twining, Agnes Jones, and Mrs Senior, together with the example of women on school boards, to bring women forward. Why so late and so few, when compared with education?

Much of the difficulty was technical. It was easy to be a school board candidate, easier indeed than to be a school board voter. Any woman, with or without property, husband, or home, could stand for any school board. The electoral system of unwarded towns, multi-member divisions, an

---

[16] *WS* 11 July 1895: M. Smedley, ed., *Boarding Out and Pauper Schools: Mrs. Senior's Report*, 1875, p. 73; *Spectator*, 7 Apr. 1877. Stansfeld described Mrs Senior as 'a woman of great intelligence, goodness and charm, and favourably known to society and the press'. The appointment met with little opposition, possibly because she was a friend of the Chancellor of the Exchequer, Robert Lowe, who thought her salary 'not half enough'; *WS* 11 July 1895. She was also a friend of Mrs Barnett with whom she worked in Whitechapel. One night the children deserted the class to watch a fight. Mrs Senior stood on the steps and sang a hymn to draw them back. Said Mrs Barnett, 'As she stood in the dark court with a background of flaring gas light which turned her flaxen hair into a halo, she seemed to some of us to be one of the angels of whom she sang'. H. Barnett, *Canon Barnett, His Life, Work and Friends, By his Wife*, 1921, p. 43. Mrs Senior's committee of helpers went on to become MABYS.

unweighted franchise (all ratepayers had the same number of votes), and the cumulative vote, could not have been better designed from a woman's point of view. The exact opposite was true for poor law work. Poor law guardians, from time immemorial, had not only to be ratepayers but substantial ratepayers, rated to a minimum of £15 outside London, and up to £40 in London. Those with most experience, married women, and with most leisure, their daughters, were effectively disqualified from standing as a guardian. Only one householder in six was a woman. Of these, very many were elderly, or widowed with dependent children; others did not always possess a sufficiently high property qualification; and yet others lived some distance from where they did their public work. It was almost impossible to find a qualified woman ratepayer in the East End of London, for example— Miss Eva Muller, who lived in the West End, had to buy a spare house in Lambeth. It was not clear whether a married woman owning property under the Married Women's Property Act of 1870 was eligible. When Miss Helena Downing married, their London home was in her name. She tried to pay the rates. They were refused. When, to protect his wife's rights, her husband refused to pay the rates, he was threatened with distraint. In Gateshead, Robert Spence Watson, 'a strong advocate for giving married women rights which were equal to their husbands', had made over the family home to his wife's trustees. For fourteen years she paid the rates out of her own money: but his name had been entered in the ratebook. In 1889 husband and wife agreed that she should stand as a guardian. While the Clerk to the guardians was out of the office at dinner, they persuaded the young assistant overseer illicitly to erase the husband's name and place hers in the ratebook. The forgery was detected, and her nomination was declared invalid. At the next triennial elections she tried again, this time basing her qualification on the Institution for Friendless Girls to which she acted as secretary, and whose rates she paid even though she did not live there. This too was challenged, but the Local Government Board referred it back to the returning officer, and she was successfully elected. However, Mrs M'Ilquham of Tewkesbury established in 1881 that married women could be qualified to serve if they paid rates on a property which they owned other than the marital home. In her case she farmed seventy-five acres held under the Married Women's Property Act in another parish.[17]

Having got a qualification, it was not easy to keep it. Hostile returning officers, ratepayers, candidates, and party agents searched for flaws to render the nomination or the election invalid. Not all boards went as far as Plymouth which decided that only inhabitants could be guardians. Even when women were ratepayers, the clerk 'considered that the word *inhabitant* applied only to males'. Miss Merrington, the first woman to be elected in 1875, was disqualified in 1879 because the election took place while she was moving house, and a ratepayer pointed out that for these few days she was not eligible. The second woman, Miss Collett, had problems as she was a quasi-lodger of an Educational Institute. The third woman, Mrs Amelia Howell, could qualify only by paying her baker's rates along with her own. Of the five women who came forward in Salford in 1875 (two ladies, a cordwainer, an innkeeper, and a draper), none survived to election day.[18] Other women were disqualified because those nominating them lacked a valid qualification. This happened to Mrs M'Ilquham in her first election. As late as 1912 a Nottingham guardian lost her chance for the same reason. It worked both ways of course, and Lady Anstruther, Miss Frances Lord, and Eva Muller were all returned unopposed in 1882 because they had helped to disqualify the opposition. A professional election agent and expensive legal advice could be essential. Only a few women could afford either.[19]

Women would-be guardians had not only to be duly qualified and duly nominated, they had also to be duly elected. Here again, poor law women faced a much harder task than school board women. The poor law vote, unlike the school

[17] *EWR* 14 May 1887. For the Gateshead case, see LGB papers, MH 12/3099 21, 26 Mar., 3, 6 Apr. 1889; *Newcastle Daily Journal*, 3 Apr. 1889; MH 12/3105 1, 16 Apr. 1892. I owe the LGB refs. to Mr F. Manders, Newcastle Local Studies Librarian.

[18] *EWR* 15 May 1887.

[19] 6th Annual Report of the Nottingham WGS, 1912–13. Miss Peppercorn was popular enough and foolish enough to be elected by two wards in Greenwich and was thus disqualified from both. *EWR* 15 July 1888.

board vote, was a propertied one. The more property you owned, the more votes you had, to a maximum of twelve; which did not help women candidates who were feared to be generous with public money, or women rate-payers who seldom had multiple votes. The voting procedure was also complicated. At school board elections, voters walked to the polls. Poor law papers were distributed to homes, and even the tiniest error in initialling the candidate's name or signing the declaration invalidated the vote. In Brighton in 1884 13,800 voting papers were delivered, but only 4,740 valid papers were returned. In Barton Regis in Bristol, where a woman was narrowly defeated, over half of the votes actually cast were void.[20] Professional paid canvassers were needed to help ratepayers safely fill in their forms. The whole process was far less predictable and far more expensive than school board polls.

As a result, women's groups time and again confessed that they could not find a candidate. When in 1887 existing women guardians stood down in Newcastle, Sunderland, and Bradford, they could not be replaced despite several months' search. Many unions were unable to fill all their seats with men, and the Women Guardians Society kept a list of vacancies for any qualified woman that might emerge. Where there was competition, unscrupulous male candidates encouraged voters to believe that no women could legally stand, circulating fly sheets telling voters not to waste their votes. Several returning officers improperly refused women's nomination papers and few women had the money or steel to take them to court. Friendly MPs brought forward bills to remove the property qualifications. In 1893 it was lowered to £5, preparatory to abolishing it altogether, as part of the great Parish and District Councils Act of 1894. From then on, women and working men flowed on to poor law boards. One hundred and fifty-nine women were guardians in 1893; eight hundred and seventy-five were guardians in 1895.[21]

There were other obstacles to women's poor law work,

---

[20] In Bedford in 1887, a woman candidate reported that of 1,882 voting papers sent out, 60 were not delivered, 320 were not collected, 323 were not fully filled in, and a further 162 were invalid; 986 valid votes were cast. *EWR* 15 Apr. 1887.

[21] WGS Annual Report, 1893.

which if less tangible were no less real. After all, in the late 1880s Manchester and Liverpool had some nine thousand women householders, Sheffield six thousand, and Salford and Newcastle each four thousand, and a goodly minority of these must have been qualified to stand.

The first of these was the almost reflex hostility of many guardians and most local poor law officials to women joining their boards: in sharp contrast to women's experience of school work. Women were elected alongside men on to the very first school boards, and in no sense could be considered trespassers. Equally they came with an ethic of educational service which clergy and caucus alike shared. All boards saw themselves as extending education to the hitherto excluded, though whether children came to board or voluntary schools was another matter. Women might not care for the high profile of school board elections (actively chasing votes was somewhat at odds with the feminine ideal of selfless personal service) but the superior political and social standing of most board members sheltered newly elected women from overt unpleasantness. Almost all women were treated with consideration and respect, sometimes even with enthusiasm.

Poor law work was far less comfortable, and poor law boards much inferior in social status. As women guardians quickly learnt, boards were dominated by small farmers in rural areas and by small shopkeepers and business men in towns, who saw themselves primarily as guardians of the rates rather than as guardians of the poor.[22] Given the language of the 1834 Act, the pressure exerted by the Local Government Board (including Stansfeld) to reduce outdoor relief, and the self-interest of guardians and voters, this was perhaps not surprising. But it did mean that whereas on school boards women could call for greater resources and obtain the backing of the board, on poor law boards if women tried to spend more money on workhouse arrangements they met instinctive hostility. Educational expenditure might be thought an

[22] Mrs Spence Watson described the Newcastle board in 1907 as including two printers, one mining engineer, two butchers, three agents, one blacksmith, one pawnbroker, one store manager, one timber agent, one builder, one tailor, two no occupation, one mineral water manufacturer, one newsagent, two married women, one widow, one Roman Catholic priest, three clergy, and one minister. 1909 RC on Poor Laws, appendix 1c.

investment in the country's future; poor law expenditure was thought to be blighting the country's present. The best-run school boards looked with pride on an expanding service— more teachers, grants, attendances, buildings, children; and if to modern eyes they played the numbers game to the exclusion of less quantifiable considerations, and were dismissive of the views and needs of working-class families, they were at least unapologetic about the demands they made on the public purse. The best-run poor law boards prided themselves on the reverse, on shrinking their 'service', on reducing outdoor relief, on deterring paupers, and on cutting the poor rate. Woman's ethic of service was somewhat ambiguous in such a context. Some boards thought it impertinent of women to want to stand at all, and suggested that women should keep to workhouse visiting, 'however ambitious the ladies might be to play the parts of men'; as simple visitors they could be 'stars of the firmament, diffusing light and help at the same time'. Miss Elizabeth Lidgett, a St Pancras guardian, wearily explained:

It was asserted that it was not well for a woman to be so prominently placed; that she should work more privately and quietly. But it was not because the position was prominent that she sought to be put on the board but because she could be useful there. The prominence was an accident of the position, but not the essence in her regard. And if the woman sought the work for the work's sake, she would lose herself in the work just as she would find her reward in it.[23]

Poor law boards, women found, were a species of male club, at which 'the chair merely confirms every admission by a stroke of the pen, while the Committee discusses local politics or the latest war news', reported Miss Alexander, a Kensington guardian and daughter of a City merchant. Mrs Leach found that the male guardians in Yarmouth gossiped and read their newspapers when the relief cases came to the board. The women took their knitting and listened. The men in turn disliked the fussy temperance views and high moral earnestness women brought to the work. Less scrupulous members found that they were inhibited from awarding contracts to

---

[23] *Richmond and Twickenham Times*, 7 Apr. 1888; *Birmingham Daily Gazette*, 31 Mar. 1882.

each other for the supply of workhouse goods. They also disliked having to watch their language and clean up their jokes. In Plymouth the guardians said the board ought not to include women because of the language used by 'male guardians when interrogating the women who applied for poor law relief'. Women circulated their comments throughout the city, and five of them were returned at the next election.[24]

A Birmingham (King's Norton) guardian said in 1882 on hearing that women were elected on to neighbouring boards:

He should like the King's Norton Guardians to pass a resolution expressing the wish of the board not to have any female assistance. The board room should be an Eden which no Eve should enter. It was absurd to say that ladies would assist in the conduct of the board. Suppose that they got some of the innocent and virtuous females, sometimes called strong minded, and sometimes designated 'prancers' on the board, if they could not have their own way they would be contradicting the board; and if they were young and pretty, no members of the board could contradict them.[25]

Guardians said they had women's best interests at heart, for poor law work with its dirty paupers and even dirtier language was no fit work for women. It would tarnish them. When Walsall contemplated the fearful possibility of women members in 1894 they decided that good workhouse management would end if they were unfortunate enough to have ladies on the board. The mayor added that he could not think of allowing his wife or daughter to sit on a poor law board and hear the evidence they had to hear. The *Women's Signal* commented, with more irritation than wit, that he should put his female relatives in a glass case and send them to the British Museum.[26]

Countless women guardians wrote to the Women Guardians Society and the women's press describing the

[24] SE District Poor Law Conference 1901, p. 455; *London*, 10 Mar. 1898, 23 Mar. 1893; Dr Slater, WLGS Annual Meeting, 10 Mar. 1908. See also comments of Mr Birrell, Hansard, House of Commons, 6 July 1899, col. 58; and Preston Thomas, a poor law inspector, observing board meetings where 'the guardians squabbled till they almost came to blows; hid away under the chair cushions each others notes for speeches'; and played practical jokes on each other. *The Work and Play of a Government Inspector*, p. 236.

[25] *EWR* 15 Apr. 1882.   [26] *WS* 17 Nov. 1894.

rancorous hostility they met on the board. Miss Mary Hopkins, much loved in Scarborough and known for her 'self-denying effort on behalf of the poor' met 'the discomfort of feeling that my presence was not welcome on the board'. Mrs Haycraft reported from Brighton that three guardians threatened to resign if she was elected. She was and they did not. In London, the situation was handled with rather more sophistication. The men were courteous enough while women kept to their traditional womanly work. However, 'Directly the lady member becomes aggressive and shows a spirit of reform, then the men who belong to the old order of things drop their suavity at once, and neither hesitate to upbraid nor to descend to meaness and malice. A woman who hopes to do good in public life must be militant and not merely sympathetic'.[27] In 1908 women guardians were still making the same comments. A new Richmond guardian found men happy for her to work under them, but 'they were not always willing to hand over a bit of work to do alone'. When on one board, women tried to take care of all the unmarried mothers' cases, there were mutterings about being 'ruled by the ladies altogether'. They formed two of a board of twenty-two. At another poor law conference, a woman guardian felt obliged to remind the men that they sometimes forgot 'that the status of a woman guardian was equal to that of a man, and that no Local Government Board had divided the work'. The chairman of the Westminster board blandly replied to such stirrings, 'There are of course ladies and ladies' and provided they worked hard and stayed silent, 'he did not object to them'.[28]

That women came with a reputation of easy charitable impulse, of being generous with other people's money, alarmed male guardians still further. Even sympathetic men addressing the Women Guardians Society felt obliged to warn them against bringing 'too much sympathy for the paupers and too little consideration for the pocket of the ratepayers to the work'. Miss Mason, the first woman inspector of boarded-

[27] Miss Brown, Camberwell guardian, *London*, 22 Aug. 1895; *Scarborough Gazette*, 27 Mar. 1884. AGM of WGS, 1884.

[28] Mrs Nott Bower, SE District Poor Law Conference, 1908, p. 610; Mrs Simpson, North Midlands District Poor Law conference, 1908, p. 569; 19th Annual Report, WGS, 1901.

out children, was her usual brisk self. When she found an injudicious lady, as she told a poor law conference, she called upon her, talked to her, and showed her the error of her ways, adding silkily, 'That is what I always do when I meet an injudicious gentleman'.[29]

Even the progressive and enlightened circuit of the poor law conferences found it hard to come to terms with women guardians.[30] Not until twelve years after the election of the first woman did a poor law conference discuss the contribution women could make to poor law work, a subject the speaker acknowledged was 'far from popular'. When he spoke of finding educated and even-tempered women guardians, he reduced his male audience to laughter. In 1893, Miss Agatha Stacey, a King's Norton guardian, pioneering care for the simple-minded, was asked to talk about the work of lady visitors. She used the occasion to plead for women guardians instead, though had to admit that 'a body of good fussy ladies may do incalculable harm'. A Birmingham guardian commented on her paper that every board should have one or two women; one of theirs was 'as valuable as any man', the other though 'most estimable' was 'quite unsuitable' and 'adds more to the cost of the parish than anyone else'. He found that ladies were self-indulgent, and sent in long detailed reports, whereas male guardians never raised complaints. A Coventry guardian was 'blessed with or suffered from' a committee of lady visitors who seemed to think 'that the condition of the pauper should be improved'. A Rugby guardian thought women should stick to charity, and a Banbury guardian found that their ladies committee made all sorts of silly suggestions, like wanting 'to classify the inmates, and not allow the godly to associate with the ungodly'; they also tried to stop the old women having their drop of gin before they went to bed. At a well-attended conference, only a handful of men welcomed women guardians.[31]

Facing such prejudice, women counselled each other to take great care, to sit quiet, and not contribute until they were well

---

[29] *EWR* 15 July 1881; Miss Mason, Central Poor Law Conference, 1892, p. 276.
[30] See below n. 71.
[31] Cllr. Middleton, vice-chairman Salford board, NW District Poor Law Conference, 1887, pp. 191–4. Miss Stacey *et al.*, West Midlands District Poor Law Conference, 1893, pp. 226–7.

established. One woman wrote in, 'I feel so strongly that we can help the men, just because we look at things from a different standpoint. I generally say that we are like the left hand to the right, one holds the needle, the other holds the work. We were careful at first not to speak or give an opinion, so that really they were annoyed with our silence'. Even Louisa Twining, a national figure before joining the Tunbridge Wells board, took seven months before making 'sundry suggestions' about the workhouse. As they themselves noted, a much higher standard of performance was expected from women than from men. At Tunbridge Wells one male guardian attended only three meetings of twenty-five in a year. 'I could not help thinking what would have been said if a woman guardian had acted thus', wrote Miss Twining. 'Truly there is in more than one department of life a different standard for men and women!'[32]

Women for their part were both amused and baffled at the struggles of male guardians trying to decide the proper diet for small babies, and the style of pinafores for older children. No ordinary man would do it at home, but put a dozen of them together as a board, and they 'rush in heroically'. Women gleefully circulated scurrilous stories of the all-male board which spent an hour discussing what marking-ink to use on the laundry, ordered 150 yards of calico with which to make one apron, deliberated at length on the virtues of metal over china pie dishes, at even greater length on which bonnets female inmates should wear; and gave a morning to determine whether dresses should be graced with buttons, or hooks and eyes, only to find that the seamstress had decided for herself.[33]

Women faced not just hostility from many male guardians, but also from many staff. Workhouse masters were determined not to have their routines disturbed by meddlesome women. The nomination of Miss Elizabeth Smith in Hyde caused 'great consternation among the workhouse officials who declared "they don't want any ladies interfering with

[32] *EWR* 14 Mar. 1885, *Parish Councillor*, 18 Oct. 1895, *London*, 23 Mar. 1893; 11th Annual Report, WGS, 1892, p. 7; L. Twining, *Workhouses and Pauperism*, p. 153. Miss Twining missed four meetings in three years. To her chagrin, her train was late for her first meeting.

[33] *EWR* 13 Mar. 1887. When Miss Louisa Carbutt joined Leeds board, her first task was to try on 40 ulsters on behalf of her male colleagues, *WSJ* 1 May 1883.

them" '. It was complained of Miss Foster Newton that she 'poked her nose into this place and that'; another matron feared the arrival of Miss Twining, as 'everything would be ferreted out'. A woman guardian commented on the difficulties 'thrown in the way of anyone but an official making any move for the better'.[34] There *were* humane and competent masters in the better-paid London workhouses, but many others were harsh and callous, uneducated and corrupt. They resisted lady visitors; blocked Brabazon schemes for the elderly and infirm until women overruled them at the board; augmented their income by fiddling contracts for food, clothing, and supplies; in certain cases were known to be drunk, sometimes sexually promiscuous, occasionally violent, and invariably terrifying to the women, children, sick, and elderly in their care. Mrs Despard, a socialist guardian, found in 1895 that the young Lambeth master was illegally punishing elderly inmates by putting them on a bread-and-water diet. She, like Mrs Surr, called in the Local Government Board, whereupon the guardians called on her to resign. She 'laughingly' refused; kept notes on the master, and finally forced the board to denounce him for sexual laxity and to investigate his financial transactions. The master resigned. A colleague recalled she had shown 'the supreme form of courage—never to falter when faced with overwhelming opposition'.[35]

Masters, appointed to the fiction of a penal institution for able-bodied men, were often quite unsuited to run refuges for the sick and helpless. Whether they were any less competent or more unsuitable than untrained staff in other closed institutions, such as prisons, hospitals, or reformatories, is unclear; but it has been calculated that at least one master in eight left his post because he was unsuitable.[36] Their attitudes

---

[34] *EWR* 14 Apr. 1888; *Parish Councillor*, 13 Sept. 1895; Twining, *Workhouses and Pauperism*, p. 99; *Richmond and Twickenham Times*, 30 Mar. 1889.

[35] *London*, 14 Mar. 1895; Lambeth Board Minutes for 1895; A. Linklater, *An Unhusbanded Life*, 1980, pp. 80, 83.

[36] M. A. Crowther, *The Workhouse System 1824–1929*, 1981, p. 120. Dr Crowther also quotes from the 1909 Minority Report, 'The men and women who we harness to the service of the General Mixed Workhouse almost inevitably develop an all-embracing indifference—indifference to suffering which they cannot alleviate, to ignorance which they cannot enlighten, to virtue which they cannot encourage, to indolence which they cannot correct, to vice which they cannot punish', ibid., p. 113.

inevitably percolated down to other staff, such as the labour master and the gate porter. Mrs Higgs, wife of an Oldham Congregational minister, went on the tramp at the turn of the century, as did Julia Varley, a trade-unionist and Bradford guardian. Both women were dismayed by the surly insolence of male staff, who propositioned them with offers of free cups of tea in return for free sex.[37] Newly appointed nurses struggling to lift nursing standards often found their lives made intolerable when to their heavy and unpleasant work was added constant squabbling with master and wife over laundry for the beds and food for the patients. Women guardians soon understood that the workhouse was run for the convenience of staff rather than of the inmates. Matrons smashed gifts of cups and saucers or individual teapots for the elderly; common china and one urn was simpler. Another rejected different shawls for the old ladies as it was easier to launder them if they were all the same. Repression, monotony and uniformity, the extinction of individuality, were tools of discipline for the officers rather than the requirements of the law. As George Lansbury said,

Officials, receiving wards, hard forms, whitewashed walls, keys dangling at the waist of those who spoke to you, huge books for name, history, etc., searching and then being stripped and bathed in a communal tub, and the final crowning indignity of being dressed in clothes which had been worn by lots of other people, hideous to look at, ill-fitting and coarse. Everything possible was done to inflict mental and moral degradation . . . officers, both men and women, looked upon these people as a nuisance and treated them accordingly.

However, the officials did not always win. Miss Thorburn, a Liverpool guardian, heard with some amusement one woman telling a portress, 'I lives here; you are only a paid person', another man telling a group of nurses, 'If it wasn't for the likes of us the likes of you wouldn't be here'. One woman had four children born by the parish and buried by it. ' "The guardians *ought* to do something for me, Miss!" She spoke as a good customer to whom we were under an obligation.'[38]

[37] Anon. (Mrs Higgs), *Five Days and Five Nights as a Tramp among Tramps*, u.d.; anon. (Miss Varley), *Life in the Casual Ward*, u.d.

[38] Miss O. Hertz, *Why We Need Women Guardians*, 1907; G. Cuttle, *The Legacy of the*

A wise woman guardian carefully cultivated the support of the staff, came at convenient hours, and looked to their comfort. Some of the more enlightened masters and matrons in turn used women to educate the rest of the board into the needs of the House. But it was not always so. Two experienced and tough-minded women, Mrs Eva McClaren (formerly Miss Muller) and Mrs Ethel Leach of Yarmouth, both faced highly publicized contests of will with male staff. Both won, but in each case it proved a pyrrhic victory.

Eva McClaren had been a Lambeth guardian. On her marriage she moved to Bradford, and joined another woman on the Bradford board. She soon persuaded the guardians to take children out of the workhouse and put them into foster care. She also learnt that the master, by other accounts a kindly, educated, and humane man, was a chronic alcoholic; she was a leading figure in the temperance world. She pushed the guardians into an investigation. They suspended and then reinstated the master when he took the pledge. Local temperance activists were angry that workhouse nurses were sacked for a similar offence, and called in the Local Government Board who insisted that the master be dismissed. Uproar resulted. Eva McClaren's meetings were broken up, she received nasty anonymous letters, was 'talked about in every public house in Horton'. She was charged with persecuting a man who was seeking to reform; of being obsessed with temperance to the exclusion of all else; of being a foreigner who had dragged the town's name through the mud. The other woman guardian (Miss M'Turk) and Miss Lupton on the school board both deserted her, and the rest of the guardians limited her right of access to the workhouse. In the April elections she was defeated. Within a year her husband became MP for Crewe, and they both moved back to London.[39]

Rural Guardians, 1934, p. 43; G. Lansbury, *My Life*, pp. 135 ff.; Miss Thorburn to RC on Poor Laws, Q. 35693.

[39] *Bradford Observer*, 14 May 1884, 31 Mar. 1885. The women's Suffrage Society, embroiled in its battles for school board places, had been delighted to run an experienced able woman for the board. Miss Catherine M'Turk, a doctor's daughter, deserted her not 'as a personal thing, but simply as a matter of public business'. Edith Lupton objected to central government interference. *Bradford Observer*, 5, 9, 10 Apr. 1885;

Mrs Leach's struggles came a decade later when, already vice-chairman of the school board, she with two other women joined Yarmouth's poor law board. She found it run as a private club by their clerk, Palmer, a leading solicitor in the town and a deputy lieutenant of the county. In contrast to her school board meetings, there were no agenda papers, no minutes, no accounts. When she asked for a finance committee to handle accounts and tenders, she was defeated. When she asked questions, she was told by the clerk she talked too much. When she asked to see the minutes of the assessment committee,

THE CLERK: 'In my office'.

MRS LEACH: 'I am not allowed to enter that'.

THE CLERK: 'Well, you can sit on the Quay, and we will send them to you with an umbrella'. Laughter all round.[40]

When the clerk asked for a rise of £30, Mrs Leach moved that it be £3 only as they could get able and civil men for less. She kept the press well briefed and waited. Soon Palmer allowed his curious sense of humour to be diverted elsewhere. At one board meeting he circulated a cartoon he had drawn suggesting impropriety (and smoking) between medical officer and nurse. Mrs Leach now pounced. She called in the inspectorate, the public inquiry attracted a delighted audience, Palmer was severely reprimanded and suffered a nervous breakdown leading to detention in an asylum. The LGB made the guardians dismiss him, and Mrs Leach finally got the board's business in order.[41]

The Yarmouth workhouse was by any criteria unreformed. Its food and drink were lavish, its outdoor relief erratic, its medical care nonexistent, its inmates and wards dirty. With the aid of the inspector and to the disgust of the chairman— 'the poor have been roughing it all their lives. What do such people want with certificated nurses?'—Mrs Leach brought trained nurses into the wards, and began to plan a new

EWR 15 Apr. 1885. Back in London Eva McClaren put her energies into the emerging women's Liberal association.

[40] *Yarmouth Times*, 4 May 1895.

[41] *Yarmouth Times*, 6 Apr., 1 June 1895. See also Preston Thomas's reference to 'lady guardians of unlimited energy fired occasional shots', *Work and Play of a Government Inspector*, pp. 249–50.

infirmary. She and the other women (known colloquially in this seaside town as The Trinity), found children in one ward whose faces were covered with sores and who had not had a bath for six months, and forty-six old men sharing three towels and scrofula in another. The chairman explained to her that they were running a workhouse, not a philanthropic institution—'The people in the workhouse now never had such a splendid time of it since they were born'—but undeterred by such visions of splendour, she got the children out of the mixed wards and into cottage homes. Male guardians got their revenge by denying women any part in the domestic details of the cottages. When the inspector criticized some of its sillier arrangements, the women self-righteously explained that they had not been consulted. When another guardian complained that the women cost the ratepayers money and seemed to think they were 'the safety valve of the entire board', the women walked out. When they walked back in again, they obtained their finance committee, audited accounts, separate out-relief committee, organized rotas for workhouse visiting, and their right to attend official poor law conferences. With Palmer gone, the press reported, the 'trickery' of the men was no match for the tenacity of the women. But Mrs Leach, who earlier appeared 'fresh and eager' after three hours fighting on the board, found it taking its toll. Her health broke down and she retired in 1898. The *Yarmouth Mercury* noted, 'She was not long in working a revolution . . . She braved affronts; she fought against unintelligence'; and she paid the price.[42]

In a number of unpleasant incidents reported from all over the country, male guardians and male staff closed ranks against the women, kept them off the committees that counted and denied them access to the workhouse. The St Pancras board in 1877 removed women from the workhouse visiting committees—the language of paupers was suddenly unfit for their ears. Some male guardians protested and offered to make way for the women but this they would not accept. After years of voluntary visiting, Mrs Haycraft was kept off Brighton's committees. In Bedford, Mrs Edwards, who was known to do

---

[42] *Yarmouth Times*, 7 Sept. 1895, 12 Dec. 1896, *Yarmouth Mercury*, 5 July 1930.

more work than any other guardian, was kept off the workhouse visiting committee for three years. She was told that 'lady guardians were the greatest failures ever witnessed . . . it was impossible that the poor could be better looked after than they were at present'. It took all the efforts of local women's groups and temperance bodies, and the weight that came from receiving nine-tenths of the votes cast at the election, to get her admitted to the visiting committee in 1889. Southport reported that their five women were displaced from the schools and children's committee, 'their places filled by farmers who know little and cared little about the matter, and who also seem to object to cottage homes as a "new fangled notion". It is terribly disappointing after such hard work, but we still hope to succeed'. Selina Cooper, who joined the Burnley board in 1901 for Nelson, found herself kept off the visiting committee; in 1904 she was removed from her Nelson relief committee and placed on that of Padiham, a snub, she thought for coming top of the poll. 'As a working woman I was too closely in touch with working people . . .' (and was thought, presumably, to be too generous in dispensing outdoor relief). She refused to attend any Padiham meeting, the Local Government Board stepped in, and she returned to the Nelson relief committee. Even those like Miss Clifford, who were on the visiting committee, were on official rota only three weeks a year. At other times, when she found potatoes inedible and the milk 'shocking', she could do nothing, as the men officially on rota refused to act.[43]

Slowly, many women persuaded their boards to open up the workhouse to public scrutiny, to permit unlimited visiting by the guardians (this was insisted on by the Local Government Board in 1893), and to allow press coverage of board meetings. Miss Foster Newton was given special permission by her Richmond board to visit at any hour. When the master complained, the board showed unusual solidarity by giving all

[43] St Pancras Board of Guardians Minutes, 21 Apr. 1881, 19 Apr. 1883. *EWR* 15 May 1877; Bedford, *EWR* 15 Apr. 1888, 15 May 1889; Southport: 19th Annual Report, WGS, 1901, though this may have reflected Miss Ryley's current school board battles; Nelson: J. Liddington, *Life and Times of a Respectable Rebel, Selina Cooper, 1864–1946*, 1984, pp. 114, 132–3. (I would also like to thank Jill Liddington for additional references to Selina Cooper and Lady O'Hagan.) See also Mrs M'Ilquham, 'Of Women in Assemblies', *Nineteenth Century*, Nov. 1896.

its members the same privilege. In time, women guardians managed to overcome hostility by slow patient work, especially with children. The chairman of the Tisbury board wrote to the Women Guardians Society in 1901 for example, that 'no one could have been more opposed to lady guardians than I was six years ago, but knowing the excellent work done by them I have entirely altered my opinion'. A few boards took real pride in their outstanding women, such as Mary Clifford of Bristol, Agatha Stacey of Birmingham, Mary Anne Donkin of Kensington. Mrs Charles had by 1885 been placed by the Paddington board not only on the workhouse visiting committee, but on the finance committee, the boarding-out committee, the infirmary committee, and the laundry committee. In her spare time she succeeded in releasing wrongfully imprisoned lunatic paupers.[44] But women were never allowed to forget that they were interlopers; and that they had to maintain standards and justify their position in ways that no man would ask of men. And unlike school board work, poor law women were seldom able to appeal to the electorate for sympathy or to political parties for support. Psephology and ideology interlocked to isolate them.

In school board work, those who paid rates, those who voted, and those who benefited from educational services overlapped. As school standards rose and as more lower-middle-class children attended board schools, the overlap strengthened. Women had a particular affinity with working-class constituents, fought for their children, and often worked in and represented the town's poorest areas. The reverse was true within the poor law. Those who received poor relief, and the increasing number who had medical aid, were automatically disenfranchised. So the concept of accountability was a very ambiguous one. To whom? Paupers? Ratepayers? Working men? It would have been inconceivable for any poor law guardian to have taken a policy steer from a public meeting, as Helen Taylor and Annie Besant did, to lead deputations as Mrs Fenwick Miller did, to plaster the division with placards as Mrs Surr did, or to speak in open fields under the auspices of working men's associations, as the

[44] 19th Annual Report, WGS, 1901. For Mrs Charles, see Paddington Board of Guardians Minutes, 15 Apr. 1885, *WPP* 26 Jan. 1889.

gentle Mrs Webster did. Many ILP and working-class guardians, such as Mrs Pankhurst, adopted such styles after 1894, but women guardians hostile to outdoor relief were never popular with the urban poor; and as few people bothered to vote in poor law elections, they lacked any constituency to support them in their battles at the board.

Such support might have come from a standing electoral committee of the influential in the town; or from the political parties, or the churches. But unlike school boards where they had a vital interest, the churches held aloof; and the political parties kept a low profile. Until the later 1890s and the arrival of guardians bent on more generous outdoor relief, there was little policy difference between the parties (and nothing equivalent to the school board debates on voluntary and free schools). Some wards were known to be Liberal, others Tory, and candidates with the appropriate proclivity were elected; but reading the sparse newspaper reports, it is not obvious which guardians were on a party ticket and what it might be. School boards shared with town councils an unweighted franchise and identical boundaries. The party machine moved across from one contest to the other. However, the weighted voting system and very different boundaries for poor law unions, which for the towns often included outlying rural parishes, meant that political parties could not easily use poor law elections as a means 'of limbering up for constituency contests'.[45] Parties did run slates in Bradford, Leeds, and Birmingham, for example, but few guardians issued manifestos, almost none ever held a public meeting or made a public speech. Policies such as scattered cottage homes were developed locally, occasionally poached from other boards, or presented at poor law conferences. There was nothing equivalent in poor law work to the Birmingham National Education League and its successors, urging progressive educational policies that were everywhere adopted on advanced school boards. MacKay, an early historian of the poor laws, wrote in 1903 that though firmness on outdoor relief was not popular, it had little electoral impact.

In the unions where a lax policy was followed, both sets of candi-

---

[45] I owe this point, and this phrase to Dr John Garrard of Salford University.

dates have been in the habit of promising outdoor relief, and plenty of it. In the unions where the law has been administered strictly, an attempt, to some extent successful, has been made to keep the subject of outdoor relief off the party ticket. The elections in both cases were decided by the strength of the two political parties, as practically both parties put forward the same or rather no poor law programme. If one party went in for outdoor relief the other had to do so also, and it was only by mutual agreement that the appeal to the delight of outdoor pauperism could be avoided.[46]

Given little publicity, low polls, minimal political salience, the suppression of controversy, and abundance of poor law seats, poor law work was not as fiercely contested as the school board or town council. In Birmingham, the parties amiably divided the sixty seats with forty to the Liberals, twenty to the Tories. After friendly press comment, Dixon, the mayor, offered Liberal women four or five seats (even the vagueness is revealing) and called meetings to encourage them to stand. Only two widows, Mrs Ashford, wife of a former councillor, and Mrs Bracey Perry, came forward, the women's committee sadly noting 'the difficulty that has been experienced in finding ladies with proper legal qualifications to serve'. They had even thought of buying up property to create their own artificial qualifications, along the lines of the old Anti-Corn Law League. The women were elected without contest.[47]

So while school board women mounted the platform in vigorous, highly publicized school board campaigns as part of a slate with a common manifesto, making keynote speeches and expecting to account for their voting on the board, poor law women, even where they had party support, still stood on their own and stood for the most part silently, relying on one or two friendly letters in the press and perhaps a commendation from the local doctor and the local vicar. At their first election, they might circulate a public address explaining why

---

[46] T. MacKay, *History of the English Poor Laws*, 1903, vol. ii, p. 583; cf. the same point made by the COS—'the advocacy of difficult economic doctrines is unpalatable to the caucuses and their position is at all times precarious'. Candidates opposed to outdoor relief risked defeat. W. Bailward, 'Local Government and Popular Election', *COR* 18, 1905, pp. 183–94. He noted that in some London boroughs, parties organized the fight, in others candidates were on their own; only half the wards were contested.

[47] *Birmingham Daily Mail*, 21 Jan. 1882; *Birmingham Daily Post*, 27 Mar. 1882, 30 Jan. 1884, 27 Jan. 1885.

women were needed in poor law work, but once elected even that ceased, they stopped canvassing, and relied on their work to speak for itself to the few who could be bothered to vote. Mrs Charles of Paddington was the exception here as in so much else, and she issued an annual address, 'regarding this as a means of giving to her constituents an account of her stewardship'. More typical was Dr Mary Royce of Leicester, of whom it was said, 'It must have cost so modest a lady a good deal to undergo the publicity of a Guardian's election, but once elected there was no more publicity for her. She did not want to speak or move resolutions. She set herself to work among the poor paupers, and hers was a work of sympathy and love.'[48] Such women could not understand why politics should ever enter public service.

Most women made the lack of a public party label a virtue. They were being elected as women of independent judgement (and independent means) rather than as puppets on a political string. Miss Foster Newton was, along with Ethel Leach, one of the few poor law women to engage in high profile campaigning, but even she, a staunch Liberal, wished her local Liberal party would not be quite so enthusiastic about her candidature. She thanked them for their goodwill but added pointedly, 'I consider the work of a guardian quite outside politics, whether they be liberal or conservative. If elected, my sole object will be to do a woman's work in an earnest womanly fashion, without any regard to political feeling of any kind. What such work has to do with politics I am at a loss to determine, and I venture to think that the ratepayers generally will share that opinion.'[49] When Florence Davenport Hill switched from St Pancras to take over her uncle's seat in Hampstead, and it was thought she might be opposed, an all-party list of three hundred influential supporters was issued within three days, and after that no one else had the temerity to stand. At Southport, Miss Ryley helped form a branch of the Women Guardians Society, which studied the poor law, canvassed discreetly, and though itself

[48] Dr Royce, *The Wyvern*, 4 Nov. 1882; Mrs Charles, *WPP* 13 Apr. 1889.
[49] *Richmond & Twickenham Times*, 31 Mar. 1888. See Ch. 1 for Miss Foster Newton.

non-political, placed four women in wards according to their political fit. All were returned.[50]

In 1894 the property qualification for guardians was removed. Labour men stood in larger number, sometimes as part of a Progressive alliance. The Tories rallied and elections were fought on sharp political lines. In Norwich for the first time virtually every candidate ran on a party ticket; and far more items at the board went to the vote instead of being determined by consensus. Women found it harder to stand as independents. Northampton refused to run non-party women. In Newcastle and Leeds, experienced guardians were displaced by young men being groomed for higher political office, impatient to be on, up, and off. Miss Margaret Baines was first elected in the 1880s as a Liberal. At the next two triennial elections, she stood as an independent, coming in head of the poll. At her most recent election she was beaten by a party man. As she believed that outdoor relief was too generous, that trade unions and the feeble-minded were in their rather different ways too active, and that drink was the main cause of pauperism, it was not altogether surprising that she lost her seat. More surprising perhaps was that she was co-opted back on.[51]

The handwritten Minute-books of Bolton's Association for Returning Women Guardians still survive and they offer a fascinating record of their efforts to break into poor law work from 1899 on. The Association was led by Miss Sarah Reddish, a former member of the school board, a trade-unionist, and organizer of the Women's Co-operative Guild. Though by now no property qualification was necessary, the

---

[50] Florence Hill, *EWR* 15 Apr. 1889; Miss Ryley, NUWW Conference, Liverpool, 1891, p. 131. See also Mrs Henniker of Fulham, *EWR* 15 Feb., 15 Apr. 1886.

[51] Lady Knightley, NUWW Conference, Manchester, 1907, p. 104; Evidence of Miss Baines to the RC on Poor Laws, Qq. 39541, 39612. She noted that two-thirds of all guardians retired or were defeated at each triennial election. Mrs Elmy confirmed the turnover after 1894—'many of the old ones having retired or been defeated, including I think many of those who had not counted the cost'. Unlike Miss Baines she thought the quality had improved—those defeated included compulsory teetotallers and vaccinators. 'I am not sorry to see women beaten on these grounds. It is one of the unfortunate results of women's position of slavery that so many seek rather power to coerce others than to free themselves.' Mrs Elmy to Mrs M'Ilquham, 18 May 1898, BM Add. MS 47451, fo. 212.

committee could not find any woman willing to stand, so in desperation decided to stand themselves. They had first to find some seats. They wrote to Conservative and Liberal wards asking if at the next appropriate vacancy, the wards would accept a woman without contest. Their letters went unanswered. They brought in Miss Clifford of Bristol and Miss Olga Hertz of Manchester to give lectures and raise public interest. The Society had already agreed that they would not run a party woman, would not challenge a sitting member, and would not upset the political balance. They would place a Liberal woman in a Liberal ward but label her independent. Not surprisingly the parties were ambivalent, trying to calculate what was in it for them. In 1905 Miss Reddish and colleague both decided to stand for empty seats as independents. The ILP and the Liberals said they were in 'entire sympathy'. 'The conservative agent on the contrary refused all help charging the two lady candidates with opposing Conservative candidates and forcing a contest in both wards'. The women were safely elected. However, other women continued to agonize over their situation. One withdrew in 1911 rather than face a contest, the other refused help in canvassing lest this be thought politically provocative.[52] The problem was not resolvable. If the Society was to attract comprehensive support and emphasize its feminism, it could only endorse independent candidates; but many women and most men after 1894 wanted women candidates safely tucked away behind party labels. Party managers could not see why they should release a safe seat to a woman who wanted their seat and their support but would not promise them her loyalty.

Many women tried to square the circle by doing what self-respecting school women deplored, seeking co-option to casual vacancies.[53] When Miss Twining retired to Tunbridge Wells in 1893, she was asked to stand for the board. She hesitated because 'I could not run the risk of an election or a defeat; I had never taken part in one, and could not think of doing so now, and should not know how to go about it.' When she learned that the vacancy was caused by a death and that no other candidate would be standing, she consented.

[52] Minutes, 22 Mar. 1900, 20 Mar., 5 Apr. 1905, 14 Mar. 1911.
[53] *COR* 1 Nov. 1883.

At a poor law conference in 1902, a male guardian thought ladies would remain ladies only if they sought co-option and not election. Miss Geare, an Exeter guardian, found contested elections 'very unpleasant indeed' but confessed she had never had to solicit a vote; she was looked after by others. Mrs Fuller of Chippenham similarly had 'not the trouble to ask for votes'.[54] A few women were more robust. When it was thought indelicate to run another woman in Richmond to help Miss Foster Newton, because she would be challenging a sitting member which would create 'bad feeling in the town', one of the other women present was heard to mutter, 'It is not for us to turn out anybody. All the candidates go before the ratepayers and it is their choice that determines the question'. Miss Henry, member of a rural district council and ex-officio guardian, said in 1907, 'It is a much better policy to go through a contested election than to rely on being coopted. The rest of the council respect you. If you are coopted, you are absolutely in the position of a poor relation'.[55] School board women valued the responsibility and accountability that flowed from winning a seat. Poor law women, with no natural constituency to account to, seldom shared their views. After 1894, women still continued to join rural boards by negotiation; but in the cities, as party positions hardened, women had to come into line, and had both to carry a label and stand for open election.

So although the 1894 Act abolished many of the technical obstacles to women's election, women still remained reluctant to come forward knowing that they faced deep-rooted prejudice from many male guardians, disgruntlement from many male officials, indifference from most of the voting electorate, and antagonism from many of the urban poor. There was one further reason why women hesitated. Male guardians had suggested that poor law work would tarnish women; and women recognized a grain of truth in such remarks. They expected to find poor law work repulsive, 'the continuous contact with misery and vice . . . very painful',

[54] L. Twining, *Workhouses and Pauperism*, p. 130; S.W. District Poor Law Conferences, 1901–2, p. 409. Several women told the RC on Poor Laws that providing a woman did not force an unwanted election, she would usually be later co-opted to a vacancy. See Miss Harriet Lloyd of Shropshire, Q. 70601, Miss F. Joseph, 69463 ff.

[55] *Richmond & Twickenham Times*, 23 Mar. 1889; Miss Henry's speech to a reception at Lord Brassey's, 14 Feb. 1907, WLGS Papers, vol. xxxii.

said Miss Mary Hopkins of Scarborough, 'the very opposite of pleasant'. 'It was most painful,' said Mrs Evans. (Though whether it was more painful and repulsive than philanthropic visiting and rescue work in slums and brothels is doubtful.)[56]

Edith Shaw, a workhouse matron, admitted able-bodied women who needed six [sic] baths before they were clean of flies and vermin. She described the distressing habits of senile, foul-mouthed, and doubly incontinent old women. It needed three of her staff to extract from one old lady 'bits of cheese, old potatoes from the dustbin, crusts, bacon, bread, butter, mouldy pudding, all were mixed up in a foul smelling mess in the pockets of her clothes'. She found the bread and butter mixed with faeces in the bolster, meat and greens in the bed. Miss Thorburn of Liverpool told the 1909 Royal Commission of the obsessional theft that went on in the workhouse—able-bodied women stealing food from the sick, clothes from the nurses, toys from the children, money from herself—partly, she discerned, 'to have the pleasure of giving the stolen things away, and in some cases it seems to be the love of accumulating, as when they did find a hidey-hole and collect in it pieces of coal, soap . . . the result of the women having no private property of their own when in the workhouse'. Miss Clifford wrote after nine years work as a guardian, 'I don't believe anyone but a Guardian of a town workhouse knows how bad people can be, and apparently only women Guardians quite realize it'.[57] It took a special kind of courage to enter the foul wards of terminal VD cases, to risk physical abuse from the able-bodied street women who clutched at their clothes, to stomach the smell of foetid wards with open receptacles for urine, insanitary drains, and rank bodies, and to risk the multiple infections of the infirmary. Florence Hill described the casual in-and-out children of the workhouse in words redolent of physical nausea—they were 'dregs' debased and

[56] *EWR* 15 Feb. 1881; *Richmond & Twickenham Times*, 30 Mar. 1881. See F. Prochaska, *Women and Philanthropy in 19th Century England*, 1980. Those women who followed Ellice Hopkins in seeking to rescue prostitutes from brothels and drunken male clients must have believed it was 'not unlike a journey into hell itself' (p. 197). By contrast with such emotionally harrowing and personally dangerous work, the workhouse was safe and relatively sanitized.

[57] E. Shaw, 'The Workhouse from the Inside', *Contemporary Review*, Oct. 1899, p. 570. Miss Thorburn to RC on Poor Laws, Q. 35693; Miss Clifford in G. Williams, *Mary Clifford*, 1920, p. 108.

diseased, children who, 'in endless succession, like the buckets in a dredging machine, discharge their moral filth' on the other children in the workhouse. She was writing about seven- and eight-year-olds. Mrs Henrietta Barnett, used to East End life and certainly no moral weakling, felt able to suggest to married women the service of settling workhouse girls into their first jobs because such work did not entail visits to the workhouse, and meant therefore no infection would be brought back to the family home. Into the workhouse came the sick whom voluntary hospitals would not nurse; the diseased and mentally disturbed, the brazen and the depraved; the derelicts, the drunks, and the tramps. They required personal service of the most physically demanding and difficult kind, unrelieved by any belief that their clients (except possibly for the children) were redeemable. In the women's eyes, they were working with morally terminal cases. Miss Clifford, the most serene and sweet of women, approached poor law work in deeply pessimistic vein. As she wrote to her Norfolk friend, Miss Blanche Pigott, herself an Erpingham guardian:

A longish board day, it was nearly four when I got out of the board room. I generally now have a long committee after the board; but it was not disagreeable, things went right I believe, though not always exactly as I intended. And then I had some visiting, looking narrowly after the sick children, having talk and reading in the Lock Ward, and the ward where the young women with babies are—the first very forlorn and sad and humanly helpless, such wrecks . . .

Yet as Miss Clifford said, 'The painful cases . . . you see, *they have to be dealt with*'; and women guardians pleaded with their sisters that just because the work was repulsive, Christian women must take it up.[58]

School board women received the love of 'their' children, the affection of staff, the gratitude of constituents. Poor law women entered public service knowing that such rewards would be rare. Occasionally a blush from a woman guardian might shame a young promiscuous girl; orphan children might be helped to grow strong and tall; a sick bed might be

[58] F. Hill, 'The Family System for Workhouse Children', *Contemporary Review*, 1870, p. 272; H. Barnett, 'Young Women in our Workhouses', *Macmillan's Magazine*, June 1879; *London*, 22 Aug. 1895; Williams, op.cit., p. 142 (letter of 1887, u.d.)

smoothed; a respectable elderly woman be cheered by the gift of a picture. But such moments were few. Duty, dedication, strong Christian faith and compassion, had to serve instead. In the words of Miss Caroline Ashurst Biggs, editor of the *Englishwoman's Review*:

It is the most Christian work a woman can take up. The poor we have always with us, and the very poorest of the poor are in the charge of the guardians—mostly old and sick, women and infants, people to whom kindness and sympathy are almost of more value than the bread they eat . . . Ladies sometimes shrink from the work because they hear there will be much that is difficult and unpleasant in the duties, but let them once be convinced how deeply necessary it is to the physical and moral welfare of hundreds of our poorer brothers and sisters that educated and thoughtful women should undertake this task . . .[59]

Women guardians, she added, should be particularly supportive of the staff and the nurses, who could not leave the workhouse behind them as women guardians did to return to their families, who worked long hours for little salary, and whose health often broke down. Books, fresh air, pianos in a staff sitting-room, a daintier diet, would not only benefit the staff but through them the morale of the entire institution.

Given the hostility they knew they would encounter and the distastefulness of the work to be done, even qualified women were slow to come forward. The women's movement recognized that it was socially and emotionally gruelling for women to enter poor law work—it was in all respects polluting—and tried to strengthen them by demonstrating that the workhouse was truly housework. It is

especially fitted for women; for it is only domestic economy on a larger scale. Accustomed to regulate her own house, a lady has had precisely the training necessary to fit her for a poor law guardian; she has had the management of children, and looked after their health, their clothing and their education; she has ordered the household supplies and is accustomed to examine into their price and quantity; she has supervised her servants and alloted to them their employments; and finally she knows something of the requirements of the sickroom. Enlarge a household and it becomes a workhouse; multiply the servants by ten and the children by

[59] *Richmond & Twickenham Times*, 4 Feb. 1888.

hundreds and you have a workhouse school; increase the sick room and it becomes an infirmary; so that every woman who has managed her own household with wisdom and economy possesses the qualities chiefly necessary in a guardian of the poor.

It was a litany calculated to encourage women and reassure men.[60]

Arrival on the Board

Women came into poor law work, therefore, only slowly. Miss Martha Merrington, a gentlewoman who had established crèches for the infants of poorer working women in Kensington, stood there in 1875. With the help of Caroline Biggs who organized her campaign, she came in eighteenth for eighteen seats. She seems to have been welcomed by her male colleagues who piled committee work on to her. Miss Margaret Collet, COS secretary, was defeated in the same election but was returned a year later in St Pancras, where she was joined in 1877 by Mrs Amelia Howell. The Hon. Maude Stanley, aunt of Stanley of the London school board, cousin of the Dean, and active in the girls' club movement, was elected in Westminster. In 1879 two notable women won seats, Mrs Charles in Paddington, and Miss Sarah Ward Andrews in St Pancras; but by 1881 these two women were the only female guardians still holding seats. After five years it looked as though the movement to promote women guardians was faltering.

So in February 1881 a handful of committed women, Miss Ward Andrews, Miss Biggs, who edited the *Englishwoman's Review*, Mrs Ormiston Chant of the National Vigilance Association, and Miss Eva Muller, met with some friendly male guardians, Louisa Twining's blessings and Mrs Fawcett's apologies for absence, to form the Society for Promoting the Return of Women as Poor Law Guardians, or the Women Guardians Society (WGS) as it was more familiarly known.[61] They were able to place Miss Elizabeth Lidgett, a Vigilance worker, and Miss Florence Davenport

---

[60] H. Muller(?), 'The Work of Women as Poor Law Guardians', *Westminster Review*, 2 Apr. 1885, p. 387.

[61] As far as I am aware, the only surviving set of WGS Annual Reports (not fully complete) is in the Blackburn Collection.

Hill alongside Miss Andrews in St Pancras. In Kensington stood Miss Mary Anne Donkin, a COS worker, who had for years patiently attended guardians' meetings to help those girls they could not. She was elected seventh of eighteen, inheriting Miss Merrington's seat.

Buoyed by their success, the WGS called its first public meeting in July 1881 to encourage other towns with a strong women's movement to bring forward poor law candidates. Lambeth, Greenwich, Manchester, Bristol, Cheltenham, and Leeds started their search. Meanwhile a married woman, Mrs M'Ilquham of Tewkesbury, and a widow, Mrs Anne Shaw of Barnet union, were both elected during the course of 1881, probably with some behind-the-scenes help from Mr Hibbert, junior LGB minister, who would not support any legal challenge to their election. As *London* noted in 1893, 'The Local Government Board are so convinced of the value of women poor law guardians, as to be most anxious to increase their number, and have pledged themselves to uphold returning officers in treating rated married women on exactly the same footing as married men'.[62]

Harriet M'Ilquham qualified because she owned property in an adjacent parish. She told the Women's Emancipation Union Conference in October 1896:

Lady candidates had to live down a great deal of open and silent opposition, the most exaggerated ideas being prevalent as to their unsuitability for poor law work. In many cases the friends and relatives of the candidates were appealed to to stop the proceedings, and [her] husband . . . was first entreated, and then abused in a public highway, for *allowing* his wife to be placed in what, it was prognosticated, would prove for her a most disagreeable and humiliating position.[63]

Mrs M'Ilquham found her first election victory invalid because her husband had improperly signed her nomination papers. She stood again, properly nominated, and was returned unopposed. She went on to become one of the staunchest campaigners for women's work in local government, becoming a parish councillor, and a rural district

[62] *London*, 23 Mar. 1893; Mrs Elmy to Mrs M'Ilquham, 14 July 1881, BM Add. MS 47449.

[63] Mrs H. M'Ilquham, *Local Government in England and Wales*, 1896.

councillor after 1894, as well as joining a small rural school board in order to thwart the ritualist vicar. She stood but was defeated for the Gloucestershire county council in 1889. Tirelessly she travelled across the country, chairing meetings for Mrs Wolstenholme Elmy's Women's Emancipation Union, the National Union of Women Workers, the National Union of Women's Suffrage Societies, and the Women's Franchise League. She helped to finance fifteen years of Mrs Elmy's feminist campaigns, as well as writing some of the most effective pamphlets of the women's movement herself.[64] Meanwhile in Barnet, Mrs Anne Shaw, a widow of independent means, had had her nomination papers rejected by the returning officer, had appealed, and had had her election declared valid by the LGB. Miss Augusta Spottiswoode, a gentlewoman of Shere, near Guildford, was elected to her board, after issuing an address promising economy in general and expenditure in most particulars. She weathered her first hostile board meeting, as she said 'without injury to her self-possession or womanly demeanour' and for a few months was discreetly silent. She then began to take her board in hand, much as Ethel Leach and Louisa Twining were to do, finding foster homes for younger children and jobs for the older ones, mainly with her long-suffering friends. She persuaded her colleagues to build a new infirmary to which she recruited trained nurses, and had the lean-to shacks which served as casual wards rebuilt. She cut the drink bill, published the accounts, and without any more electioneering was repeatedly returned head of the poll. She seems to have worked wonders single-handedly.[65]

Over in Bristol, the efforts of the ladies committee now saw success. Mary Clifford, a vicarage daughter, had like Frances Power Cobbe and Louisa Twining, long been a workhouse visitor. Proposing to stand herself, she brought together a committee of friendly men and women, interviewed a dozen ladies, and chose three to stand with her; 'canvassed heartily'; and in April 1882 all four were triumphantly elected. Over the years Miss Clifford came to be regarded as the most inspiring of all women guardians. In her own words, she 'went

[64] *Vote*, 12 Feb. 1910.
[65] *WSJ* 1 Nov. 1881; anon., *Ten Years Work as a Lady Guardian*, Dorking, 1894.

upon the Board of Guardians to protect old women who seemed oppressed, and quickly found matters calling for attention, such as the education and training of children, classification to promote purity, nursing of the sick, emigration, legislation on new matters . . . It is rather wonderful how beginning to understand a bit enlarges our view, and brings countless needs into sight'. She was a much loved and charismatic figure in her glowing art nouveau cloaks (under which children sheltered from the rain while she told them fairy stories), deeply Christian, and blessedly free of COS authoritarianism. She could lift the sights of the most parsimonious and hard-bitten of male guardians.[66]

The WGS steadily enlarged its activities. By the mid-1880s it had branches around London (Croydon, Richmond, Bromley), Brighton had daughter branches at Eastbourne and Hastings, Bristol sheltered Bath. In Manchester Miss Margaret Ashton emerged to federate Lancashire groups into a North of England Women Guardians Society. Central WGS membership stabilized at between one hundred and one hundred and fifty, with many more in the local societies. Funds were sufficient only for a part-time secretary. However, the WGS had an invaluable resource in Caroline Biggs's *Englishwoman's Review*, one of the best of the women's periodicals, in which every setback and success was carefully chronicled. The WGS was also at the centre of a web of related women's organizations—the Association for Boarding-out Children, the Metropolitan Association for Befriending Young Servants, and the Girls Friendly Society, the Workhouse Girls Aid Society, the Workhouse Nursing Association, as well as the Brabazon committee of Lady Meath. Its branches were affiliated to the Women's Local Government Society; more obliquely, it shared overlapping membership with the women's suffrage societies, the COS, the National Vigilance Association, the anti-Contagious Diseases ladies committees of Josephine Butler, the newly growing women's Liberal organizations, and the increasingly important National Union of Women Workers. All of which gave the Society an influence and a range of contacts far beyond its

[66] M. Clifford, *The Relation of Women to the State*, 1894.

immediate membership. As *London* said, the WGS 'has influential supporters among the leading members on both sides in the two Houses of Parliament, and from its quiet little office in Dean's Yard, Westminster, has already done a vast amount of work in securing the candidature of efficient women in all parts of the kingdom'.[67]

Its main task was to help form 'a good local committee' which could bring forward women candidates, preferably two or three at a time, to give each other support. The WGS would give advice on technical questions of qualification, and provide speakers and cheap literature. It also began to host conferences, bringing women to London to discuss some of the more difficult questions facing them, such as cottage homes versus boarding out, the scale of outdoor relief, the propriety of detaining unmarried mothers in the workhouse against their will.[68] Such informal conferences continued into the 1890s when women were more fully admitted into the official poor law circuit. The WGS built its own library, with help from Louisa Twining (in kind), and along with the Women's Local Government Society tried to keep an eye on the parliamentary scene.[69]

For its first decade, therefore, the WGS with its meetings, conferences, pamphlets, and reports, was a lifeline of mutuality and support for lonely and isolated women in poor law service. But from 1894 it dwindled in significance. The 1894 Act removed most of the formal barriers inhibiting women's candidature; official poor law and NUWW conferences meant that policy initiatives were being discussed elsewhere; the Women's Local Government Society (WLGS) was widening its brief to include poor law activity; and the politicization of poor law work with the emergence of Labour guardians meant

[67] *London*, 23 Mar. 1893. Miss Caroline Ashurst Biggs, member of a radical Leicester family, former secretary of the London Suffrage Society, pivotal figure in the WGS. For the WGS, see their Annual Reports, and *EWR*.

[68] 9th Annual Report, WGS, 1890. Bristol, Southport, Leicester, Nottingham, Chorlton, and Plymouth all successfully ran teams of three to five women in the late 1880s and 1890s. Miss Foster Newton noted the value of conferences in describing her first year in office, 'chiefly a learning time. She had been to meetings in London where the poor law was discussed, she had had correspondence with women guardians, had gone over other Unions, and had studied the poor law in all ways that were open to her.' *Richmond & Twickenham Times*, 30 Mar. 1889.

[69] *EWR* 15 Aug. 1882.

that more and more women were choosing to stand as party rather than as WGS-sponsored independent candidates. Before 1894 the WGS could reach almost every woman guardian. But by 1897 its Annual Report sadly noted that only a hundred of the country's nine hundred women guardians were members of the society. It lacked the money to contact the hundreds of newly elected women in the provinces. It had become confined to London. In 1904 the Society folded. Fully in character, its library went to the COS, its leaflets to the WLGS, and its premises back to the Church Temperance Society.[70]

Apart from the WLGS, two forums took its place. The first of these were the official poor law conferences, started in the late 1860s to promote best practice and encouraged by the Local Government Board. Their secretary, Sir William Chance, started regional district conferences to which all unions were encouraged to send a couple of members, and suggested topics for discussion and speakers for their attention. Its papers and reports were widely circulated. Chance himself held orthodox COS views on outdoor relief and was an enlightened campaigner for workhouse children. On both counts, he welcomed women guardians, bringing them on as speakers and committee members. By the late 1880s women guardians were beginning to address conferences and a few women to attend them. By the 1890s, leading women guardians were on the circuit, and women were forming around a quarter of conference audiences. Some unions were now represented only by women, partly because they had the time to attend. Another breakthrough came when in 1890 Florence Davenport Hill went on to the central committee, the blue ribbon, said Sophia Lonsdale, of poor law workers, while many others served on district committees, where they helped to shape the agenda of poor law policy.[71]

The second forum was the National Union of Women Workers, later the National Council of Women, which from 1888 brought together workers in philanthropic, religious, and

[70] 18th Annual Report, WGS, 1900.

[71] Most women speakers kept to womanly subjects (children, unmarried mothers, etc.). Mrs Pankhurst was the first to depart from the domestic brief when she spoke on unemployment in 1895. Miss Clifford, Miss Stacey, Mrs Chapman, and Miss Lonsdale all served on the central committee.

temperance societies, but which increasingly sought to promote women's work for women within local government. Its annual conferences became the women's movement in council. By 1914 its 1,500 affiliated societies and local government branches were, like the WGS before it, seeking out and supporting women candidates. Every annual conference, report, or occasional paper brought news of further advance. Typically, for example, in 1905 NUWW women wrote to all Kent clergy 'in 14 Unions unrepresented by a woman guardian' asking them to support women candidates. While the official poor law conferences brought women into the formal male poor law world, the NUWW brought behind women guardians all the moral energy and practical support of organized female philanthropy.[72]

When the WGS had been formed in 1881, only two women were guardians (both incidentally elected within Elizabeth Garrett's old school board seat of Marylebone, currently held by Alice Westlake with the best women's machine in London). Over the next few years, women gained seats in other north-west and west end unions, Kensington, Islington, St George's, the Strand. During the later 1880s, women found seats south of the river, Lambeth, Fulham, Greenwich, Wandsworth, and out to Woolwich. As the WGS reported, 'Nearly all the Metropolitan West End Boards now include one or more ladies among their members, and the fact that the poorer districts possess fewer women guardians is not because women will not undertake the work there, but because of the difficulty they experience in obtaining the required qualification.'[73] By 1893 all the London unions had women guardians, some forty in all, except the City, and the very poorest unions, Southwark, Poplar, Whitechapel, Bethnal Green, and Shoreditch. The 1894 Act meant that the numbers of women guardians in London doubled to eighty-six by 1901 and they were on most East End boards but had still not found seats in Southwark and Shoreditch.

[72] *NUWW Handbook*, 1905, p. 93. More generally, see Ch. 1. Indeed, such was the degree of networking and overlapping that Miss Mason and Miss Clifford suggested that committees of the WGS, NUWW, and WLGS should 'cooperate in forming a machinery for the selection of suitable persons' as guardians. NUWW Central Conference Committee Minutes, 16 Nov. 1900.

[73] 10th Annual Report, WGS, 1891.

It was all very different from school boards. Because most poor law women got their property qualification from their home, they would only be eligible to stand where they lived—inevitably the more prosperous and conservative parts of London. They therefore did not forge that alliance with progressive working men that was so notable for London school board women who represented Finsbury, Tower Hamlets, Southwark, and Hackney. Emily Davies and Alice Westlake were both radicalized by their working-class voters. Poor law women were not only geographically distanced from working-class poverty, they were further insulated by the fact that recipients of poor relief lost their vote and therefore any leverage they might exert on guardians. So, usually single, sieved by property qualification, toughened by the COS, neglected by party, the class, cultural, and charitable preconceptions that London lady guardians brought to their work remained largely undisturbed. In so far as they came on to poor law boards determined to withdraw children from and bring nurses into the workhouse, they were progressives, and educated many of their male colleagues to their way of thinking. In so far as they continued to read pauperism as moral degeneracy rather than as economic or social dislocation (which under COS influence they largely did), then they became on matters of out-relief among the most conservative members of their boards, the most judgemental, and the most unpopular.

One such moment indicating the distance between women guardians and working men, may stand for many. Mrs Shaen, a Kensington poor law guardian, was addressing the Eastern District poor law conference in 1898. She described paupers as 'a vast invertebrate class among our people, living as a parasitic growth upon the industry and thrift of the Commonwealth'. Their lack of prudence, love of drink, and willingness to 'sponge off' others was undermining the nation's moral fibre. George Edwards, the farm worker who was rebuilding Norfolk's agricultural trade-unionism, rose to his feet, and asked her how agricultural labourers were expected to save for their old age on 12s. a week? Pauperism and poverty, he said, had one root, and that was low wages. Some of the newer women guardians elected after 1894 aligned themselves with

George Edwards. With exceptions that can be counted on one hand, older women guardians never did.[74]

To some extent, the London pattern can be read across the country. During 1881 as we have seen, Mrs M'Ilquham, Mrs Shaw, and Miss Spottiswoode were elected. As the 1880s progressed, women stood as guardians in three sorts of town. Predictably, they came forward in towns with a strong women's movement based on philanthropic and suffrage work—Bristol, Birmingham, Leeds, and Bradford by 1883 had women guardians, Manchester not until 1889. Women guardians also stood in those cathedral, resort, and spa towns rich in strong-minded, single, and affluent ladies, such as Brighton, Oxford, Chester, Eastbourne, Bath, Cheltenham, Southport, Worthing, Hastings, and Winchester, often those same towns that had already returned strong-minded, single, and affluent ladies to their school board. The third cluster of women guardians were to be found in those fashionable quasi-dormitory towns around London such as Guildford, Wimbledon, Croydon, Richmond, Lewisham, Kingston, Bromley, West Ham, and, of course, Barnet. Almost no women were elected by rural parishes, in contrast to the way they joined rural school boards, even though many women must have owned landed property. Mrs M'Ilquham, Miss Spottiswoode, and Miss Emily Siddon were the exceptions, and of Miss Siddon, a substantial Honley landowner, it was asked, could Honley not find a man to represent it?[75] By 1885 there were 21 guardians outside London, by 1889 43, by 1893 119. A team of five stood in Plymouth in 1892 (all of them promptly defeated two years later when they stopped workhouse children attending the pantomime),[76] and others were elected in Nottingham, Newcastle, and Liverpool. As in school board work, the ports of Southampton, Hull, and Portsmouth resolutely refused to elect women to anything at all. More surprisingly, radical cities like Norwich and Sheffield still had no women guardians by 1893.

---

[74] *Norwich Mercury*, 26 Nov. 1898. For a fuller discussion of this point, see pp. 286 ff.

[75] In 1898 Miss Siddon refused the chair of her board, to the dismay of feminists. 'Many women would have been helped by her acceptance', complained Mrs Elmy to Mrs M'Ilquham, 13 May 1900, BM Add. MS 47452. A parish would have five places on its school board, but elect only one guardian.

[76] *EWR* 15 Jan. 1895.

Poor law women therefore were to be found in most of the same towns as school board women, and shared a similar social and philanthropic background. They were district visitors, parish workers, active in the Girls Friendly Society, in temperance or in local nursing schemes. They included Miss Anne Ely, a Croydon guardian who travelled into darkest Lambeth for years to work with the poor; Mrs Copp, a Barnstable guardian, wife of a sheep dealer, whose spare time went on making garments for the poor, and running the North Devon Infirmary, Dispensary and Nursing Association; Miss Fullagar, at the centre of Leicester's NSPCC; Miss Blanche Pigott, friend of Mary Clifford and a Cromer guardian, equally busy as a Bible missionary at Norfolk fairs and fund raiser for Armenian refugees; or Mrs Shaen herself, for twenty years a workhouse visitor, manager of five board schools and several refuges for unmarried mothers. And like school board women before them and council women after them, they used their philanthropic networks to imaginative effect.

Poor law women were, however, far more likely to have trained with the COS or worked for a social purity cause, school board women to have worked for girls' higher education and for women's suffrage. School board women were usually Chapel but said little about it. Quite a few poor law women were Church, and most of them clearly drew strength from a living faith and its image of the Good Samaritan. School board women were almost always staunchly Liberal even when they stood as independents, and most of them abandoned any pretence of neutrality when expediency demanded. Poor law women do appear to have been largely indifferent to politics (just as, or perhaps because, the political parties were largely indifferent to poor law work); and many of them were deeply upset when battles over out-relief after 1894 politicized the boards, and their service. As we have seen, the apolitical nature of poor law work together with the property qualification meant that women guardians perceived working-class men and women as inmates, recipients, dependants, paupers; whereas for school board women they were voters, parents, party workers, lobbyists, and allies. So though these 'ladies elect' shared a similar social class and regional background, property, politics, and ideology interlocked to pull them sharply apart. Interestingly, only a handful of women, and those mainly in rural areas (Miss Avery in Barnstable, Miss Kate Spiller in Bridgwater) were ever both guardians and school board members before 1894; though in a few families (the Hills, the Mullers, the Guilfords), one sister went on to the school board, the other on to the poor law board. Normally their paths did not cross, except through the good offices of the NUWW. There is not much evidence that they worked together (on matters of school fees, for example) and some evidence that they did not, coming to metaphorical blows, for instance, over their mutual responsibility for handicapped children. The choice of school or poor law work seems also to have reflected a deeper personality bent. There was a spring to the school board woman's step, an obvious zest and relish for the work, a buoyancy about the battles ahead. Poor law women had much the harder task of it. Theirs was gruelling, unrewarding, and personally harrowing work, and if they protected their own emotional health by distancing themselves from clients, becoming case-hardened in the process, this is perhaps understandable and was perhaps necessary. Those who remained emotionally generous and gave without stint often broke and had to retire. They were mostly on their own, with less family, party, or board support than school board women enjoyed, and their presence was bitterly resented. It took reservoirs of moral courage to stay the course.

In 1894 the property qualification was scrapped. The effects were dramatic. From 40 women guardians in London, 116 in the English provinces, and 3 in Wales, numbers rose to 86 in London, 716 in the English provinces, and 73 in Wales. A further 300 women stood without success. Many headed the polls. Most of them were married women. Increasingly, they were party women.[77]

[77] *Women guardians before and after 1894*

1893	London		Provinces	
Single	27		68	
Married/widowed	13 (40)		51 (119)	
1895	Re-elected	Elected	Re-elected	Elected
Single	16	20	54	190
Married/widowed	6	44 (86)	43	502 (789)

Sources: WLGS, WGS Annual Reports.

The biggest advance was in the industrial North—over seventy women were elected in Lancashire, over eighty in Yorkshire (in Blackburn, Bolton, Dewesbury, Halifax, and Wigan)—and in the rural counties. Cambridgeshire, Huntingdonshire, and Rutland still had the distinction of returning no women guardians, but Devon, Norfolk, Somerset, and Lincolnshire each returned around thirty women, usually for small townships such as Bovey Tracey and Cromer. Around a hundred and forty of the rural guardians were also rural district councillors, especially in Norfolk. Others combined it with parish council work.[78] Eleven women joined Newton Abbot union, where previously there were none. Eight came on to Exeter, to Huddersfield and to Kingston. Over half of all unions now had at least one woman member, though, as the WGS pointed out, 875 women among some 29,000 guardians was still a mere 2½ per cent. By 1898 numbers had risen to nearly a thousand, though in the sparsely populated large rural unions women were often deterred from standing because they faced distances of up to twenty miles to meetings without benefit of bus, car, train, or carriage.[79] Here as elsewhere, the bicycle was to liberate women from immobility.

Some poor law guardians were for the first time working-class women, the wives of artisans, railway drivers, and small shopkeepers. In the North, a few were textile operatives. The Women's Co-operative Guild, in conjunction with the Labour party, tried to encourage working-class women to come forward.

The Guild had been founded in 1883 to bring women into the co-operative movement. It started as a social club which began to educate women into their power as consumers; and from there to campaign for non-polluted milk supplies, maternity clinics and benefits, and better housing. By 1900 it had

[78] Cambridge was notorious. Mrs Anna Bateson was elected in 1884, as a progressive, but was defeated two years later by nine votes when the Tories ran the vicar against her. Clara Rackham stood in 1904, Florence Keynes in 1907.

[79] The *EWR* calculated 15 Jan. 1895 that ¾ of all women candidates were elected, 20 per cent at the head of the poll, 10 per cent without a contest. For RDC work, see below, ch. 7.

strengthened in size under its able secretary, Margaret Llewellyn Davies, to 350 branches and 18,000 members, largely working-class wives.[80]

With the removal of the property qualification, Mrs Edith Abbot, a leading guildswoman from Tunbridge Wells, urged guildswomen to enter poor law work.

Form yourself into a committee and get others to join. Speak to your minister and doctor, and see if they know of any other men and women in the town who will help you, and don't be discouraged if they don't. Then choose the right woman, and get the town or village canvassed in her favour ... Tell your neighbours why a woman should be elected. Would they like their homes and their children to be managed by men only, even if they were the best of men? Would they like men only to choose their clothes and their food, their education and their nurses, to inspect their sickrooms and their nurseries, to cross question them when they have fallen? Should we trust to men the work that centuries of inherited training have given us the best right to do? ... I am not one of those who think men should have no place in this world, but be he the best possible, there are things *I* know, and that he can never know—a baby's needs; a little girl-child's troubles when she is growing up to womanhood; a woman's misery when shame and trouble come upon her; an old wife's pain when told she must spend the remaining few years of her life apart from the husband who has been half her existence for perhaps fifty years of her life.

Won't you help to put a woman where she will have the power and authority to soften these troubles—and you can do it, if you will put in action the magic strength of the force of *Cooperation*.[81]

In December 1894, forty-five guildswomen stood for poor law boards, twenty-two of whom were successful; but these were mostly the middle- and upper-class patrons of local guilds, such as Mrs Toynbee in Oxford. In Burnley, the two guildswomen guardians were doctors' wives, a third an advanced liberal (Lady O'Hagan), the fourth Selina Cooper,

[80] M. L. Davies, *The Women's Cooperative Guild, 1888–1904*, 1904, p. 57; P. Moulder 'What the Women's Cooperative Guild is Doing for Working Women', *Englishwomen*, 22, 1914, pp. 32 ff.

[81] Mrs Abbott, 'How as Guildswomen shall we most fully use our powers under the new Local Government Act?', paper given at Doncaster, June 1894. Mrs Knott in *WH* 6 Aug. 1892.

the only working-class woman of the quartet and also the
SDF/ILP candidate. In Cambridge Mrs Clara Rackham,
herself a Newnham graduate and wife of a Cambridge don,
started her local branch in 1902. As a guardian and then a
councillor she battled for municipal milk and abattoirs,
nursery schools, clear footpaths, and clean market stalls.[82]
Sarah Reddish, the Guild's organizer, toured the country
urging working-class women to stand for public office, to 'take
a share in the larger family of the store, the municipality
and the State'. Leaflets were distributed by the thousand; and
some existing guardians went out of their way to encourage
them. Mary Clifford described one occasion to her brother in
1899: 'I spent the afternoon giving a popular lecture on our
doings for the Workhouse children to 150 charming poor
women, members of the seven local branches of the Co-
operative Guild. Their interest and enthusiasm were quite
inspiring. They could hardly sit still, and at the conclusion
rose up and made excellent little speeches, requesting visits to
the Homes, asking that the children should no longer wear
workhouse dress, offering to foster "for a nominal sum" and
left rejoicing that they were helping in their Poor Rate to pay
for neglected children to be well brought up'.[83] At branch
level, guildswomen prepared themselves for public life quite
self-consciously, practising public speaking, learning committee methods, and discussing policy issues such as child-care,
outdoor relief, and unemployment. From the late 1890s
between thirty and fifty guildswomen were elected as
guardians each year and, as the NUWW recognized, 'The
Guild has done much to fit them for such responsible
positions'.[84]

It was not easy. Working-class women struggled to organize
their poor law work around the constant washing, cooking,

[82] An advanced liberal, she joined the Labour party, became a guardian, JP,
factory inspector, city alderman, and county councillor—and, somewhat frail, joined
CND marches in the 1960s. *Cambridge Independent Press*, 11 Apr. 1930, *Cambridge News*,
2 Dec. 1965; M. Stocks, *My Commonplace Book*, 1970, p. 77.

[83] G. Williams, *Mary Clifford*, p. 147, dated 2 Mar. 1899; Sarah Reddish, 11th
Annual Report, Women's Co-op. Guild, 1893–4.

[84] NUWW Conference, Croydon, 1897, p. 148. By 1922, there were 162 guardians
and 66 urban, borough, and county councillors. See Women's Co-op. Guild Annual
Reports.

cleaning, and mealtimes of family life. Hannah Mitchell who took over the Ashton seat coped only because a kindly neighbour looked after her home in her absence. Another guildswoman wryly noted that she bought domestic peace by first making pies—jam pies, bacon pies, rabbit pies, and meat pies for the men's dinner, and then by making 'cake for breakfast, cake for their dinner, cake for their tea. They'll eat cake nearly always'. She could then set off for board meetings. Few working-class men made any concession to their home comforts should their wives scrape up the courage to take on public work. Every hour of the day was planned, squeezed, and stretched, to fit workhouse visiting in between the washing and the ironing.[85]

In so far as working-class guildswomen were Labour candidates, they encountered political and class prejudice as well. Selina Cooper joined Burnley's board and managed to get some modest concessions for the elderly, such as allowing them to move freely around the grounds during the day, but could not get workhouse dress abolished, the casual wards reformed, nor outdoor relief made more generous. Her daughter recalled, 'When she became a Guardian, oh, she used to have such rows. They had to literally fight—the Tories used to fight rather than give sixpence.' At more heated moments her ILP comrade used to take off his wooden leg, with which he belaboured the opposition.[86] The Rotherham Guild in 1897 adopted a candidate who was also a member of the Women's Liberal Association, whom they considered very suitable in every way. 'But we are told', said a member of the Guild, 'that it must be "a lady", and not a "woman", according to the Gentleman's Liberal Association'. The woman withdrew. Mrs Bury, a former mill worker, noted that the men 'bitterly resented this advent of the women in their special preserve' but added that this was soon lived down, and within a short while wards decided they could not do without their women guardians. 'One of the Guardians is rather deaf', she said, and found it difficult to follow the proceedings, but he got round the problem he claimed, by watching 'the women and see how they do, and I'm never far wrong'. She added that

---

[85] H. Mitchell, *The Hard Way Up*, 1977 edn., p. 125; Davies, op. cit., pp. 152 ff.
[86] Liddington, *Respectable Rebel*, pp. 116–17.

this was the greater compliment because 'he holds political views opposite to our own!'[87]

The number of guildswomen was never very large, the number of working-class women even less, but they brought a first-hand knowledge of the family economy to the board table and personal experience of the havoc wrought by sickness and unemployment to precarious and hard-won respectability.

[87] 15th Annual Report, Women's Co-op. Guild, 1897–8, p. 8; Mrs Bury in Davies, op. cit., p. 57.

5

The Work of Women Guardians 1875–1914

Managing the Workhouse[1]

THE movement to bring women on to school and poor law boards, said a friendly Birmingham clergyman in 1889, 'was not as some people seemed to think, an ambitious scheme for putting women into positions for which they were not intended by Providence. On the contrary, it was intended to put them in exactly the place for which their special faculties and their womanly instincts fitted them'. That place was of course the home. Just as Miss Muller had argued in the *Westminster Review* that poor law work was simply 'domestic economy on a larger scale', so Caroline Biggs called women to be 'state housekeepers for those who are too poor, too ignorant or too helpless to keep houses of their own'. Olga Hertz, a Chorlton guardian, was still writing about 'wholescale housekeeping' in the years before the War.[2]

As at home, this meant keeping a watchful eye on the moral and physical welfare of all members of the 'family', as well as devoting meticulous attention to clothing, food, and sanitation. Armed with check-lists of helpful hints provided by the Women Guardians Society and Louisa Twining, scores of women guardians introduced underclothes, nightclothes, winter clothes, and indeed clean clothes for the first time to

---

[1] This account is largely confined to urban boards and workhouses; for the rural poor law, see below, ch. 7. For an excellent account of the workhouse, see M. A. Crowther, *The Workhouse System 1834–1929*, 1981. This chapter focuses on the concerns of women guardians and does not seek to be a comprehensive description of the poor law.

[2] *EWR* 15 Feb. 1889, 14 Mar. 1885; see also *COR* 19 June 1882; O. Hertz, *Why We Need Women Guardians*, 1907, p. 9. There were 194,650 indoor paupers in 1880. 11,472 were able-bodied men; 15,519 were able-bodied women, and there were 19,551 dependent children. 53,281 were non-able-bodied men, 34,508 non-able-bodied women, 38,579 dependent children. 16,394 men, women, and dependent children were classified as insane. 26th Annual Report, LGB, 1986–7, Appendix E. For the outdoor poor, see n. 68 below.

hundreds of inmates. One woman guardian met the children lined up before her, neatly dressed, clean necks to the collar line and months of grime below it. Shrewdly she had them take off their boots, and found them all wearing socks without feet. Some women inmates were still without drawers years later, at the cost of hygiene as well as decency since they cleaned floors on their hands and knees. Sheets and shifts were not changed for months on end. Mrs Evans was a businesslike widow with young children who kept a photographic shop. She went on to the Strand board in 1887, she confessed, not from any sense of service but because she suspected extravagance and wanted to get her rates down. She made her first duty visit on a cold April morning, and found girls shivering in thin short-sleeved cotton dresses as they endlessly scrubbed damp stone corridors. Their petticoats were as flimsy as cobwebs, their eyes were sticky, their noses running, their bodies covered with ringworm. Going home in the train, 'very sad', she decided the girls must have warm serge frocks. She got these and their handkerchiefs after asking male guardians whether they would like their afternoon tea served by a young servant wiping her nose on her skirts. She became the board's most determined reformer.[3]

Women guardians tasted the food for themselves, though male guardians thought this was overdoing it. One of Miss Spottiswoode's first changes at Guildford was to introduce a trained cook (rather than having the cooking left to paupers), to stop the tea being stewed with soda, and the potatoes boiled away. Louisa Twining found tea and cabbage boiled in the same cauldron at Tunbridge Wells, though admittedly not at the same time. One inspector found two coppers without lids standing on the oven three inches apart, with clothes being boiled in one, soup in another. 'When the soup boiled over into the clothes I raised no objection, but when the clothes boiled over into the soup, I said I would not stay to dinner'. Babies, sick children, maternity cases, the able-bodied, and the elderly were all getting the same food, solid chunks of stale bread, whatever their age, health, or teeth. Women guardians had it sliced. Mrs Despard found in Lambeth in the late 1890s

---

[3] Mrs Evans, *Richmond & Twickenham Times*, 6 Apr. 1889.

that elderly women were being served rotten potatoes, half-cooked meat, and thin gruel. The master and the contractor were both corrupt. Any complaints were met with the punishment of bread-and-water diet. Mrs Despard had the master sacked. Other women found that babies were being fed from dirty bottles, and dying of summer diarrhoea. Adults were sharing one mug between them. The sick had their food in bed without the luxury of plates. Miss Augusta Brown spent some time in Camberwell trying to improve the official soup, of water, onions, and grease. In desperation, she brought a bowlful of it to the board, but could persuade no guardian to try it. As the same soup had been turned down by both her cat and her dog, she thought that wise. In some workhouses it was poor ingredients, in others it was downright corruption. Women guardians placed themselves on the contracts subcommittee where they argued fiercely that the board should not accept the lowest tender irrespective of quality and adulteration. They checked the specifications for themselves.[4]

Women recognized that inmates had little to do but sit around awaiting meals, and that they therefore suffered almost as much deprivation from the food's monotony as they did from its abysmal quality. The workhouse diet provided bread, cheese, boiled meat, and soup, with none of the pickles, onions, bloaters, and cockles that gave relish to the food even of the very poor. Women introduced roast as well as boiled meat, fish dinners (one of their less popular innovations), more fruit and vegetables, cocoa for the children, tea and a mincing machine for the elderly. Some of the inexhaustible supplies of stale bread were made palatable with margarine or baked with milk into a pudding. Even so, observers agreed that convicts continued to have a better and more varied diet than paupers. Prostitutes were known to break windows to go to prison rather than enter the workhouse.[5]

Women nosed around (often literally so) those sanitary facilities and 'parts of the hospitals and workhouses which the gentlemen very rarely visit', said Miss Thorburn of Liverpool. Mrs Evans suspected that the ophthalmia and ringworm she

[4] H. Preston Thomas, *The Work and Play of a Government Inspector*, 1909, p. 293; *Parish Councillor*, 1896, p. 512.
[5] *Eastbourne Chronicle*, 28 Feb. 1885.

noted spread because fifty-six girls bathed in one tub of water, and shared half a dozen towels, five dirty brushes, and two and a half combs between them. Another workhouse, they found, had two small hand basins for 120 women, and WCs, without paper, which were locked at night. Some infirmaries had no lavatories at all, and the sick had to go downstairs and outdoors to 'a sort of shed' on winter nights. Rural unions were even more backward. Dunmow in 1904 swarmed with rats, Billericay in the 1890s had no lavatories at all, its sick wards were 'dark, stinking and dirty'; its sick in 1905 were still sleeping on the floor. Louisa Twining found euphemistically described dust heaps, and stained and dirty linen lying around the wards. Women introduced toothbrushes, sufficient towels, decency and privacy in the lavatory arrangements. They inspected the kitchen and the laundry for safety and cleanliness, and brought in new equipment, such as washing machines, to ease the life of the staff.[6]

As they visited the wards, they prodded the beds, had the mattresses picked over to remove the lumps, and changed the linen. Mrs Despard and Mrs Pankhurst were just two of many who threw out the hard forms on which the elderly sat out their lives, and replaced them with comfortable windsor chairs. They added lockers for personal bits and pieces, put up curtains and put down rugs to reduce the physical and emotional chill. Wards were mind-numbingly monochromatic with whitewashed walls, grey stone floors, grey dresses. More sensitive guardians realized that inmates were starved of colour and decoration. Mrs Fordham of Royston brought in red flannel shawls, red twill curtains, brightly coloured cushions, and scarlet geraniums. She had the windows lowered so that inmates could see out of them. In Norwich, Mrs Rump and Mrs Mitchell provided personal drawstring bags for every inmate, each with its own brush and comb, to hang on the beds, slippers for the infirm, hoists, lifts, and bath chairs for the sick, china rather than tin ware, and pocket handkerchiefs for all. One working-class member of the workhouse visiting committee, Mrs Emma Winter, unable to

---

[6] Miss Thorburn, Liverpool, RC on Poor Laws, Q. 35693; Evidence of the WLGS to the RC on Poor Laws, 1909, appendix lxxxv; *Parish Councillor*, 30 Aug. 1895; G. Cuttle, *The Legacy of the Rural Guardians*, 1934, p. 18.

afford to buy 'comforts', would hang around Norwich shoe factories to pick up cigarette ends. These she would bear home and painstakingly unpick, bringing the tobacco into the workhouse as cheer for the tramp wards. Everywhere women came bearing books and newspapers, pictures, flowers and toys, tobacco and snuff for the old men, twills of tea and sugar for the old women.

It was a culture of domesticity and emotional warmth, rooted in an endless attention to small details. As standards of comfort for all social classes slowly rose outside, women guardians tried to ensure that workhouses were not left too far behind.[7] What mattered most was the quality of the staff, as their kindness, competence or lack of it determined the weal and woe of the rest. Though women guardians were able to do something about the nursing staff, they were unable to do much about the others, hardened as they were into indifference.

The Care of Children

Most women, however, came into poor law work less to be state housekeepers than to respond to Mrs Senior's plea that they should mother workhouse children. Some fifty-five thousand children were in workhouses during the 1870s, up to a third of all their inmates. Of these, 40 per cent were in permanent care, having been orphaned or deserted by their parents. Another 15 per cent were the children of inmates, widows, or prisoners, who were in the workhouse for several months at a time. The rest were Florence Hill's 'dregs', the in-and-out children of vagrants, hawkers, and tramps, dirty and unsettling, who might enter and leave a workhouse twelve times in as many months.

[7] *Parish Councillor*, 31 July 1896; E. Pankhurst, *The Making of a Militant*, p. 25; *London*, 6 June 1895; A. Linklater, *An Unhusbanded Life: Charlotte Despard, Suffragette, Socialist and Sinn Feiner*, 1980, *passim*; see also Crowther, op. cit., p. 82. From the early 1890s, comforts could be provided on the rates. Mary Stocks used to accompany an aunt to St Pancras workhouse before the Great War, and recalled the 'vast bleak day rooms', *My Commonplace Book*, 1970, pp. 47, 81. On Poplar board, a Tory board without women, George Lansbury performed much the same role, getting the diet improved, clothing individualized, the Brabazon scheme, separate rooms for aged couples, newspaper and tobacco, and the children sent to ordinary schools. G. Lansbury, *My Life*, 1928, *passim*. For Norwich, see Board of Guardians Minutes, workhouse visiting committee 27 Feb., 10 Apr. 1895.

After 1834 there is little evidence of deliberate cruelty to workhouse children, but much evidence of damage to their moral and physical health from being 'barracked and warehoused' within workhouse walls, emerging only on a Sunday to walk in file to church. Poor diet, endemic infection, and little fresh air or exercise meant children were sickly and stunted; mixing with paupers, often cared for by pauper lunatics, they picked up the so-called pauper taint and fared badly outside. Reformers of the 1840s, headed by the senior Education civil servant, Kay Shuttleworth, tried to limit the damage done to children by removing those in permanent care to large district boarding-schools where it was hoped that a child-centred regime, a sound education, healthy surroundings, and industrial training, would give children a better chance at independent living.[8] In much the same way, Mary Carpenter was removing delinquent children from prison in the 1850s to smaller family-style reformatories.

But over the years, the new large district poor law schools were to replicate the worst faults of the old workhouses; and mounting concern about children's health from lady workhouse visitors and the Hill sisters had led James Stansfeld to ask Mrs Senior to report on them. Meanwhile, Miss Annette Preusser of Windermere had taken matters into her own hands by boarding some Bethnal Green children in the Lake District at her own expense. The LGB reluctantly agreed to permit boarding out where children were properly supervised, hesitant that it would make them more privileged than labourers' children; by 1876 some forty voluntary committees were quietly supervising three hundred children in foster care.

Mrs Senior's *Report* of 1873 was devastating. She found that formal educational standards in district schools were high, if mechanical, which meant they were rather better than the average board school, but in almost all other ways large

---

[8] 'Barracked and warehoused' is Ellice Hopkins's phrase, *EWR* 14 Nov. 1885. For a horrifying description of a large pauper school in mid-century, see J. Buckley, *A Village Politician*, 1897, ch. 2. Crowther argues that there is no clear evidence that either a significant proportion of workhouse children later became paupers, or that many adult paupers had been brought up in the workhouse, given that the children came from the most deprived backgrounds. However, the quality of their childhood was undeniably poor (op. cit., pp. 220–1).

district schools were as damaging as the workhouse itself. The girls especially were sullen, apathetic, lacking any spontaneity, liveliness or intelligence. They cared about nothing or nobody, though would sometimes fly into unpredictable and ungovernable rages before subsiding again into lethargy. 'The most absolute monotony, both of sight and sound—only it is called order and regularity—and there is intense dullness under the name of method and discipline. White walls, dull rooms, the same dinner (and the same quantity) to eat on a given day of the week'. When there were fireworks in one nearby garden, 'the blinds were drawn down for fear the children would be disturbed . . .'. And added Miss Synnot, who accompanied Mrs Senior, and was later a guardian herself, 'there are young children who are really old men and women. We find them by hundreds in our monster schools'.[9]

Starved of play, toys, treats, variety, the children were 'wooden', 'flattened', 'mechanical'. Starved of colour and ornaments, the girls spent all their money when they became domestic servants on decoration, coral beads, and large feathered hats, to the puzzlement of their employers. One child cut off the ribbons of her mistress's bonnet to make dolls' clothes. Starved of mothering, they had little or no moral sense, stealing and lying with equal unconcern. Mrs Barnett became manager of the Forest Gate district school and found it without toys, books, flowers, pets, or pictures. Meals were eaten in silence, the girls' hair was cropped to keep it free of lice, the whole was hideously clean. No child knew how to play. She wept for them. George Lansbury, who knew the school well, said, 'The place was organized and controlled as a barracks'.[10] When fire and then poisoned meat ravaged the Forest Gate children, Mrs Barnett determined to launch the State Children's Aid Association to lobby on their behalf. Sophia Lonsdale found the Lichfield children, placed in West Bromwich school, in 'wretched' condition, though the local guardians and the local inspector seemed well satisfied. She

[9] H. Synnot, 'Little Paupers', *Contemporary Review*, 1874, pp. 970–1. See also Buckley, op. cit., p. 253.

[10] H. Barnett, 'Young women in our workhouses', *Macmillan's Magazine*, 1879, pp. 133–9; M. B. Smedley, 'Workhouse Schools for Girls', *Macmillan's Magazine*, 1874, pp. 27–36; Lansbury, *My Life*, p. 149.

went to London over their heads, and warned that if nothing was done she would place a question in the Lords, a letter in *The Times*, and 'I'll throw that school to Mrs. Barnett to tear to bits'. The managers threatened her with a libel action, but she got new staff, had the school spring-cleaned, and the children began to thrive. One of the first Nottingham lady guardians found children sitting listlessly on benches in a room without toys. She took them outside, and found 'they did not know what grass was or how to run'. Louisa Twining discovered nearly forty small children in a bleak room at Tunbridge Wells in 1893 without a single plaything between them.[11]

Newly elected women guardians were determined that children must neither rot into pauperism within the workhouse nor be damaged by the institutionalizing effect of large district schools. As Mrs Henry Kingsley, a Richmond guardian, said to the Central Poor Law conference in 1887:

I unhesitatingly say that if I had the power as I have the will, not a single child of any kind would ever be in the workhouse, notwithstanding the great kindness of any master or mistress, notwithstanding the greatest care of the Guardians . . . We take everything that is right and good away from them; we take fatherhood and motherhood away. We certainly feed and clothe them but that is not all. We want something higher and better if we want to raise the tone of the whole of the coming nation.

That something 'higher and better' was family life. Every day within a family, said Mrs Rye to the NUWW conference of 1897, 'children have small sacrifices to make for the sake of the comfort of some other member of the family, denials which become gifts . . . In a family, children learn something by the moving, the turnings out, the cleanings, the washing days; and more deeply, the sicknesses, the accidents, the deaths, births and separations, and losses and gains of income—in a word, the change of all sorts . . . These make the individual ready and adaptive, create or quicken the sympathies, and give backbone to conduct . . .'.[12]

[11] V. Martineau, *Recollections of Miss Sophia Lonsdale*, 1936, pp. 171–3; N. Midlands Poor Law Conference, 1908, p. 545; L. Twining, *Workhouses and Pauperism*, p. 139. See also Miss Clifford, N. Midlands Poor Law Conference, 1892, p. 73.

[12] Mrs Kingsley, Central Poor Law Conference, 1887, p. 287; Mrs Rye, NUWW Conference, Croydon, 1897.

Guardians were only willing to board out long-stay children, such as orphans, the deserted, some handicapped, who were over the age of two and under the age of ten, fearing baby farms on the one hand and domestic drudgery on the other. Though well-intentioned, this meant that small babies continued to die at the hands of retarded pauper nurses, and older girls become unfit for family domestic life. The women's Boarding Out Association and Mrs Barnett's State Children's Association both tried hard to widen the categories of children eligible for fostering. Though in many European countries and in Scotland most pauper children were boarded out, less than 10 per cent of English pauper children were in foster homes by 1895.

Boards normally fostered children with local families, often relatives. Women members, obsessed by the notion of taint, preferred to send them some distance away where they could make a fresh start and not be reclaimed by grasping relatives the day they could start to earn. There were occasional problems. Some children were not properly supervised. Miss Lidgett, always doubtful about the value of boarding out, had cases where the children were taken only for the money; and became lodgers and drudges. One Billericay guardian remarked 'they take the children for what they get, not for love, just the same as I grow dahlias. I used to do it for love, but I don't now'. Supervision was essential. Because each local committee was confined to a patch of four to six miles, many former workhouse children were 'unhealthily' concentrated into one or two villages; boarding out, guardians worried, became the village staple industry. The boys would later find it hard to get local jobs. More generally, women guardians could not find enough homes with working-class living standards but middle-class values. When one male working-class guardian from Lewisham ventured to suggest at a conference that he personally knew of many suitable homes, the lady guardians shouted him down.[13]

It was undeniable, however, that boarding out was good for the child and cheap for the ratepayer. Of a hundred and two

[13] Miss Lidgett, Central Poor Law Conference, 1896; Cuttle, op. cit., p. 126; NUWW Conference, Manchester, 1896, p. 91; Miss Lonsdale, N. Midlands Poor Law Conference, 1898.

children placed by Joanna Hill in Birmingham, eighty-five were thought to be thoroughly satisfactory. Almost as satisfactory was the cost. Payments to foster parents averaged 4s. a week; the cost of a child in a district school, where there was no cheap pauper labour, came to 8s. to 10s. a week. Boarded-out children often joined a family and received its affection for life, returning home for holidays and Christmas long after payments had ceased. Visitors found foster-parents giving their children birthday parties and nursing them devotedly through epidemics, the children in turn spending their first wage packet on presents for them. One little girl who was boarded out was asked by Miss Preusser if she wished to return to a district school. She refused, because there 'I belong to nobody'.[14]

Women guardians made child placements their especial care. Miss Brodie Hall was elected at Eastbourne in 1883. Within six months she had built a small ladies committee and they placed every eligible child in a suitable home. The women then turned to the difficulties facing older girls, started a small training home, and helped them through the trauma of their first situations. Miss Spottiswoode found fifty children in her workhouse. She boarded out every one, and over ten years had not a single child return to the workhouse. Mrs Charles pointed out to the WGS the difference boarding-out policies could make; Fulham had seven hundred children in district schools, she had brought Paddington's figure down to just forty-six. Miss M'Turk, a doctor's daughter, had started a private orphanage to rescue babies from the workhouse; as a Bradford guardian from 1883 she managed to board out most of the older children. At Croydon, Miss Anna Ely and Miss Lucy Morland boarded out eighty children 'with great success and at half the cost'. Louisa Twining, who was not much interested in children, none the less found one hundred and twenty Tunbridge children in the workhouse and placed forty of them in foster care. In Bristol, Miss Clifford devoted herself to the children left behind, taking them on treats to Weston-

[14] Miss J. Hill, W. Midlands Poor Law Conference, 1883, p. 69; *COR* 1 Feb. 1885, p. 64. Another factor encouraging boarding out was the pressure on space within the workhouse itself, especially when the LGB began to insist on proper infirmary wards. Billericay chose to board out their children rather than build a new infirmary. Cuttle, op. cit., p. 126.

Super-Mare where she allowed them to explore the town on their own; or for walks on the Downs. When it rained she turned her green cloak into a wigwam under which her children huddled, 'only their legs appearing', and told them magical desert island stories. 'I feel one ought to go freely into the impossible', she confided to her friends.[15] In a more formal capacity, she took case notes on every child, accompanied Miss Mason, the inspector, in searches of suitable homes, and twenty years later was still writing to her boys now fighting in the Boer War.

Many women guardians after 1883 persuaded their boards to emigrate the best children to Canada. They appear almost always to have done well. One exception to this is perhaps worth quoting. Miss Brodie Hall had helped a fourteen-year-old boy to emigrate from Eastbourne. At eighteen he was found guilty of murdering a member of his employer's family. Miss Hall was at the other end of Europe, but dashed home, checked her records and his letters, found 'evidence of irresponsibility and insanity' in the family, wired the material, and the boy's death sentence was commuted to ten years. On his discharge from prison, Miss Hall placed him with a builder in Saskatchewan and in time he set up as a mechanic in his own right. As of 1904 he was doing well. She gave another kind of help to a lad who was placed with a master and three nice daughters, whom he studiously avoided as they were 'radicals' and were rude about the Queen. Miss Hall sent him a biography of Her Majesty and the boy reported that they were now all 'converted'. Miss Spottiswoode and Mrs Despard were only two of several women guardians who travelled to Canada at their own expense to check on their children and found them well.[16]

The arrival of working-class guardians after 1894, however, challenged assumptions here as in so many other fields. When a Winchester woman guardian recommended emigration for bright healthy children, Will Crooks, a Poplar guardian,

[15] Miss Brodie Hall, *Eastbourne Chronicle*, 8 Dec. 1883, 1 Mar., 21 June 1884. Miss Brodie Hall became hon. sec. of the National Association for the Advancement of Boarding Out, *COR* 1 Feb. 1888. Anon., *Ten Years Work as a Lady Guardian*, 1896; London, 22 Apr. 1897; 15th Annual Report, WGS, 1897, Twining, *Workhouses and Pauperism*, p. 148; G. Williams, *Mary Clifford*, 1920, p. 146.

[16] Miss Brodie Hall, Central Poor Law Conference, 1904, p. 695.

crisply commented that these were just the children who did
well at home. Indeed efforts by childless women guardians to
break the links between child and erring parents seem to have
bordered on the punitive and the obsessional. Miss Smedley,
along with Miss Synnot one of Mrs Senior's assistants, had
written in 1874 that even 'the lowest form of family life' awoke
affection and conscience in the child, so that it grew 'far
healthier, far more capable, far more human, than the items in
this vast unnatural conglomerate' of district schools.[17] School
board women knew this instinctively and fought hard to
stop board children being sent to distant reformatories, much
preferring that they should attend local day industrial schools.
Poor law women seldom did. Mary Clifford, in many ways
one of the most sensitive of guardians, was quite ruthless
about denying parental rights. She went to immense trouble
to ensure 'that the parents should have no clues as to where
the children are'. Miss Hopkins of Scarborough found little
pleasure in poor law work but she did describe to the *British
Women's Temperance Journal* one of her moments of reward.

A day or two ago I was asked by a most degraded, wretched woman,
to walk into her filthy home, to read a letter, and to look at the
photograph recently received from her little daughter, who had been
sent to Canada by Dr Barnardo. When I contrasted the mother and
child I could but give thanks that I had been the instrument of
sending that bright little maiden safe out of the reach of the miserable woman when they were both in the workhouse . . .[18]

There was little compassion for a mother who clearly loved
her child. Miss Mason, the able, hard-working, LGB inspector of boarded-out children, who left a trail of sugar mice
behind her on her visits, was unhampered by any concept of
bonding and was quite unsentimental about removing
children from foster homes she thought grubby or unsuitable.
Miss Clifford, who often went with her, found homes where
there was 'evidence of abundance of kindness, the children
were fairly clean and quite well fed, yet it was clear the habits
and tone of the people were not high enough to make a strong

---

[17] *Macmillan's Magazine*, 1874, pp. 27–36; Miss Bromfield and Will Crooks,
SE Poor Law Conference, 1898–9, pp. 535 ff., 551.
[18] Quoted *Richmond & Twickenham Times*, 26 Jan. 1888.

example and education for the boys'. They were removed. She explained why.

> It is not the home which is the most loving and tender, nor the one that incorporates the child most thoroughly into its midst, that is the best, but the one in which the foster mother can instil character and backbone into the child, and this can scarcely be done unless the mother has some strength of character herself.[19]

The pauper child was not as other children. It needed a double dose of fibre if it was to overcome its parents' taint, its acquired or inherited trait of being born tired.

Women guardians certainly needed to protect children from natural parents who battered and brutalized them, who tried to entice them into brothel life, public house work, street hawking, or begging. Clearly also they were watchful of foster-parents who took children for their money or their unpaid labour. But their judgements went beyond that. Even the COS, a loyal supporter of boarding out, began to hesitate at the ruthless severing of family ties.[20] The treatment of respectable widows seemed particularly harsh. They could come into the workhouse; or, under the LGB guidance of 1871, could put one or two children into the workhouse while they struggled to keep some semblance of family life going for the rest. Having lost their husband, they had now to lose some of their children (and the agony of that choice), rather than be given adequate and less costly outdoor relief. Mrs Charles of Paddington persuaded her board to restore outdoor relief to such women and in 1888 the LGB modified its own guidance. Not surprisingly, newly elected working-class guardians grew restless at the moral and social superiority evinced by affluent spinster guardians, the fracturing of family ties, the punishment of poverty, the denigration of working-class culture and life-style, the easy assumption that it was all about drink and never about wages. For their part, women guardians felt they had little option. Without any professional salaried social workers to help them, they could not return children at risk to inadequate families, even if guardians had been willing to

[19] Miss King, Northern Poor Law Conference, 1902, p. 113; N. Midlands Poor Law Conference, 1892, p. 76.
[20] *COR* 1 Jan. 1889.

'reward' such families by offering financial and practical support. In their eyes, it was removal or nothing.

Slowly, the number of boarded-out children rose. By 1886, 3,000 children were fostered, largely within their own union, by 1896 nearly 6,000 and by 1914 nearly 12,000. A similar number of older children were in small homes training for domestic service, or were handicapped and in special certified schools. What of the rest, those children only temporarily in care, illegitimate children, and the 'casuals' for whom long-stay fostering seemed inappropriate?

For the most part, these children remained within the workhouse and its grounds, though often in special cottages with their own housekeepers. Miss Varley in the 1880s improved the food and abolished workhouse dress for Islington's children, started a nursery for the babies, and herself took the older children to the local swimming-baths, exhibited their school work, and tried to enrich their range of experience. Miss Spottiswoode and Miss Twining both closed their workhouse schools and sent the children out to mix with others at the local elementary school, a policy particularly popular with guardians once school fees had been abolished in 1891. By 1908 only one child in a hundred still went to an exclusively workhouse school. The rest were beginning to experience a more normal childhood. Mrs Walter, a Norfolk guardian, wrote in 1895, 'To my delight the children are now dressed as ordinary school children, so that meeting them out for their walk you would not know they belonged to a workhouse. Now too the children over five go to a school at Docking—another good step I think. Surely mixing with other children must be right and best for them'.[21] As the children were clean, punctual, polite, and invariably regular in their attendance, they were popular with village teachers, earning them high grant. There was still the problem of weekends and school holidays—inspectors encouraged boards to find playing-fields, though one inspector thought integration into the local community had gone a bit far when he found the guardians allowing boys to 'turn cartwheels in the street in order to attract the pence of tourists on excursion cars!'[22]

[21] Mrs Walter, *Parish Councillor*, 4 Oct. 1895; Miss Varley, *London*, 24 Nov. 1898.
[22] Preston Thomas, op. cit., p. 233.

One kindly lady guardian in Chelmsford arranged pocket-money for the children—1*d*. a week for the older ones, ½*d*. for the young ones—but this was banned by the LGB. In Maldon, to take one example of hundreds, guardians took children off to fêtes, flower shows, and the seaside. Many workhouses appointed matrons who were devoted to their children. During the 1890s even those children who remained behind seem to have had a much better time of it, owing largely, said McKay, 'to the efforts of lady guardians'.[23]

Many progressive boards, anxious to take children out of the workhouse entirely but unable to place them in foster care, developed children's colonies, a purpose-built cluster of large cottages each with twenty to thirty children, cared for by houseparents, each village with its school, infirmary, playing-fields, and baths. These had been pioneered in Germany, copied by Dr Barnardo and other enlightened church orphanages during the 1870s, and now filtered into poor law policy. Two of the best known were at Banstead, built by Kensington union as an alternative to large district schools, and at Marston Green, near Birmingham. Advocates of boarding out, such as Joanna Hill, thought these villages misconceived. They were still far too large, with several hundred children, their cost was high, the health of the children poor, and their adult life unpromising.[24] In Sheffield, Mrs Wycliffe Wilson's husband suggested a modified version of the cottage home by introducing scattered homes, as Barnardo and the Waifs and Strays Society were doing. The board purchased ordinary large cottages scattered throughout Sheffield and placed a dozen to twenty children in each with houseparents. The children attended the local school and were absorbed into the neighbourhood, making their own friends, visiting other children. Even here, women guardians felt that these scattered homes were too large and impersonal. Foster-mothers were overworked, older girls were press-ganged into constant cleaning, and there was no emotional space for toys and the kitten

---

[23] Cuttle, op. cit., p. 106; T. Mackay, *History of the English Poor Laws*, 1903, vol. ii, p. 432.

[24] *Birmingham Daily Mail*, 1, 3 Feb. 1887; *Local Government Chronicle*, 29 Jan., 5 Feb. 1887; *Weekly Post*, 5 Feb. 1887. Her criticism produced 'consternation'; attacks and counter-attacks continued for some time in the press.

before the fire. It was also difficult to find competent foster-mothers who could be trusted to care for so many children without the close supervision of the grouped homes arrangement. Nine of Sheffield's first twenty-three foster mothers were sacked, in some cases because they were on the game.[25]

Two Bradford women guardians, Mrs Stott and Mrs Moser, went over to Sheffield to see the cottages for themselves. They persuaded their board to acquire scattered homes for family groups of no more than six or eight children, keeping brothers and sisters together. By 1895 Mrs Moser had acquired four cottages. Miss Margaret Baines persuaded her Leeds board to follow suit, and in a few years was supervising eleven small scattered homes which offered a 'simple, free, natural and economical' family life for knots of youngsters in their care. They were judged very successful. In the Huddersfield area, Miss Siddon bought four scattered cottages and acted as their superintendent. In Grimsby, each woman guardian took responsibility for mothering a handful of children, boarding out those in permanent care, placing others in cottage homes, finding older children their first jobs, and befriending them through its trials. In Oldham, Marjorie Lees visited each of her scattered homes every two or three days, settling squabbles, inspecting clothing, checking on sick children, ordering supplies, studying the foster-mother's punishment book; and recorded every detail in her Memorandum Book. Norwich had long had a separate boys home. With the arrival of women guardians it opened a thirty-bed girls home. It proved a success. Not until after 1910 did Norwich start systematically to board out children. Then it carefully instructed foster-parents 'to manage your home in as nearly as possible the same manner as if the children entrusted to your care were members of your own family'. Foster-mothers were to make the children's clothes, take them out and about, and not use corporal punishment.[26]

[25] SW Poor Law Conference, 1900, p. 127.

[26] *Bradford Observer*, 25 July, 19 Sept. 1895, 20 Feb. 1896; Miss Baines, RC on Poor Laws, Q. 39550; *Yorkshire Post*, 21 July 1898; 2 Jan. 1902, 16 July 1903; Miss Siddon, RC on Poor Laws, Q. 40583; Miss Lees, *Memorandum Book*, Lees Collection item 172; Norwich Board of Guardians Minutes, 27 Nov. 1895; *City of Norwich Union Instructions*, 1912 (in possession of the author). See also Mrs Ellis of Leicester, who with her

By 1909, of the 75,000 pauper children in England and Wales, 13,000 of them were still attending district or separate workhouse schools; 11,500 were in cottage homes, over 8,000 were in scattered homes, a similar number were boarded out, and 9,500 were in special care, such as homes for the handicapped. The rest, around a third, still lived within the workhouse; they were mainly rural children attending the village school and very much part of village life. Finally, in 1913 the LGB forbade children between the ages of three and sixteen to remain in a general mixed workhouse for more than six weeks. Young children were to stay with their mother.[27] Thirty years after Mrs Senior's death, too many children still remained within the large district schools and workhouses that she had denounced; but it was undeniable that even for these children their quality of life had immeasurably improved. No one doubted that this was the work largely of women.

The Training of Girls

What, however, would happen to a child when at thirteen or fourteen it left the day-to-day care of the guardians? Possibilities were limited. Without families of their own, the children had to seek living-in jobs. Boys might enter the Army or Navy, the lucky few might become a living-in apprentice, others might survive in lodgings as a factory hand. For girls, the only realistic option was domestic service, entering someone else's home, living under someone else's protection. So far workhouse girls were no different from the majority of young rural girls who at twelve or thirteen were going into their petty place. Only slowly were other jobs, in shops, clothing factories, and laundries, opening up for working-class girls. Poor law girls were disadvantaged, however, because

two women colleagues toured their scattered homes and checked on foster-parents in a wagonette. I. Ellis, *Records of Nineteenth Century Leicester*, 1935, *passim*.

[27] NUWW Conference, Portsmouth, 1909, pp. 17–18. Cuttle's study of Essex guardians shows that neither Maldon nor Dunmow started to board out children until shortly before the War. However, in both unions the children appear well cared for, with good food, clothing, and treats, in Maldon because the clerk was child-centred, in Dunmow because the master was devoted to their welfare (op. cit., p. 118). The problem of children who were frequent 'ins and outs' was especially difficult in small rural unions.

they entered domestic life unsuited and unfitted for it.
Henriette Synnot described dinner at a district school;

A large hall capable of holding 1,500 children. A smell as of laundry
and kitchen combined, and a great deal of steam. Enter four men
bearing two enormous wooden washing tubs, containing a stew or
soup of Australian meat. That is ladled out into small basins, and
placed by boys on the tables. The sound of martial music is heard in
the distance, and presently a thousand little paupers, headed by
their brass band file in and take their places. After singing a grace,
they fall to . . . When the girl who has assisted at this meal goes to
service, she has not, as far as I know, to serve out soup, from a
washing tub, or to *sing* a grace, or to throw away what is left on the
plates . . . nearly everything which a girl sees and does at school is
absolutely opposed to what she will see and do at service'.

Twenty years later, Florence Davenport Hill was still describing the 'never-ending scrubbing of vast dormitories and
dreary corridors, the preparation of vegetables by the bushel,
the mending of clothes by the hundred' as the only experience
a girl brought with her to domestic service.[28] She would not
know the name of the commonest piece of furniture she had to
polish.

Pauper girls went to their first situation not only untrained,
but virtually untrainable. Lethargic, slow-witted, sullen, and
often sickly, they were unable to obtain good situations and
were exploited or sexually abused in bad ones. They seldom
stayed long in any job—Miss Brodie Hall had sent seven girls
out to service in five years from Eastbourne union and her
ladies committee had had to find thirty-seven placements for
them. They were not bad children, she considered, but what
would have happened to them, she asked rhetorically at a
poor law conference, if they 'had not had kind ladies to work
for them?' Miss Clifford answered her. She knew one young
woman who had left the workhouse at thirteen. She was now
twenty-four years old. She had not spent 'any part of her life
out of gaols, workhouse, hospitals, home or the street'.[29]
Pauper girls had no one to turn to and nowhere to go. If they
became pregnant, they came back to the workhouse; if

---

[28] Synnot, 'Little Paupers', p. 966; F. D. Hill, 'The Boarding out of Pauper
Children', *Economic Journal*, 1893, pp. 62–73.
[29] Central Poor Law Conference, 1895, pp. 429–31.

diseased, went up to the VD Lock (or 'foul') wards; if delinquent, to the reformatories. Boarding out, it was hoped, would break the circle. Growing up in families would prepare girls for domestic work with other families. Those too old or ineligible for boarding out, however, needed compensatory care and training.

Norwich, back in the 1850s, had pioneered a small training home for its older girls, although it closed in 1874. No other board seems to have started a similar one. Louisa Twining, with money from Baroness Burdett-Coutts, had opened a training home in London in 1861, and over the years some 750 girls were given a few months' preparation for domestic service. Mrs Senior, fifteen years later, had started the Metropolitan Association for Befriending Young Servants (MABYS—or Mind and Behave Yourself, as it was known familiarly to its girls). In the provinces the Anglican Girls Friendly Society (GFS) quickly built up a network of support and practical help for young women, helping them find decent situations, rescuing them from bad ones, visiting them in their new places, arranging recreational evenings, and providing safe lodgings between jobs.[30] By 1898 there were over two hundred training homes in England and Wales and every self-respecting woman guardian was supposed to have the list in her head. Many of them were run by the GFS, to which guardians on the advice of their women members would send older girls for a few months at a time. By then, MABYS was working as a semi-official after-care organization in almost all of the London unions and the GFS had workers and associates in some four hundred and fifty provincial unions, stubbornly befriending lonely and isolated young girls through a succession of employers, boyfriends, and crises. By the later 1890s workhouse girls were being sought as servants, unlike twenty years before. They were clean, trained, and obedient and could therefore get better homes.

Women guardians naturally made such work their own

[30] For MABYS, see E. Ross, 'Women in Poor Law Administration', London MA, 1955, pp. 127 ff.; for the GFS, B. H. Harrison 'For Church, Queen and Family: the Girls Friendly Society, 1874–1920', *Past and Present*, 61, pp. 107–38. The GFS was not as effective as the London-based MABYS, partly because many guardians resented its denominational character, and preferred to rely on non-sectarian ladies committees.

field. Some indeed had first started public work as a MABYS or GFS visitor, such as Miss Bramston, six years a full-time MABYS organizer before becoming a London guardian. Miss Donkin helped to start her branch in Kensington, Miss Cadbury on becoming a Birmingham guardian in 1884 established a ladies after-care committee for the girls from Marston Green cottage homes. For twenty years she acted as its secretary. Mrs Henrietta Barnett trained 135 pauper girls in her own home, three at a time, at some cost to her delicate furnishings, before placing them in suitable jobs. Miss Mason, the LGB inspector, had started her professional life as a GFS organizer, and when Miss Clifford found Bristol guardians rather contemptuous of ladies who busied themselves with little servant girls, she brought Miss Mason down to enrol their wives into the GFS, so that they might know better. In Oxford Mrs Toynbee specialized in the after-care of Cowley poor law children. In Southport the ladies ran their own classes in 'simple cookery, laundry work and the cutting and making of clothes' for their older girls, and personally inspected every situation before they released a girl from their tutelage.[31]

With memories of Mrs Senior's Report in mind, women guardians feared that only painstaking after-care stood between a workhouse girl and the streets. If a girl in her first situation rejected their affection and attention, what, the WGS agonized, could they do? The question was given even greater saliency by W. T. Stead's revelations in *A Maiden Tribute to Modern Babylon*, when he bought a thirteen-year-old, Eliza Armstrong, from her mother for £5. Following his exemplary prison sentence, the age of consent was raised from thirteen to sixteen in 1885. The LGB responded to a deputation of women guardians in 1887 by similarly extending the powers that guardians might retain over workhouse children.

[31] M. C. Cadbury, *The Making of a Woman Guardian*, 1937, p. 13; Mrs Barnett, NUWW Conference, Manchester, 1895, p. 91; Williams, *Mary Clifford*, p. 114; for Southport, NUWW Conference, Liverpool, 1891, p. 131. However, Mrs Paul, a Truro guardian, was still finding girls going straight into domestic service without any training or support in 1908; if their first mistress failed to care for them, their future was bleak. SW Poor Law Conference, 1908, pp. 422 ff. See also the tribute paid to Mrs Elizabeth Shanks of Croydon on her death for her work for girls: it 'would be difficult indeed to supply her place'. Chairman, Croydon Guardian Minutes, 12 Mar. 1895. (I owe this reference to the kindness of Croydon library staff.)

If guardians adopted their children, they could continue to supervise girls until they were eighteen, boys until they were sixteen. The Act was permissive but women guardians everywhere pressed their boards to use it to ensure continued protection for children at risk.[32]

There was one other way of removing the pauper taint from pauper children, and that was to remove them from the care of the poor law altogether. As so many poor law children now attended elementary schools, and as school boards were pioneering classes for mentally and physically handicapped children, Mrs Barnett and her State Children's Association argued that the best way of removing workhouse stigma was to transfer the children to the care of the Education Department and local school boards. A Government Select committee of 1896 favoured the idea, and a bill was introduced for London. Women guardians rose in sorrow and in anger. Nine hundred women, they said, had become guardians, 'in many cases especially for the children—I think we have hardly had a fair chance yet'. School board and poor law women fought over their territorial responsibilities, lobbying was intense; but, said Mrs Barnett, the children's well-being was sacrificed to provide interesting work for women guardians.[33] The bill was withdrawn.

The Rescue of Unmarried Mothers

After-care for workhouse girls often and sadly became after-care for unmarried mothers: 'to lift up those who are down'.

One child in twenty, it was calculated in 1880, was illegitimate and many of these would be born in the workhouse. This presented problems both moral and medical. Many girls were pregnant for the first time, often respectable young servants who had been seduced and abandoned. Women guardians tried to protect them by having their cases reviewed by a small subcommittee away from the jokes and sniggers of the main board. When Sophia Lonsdale joined Lichfield board she found 'some of the farmers used to make coarse jokes and

[32] F. D. Hill, Yorkshire Poor Law Conference, 1889; J. Hill, W. Midlands Poor Law Conference, 1890. There is a vast literature on prostitution and sexual standards but see F. Finnegan, *Poverty and Prostitution: A Study of Victorian Prostitution in York*, 1979, for the connections with simple-minded and pauper girls.

[33] Hastings branch of the WGS, 17th Annual Report, 1899.

nudge each other when a girl in this condition appeared. For some time I endured this, but at last I determined to make a stand'. The cases were in future dealt with privately. One newly elected woman wrote to the *Englishwomen's Review* that reciting their story at full board was bad for the girls—it hardened them; but it was also bad for the guardians 'many of whose faces are not very good or pleasant to see as they turn their heads to watch her out'. At N———— (Newcastle or Nottingham from the date), women members were horrified at the salacious interest exhibited by the men in the girls' sexual histories. As girls had only the haziest notion of physiology, 'very *pointed* questions' had to be put to them before the date of the birth could be calculated. Women guardians usually managed to shame their male colleagues into agreeing to form a small subcommittee of women and medical men to handle all such cases. The girls needed consideration, privacy, and courtesy, if they were to hold on to their fragile self-respect.[34]

Women members would, if at all possible, turn to their philanthropic networks to place girls in small homes for a few months while they had the baby and rebuilt their moral and physical strength. If they came into the workhouse for their confinement, women tried to keep them apart from the more brazen sinners, for otherwise, said a Birmingham woman guardian, 'they came out worse than they went in'. The Workhouse Girls Aid Society was founded in 1880 by London women guardians to help girls through their troubles. Elsewhere, Mrs Barnett, Miss Clifford of Bristol, and Mrs Haycraft in Brighton all set up ladies committees to help the girls find foster-parents for their child and a suitable living-in job for themselves, whereby they might re-enter respectable life and not be forced back on to the streets, which would otherwise be their fate.[35] Many continued to slip through their net, but in the words of Miss Augusta Brown, a Camberwell guardian, women members 'could help one here and there by going to their friends, or standing by their side in the Police court, or

---

[34] Martineau, *Sophia Lonsdale*, p. 170; *EWR* 15 Aug. 1888; *Parish Councillor*, 4 Oct. 1895: WLGS evidence to RC on Poor Laws, appendix lxxxv.

[35] Barnett, 'Young Women in our Workhouse', *Macmillan's Magazine*, June 1879; SE Poor Law Conference, 1889, p. 290; W. Midlands Poor Law Conference, 1906–7, p. 20. Penniless girls could not both keep their babies and go 'straight'.

getting them situations, or inducing the men to marry them. The work is very arduous and wearying, but it is well worth doing, for it frequently means the making of a life for all time that would otherwise have been a wreck for all time'.[36] One woman guardian took a girl into her home who had had three children by different men; now, the girl 'is living respectably and giving satisfaction to a good mistress'. As Miss Thorburn put it, 'a girl gets more than one chance before we give up trying to help her'.[37] In this field more than any other, only women could aid women, could bring to the board the delicacy of taste, the compassion, and moral insight, 'a right tone, born of the firm conviction that every girl, however poor, lowborn, defective, unprotected, is precious in the sight of God; that it is no kindness to any individual man or to society in general to pass over a man's sin lightly because he is a man; and place all the irremediable shame and stigma on a woman's sin because she is a woman.'[38]

Women guardians, however, differed, and were harsher, on girls in for their second and subsequent confinements. Marylebone, according to Louisa Twining, had one hundred and seventy-six births in its workhouse in 1888, and of these forty-seven were second confinements. Miss Donkin had two cases of women having their eleventh illegitimate child—on the rates. Perhaps a third of the girls were feeble-minded and in need of protection; the rest were thought to be morally delinquent and in need of correction. Some women guardians (Louisa Twining, Agatha Stacey, Mrs M'Ilquham) wanted powers of compulsory detention from the LGB; that way they would get maintenance money from the man and a bit more remorse from the girl. Other women would have none of this. It smacked too much of the arbitrary anti-women clauses of the Contagious Diseases Acts they had spent the last ten years fighting. Mrs Charles of Paddington—eccentric as ever—robustly argued that the whole proposal was irrelevant; illegitimacy rates would fall only when wages and prosperity

---

[36] *London*, 22 Aug. 1895. Two Liverpool women guardians visited 59 cases in 12 months in their maternity ward. Forty of the girls subsequently did well, three of them marrying the babies' fathers. They also received 70 letters from young women they had helped to start on a new life. *Eastern Weekly Leader*, 20 Dec. 1894.

[37] *Parish Councillor*, 4 Oct. 1895; Evidence to RC on Poor Laws, Q. 35693.

[38] *WS* 28 June 1894.

rose. In the end, the Women Guardians Society produced a fudged motion on the subject to send to the central poor law conference, which gathered up most shades of their opinion.[39]

The female able-bodied wards to which such women were admitted were the most unmanageable and refractory in the entire workhouse. Josephine Butler with her extraordinary gift for empathizing with the roughest and most degraded women was one of the few able to reach out and reclaim them, as she picked oakum alongside them in the sheds and cellars of the workhouse, or sheltered street girls in her own home.[40] Even women guardians were seldom able to rehabilitate the most hardened female paupers who stubbornly resisted being 'rescued'; VD cases discharged themselves as soon as they could. But able-bodied women paupers were declining in number. By the turn of the century rural unions were complaining that they had no able-bodied women in the workhouse left to do housework. All would be forgiven the woman with her fifth illegitimate child if she could only cook.

For the few respectable married women in the workhouse, the company of the rest with their foul language, lewd stories, and violent behaviour must have been distressing. A married woman would find herself in the workhouse if her husband accepted indoor relief; she was held there at her husband's pleasure; she could not leave without his consent to find rooms or work, and was forced to leave when he chose. She had no legal rights of her own. Even worse could be the plight of the deserted or newly widowed woman. 'To appear before a Relief Committee, much as though you were a prisoner in the dock, is a veritable ordeal; but to be cross examined by a relieving officer as though you were a criminal or a fraud, to have the validity of your marriage questioned and the illegitimacy of your children assumed, is an insult that respectable women will not submit to, save in the last extremity'. The crisis of

[39] NUWW Conference, Birmingham, 1890, p. 139; Report of WGS Conference, *EWR*, 15 July 1885. Their fudged motion called for the detention of 'able bodied paupers of bad character of both sexes, including women who came into the House for their second confinement'.

[40] One young girl stumbled into Mrs Butler's home, dying of TB, saying, 'God has given me to you that you may never despair of any'. Mrs Butler filled her open coffin with white camellias. Mrs Fawcett wondered how Mrs Butler, the wife of a school teacher, could afford such flowers in March—and then felt herself the Judas in wondering. A. S. G. Butler, *Portrait of Josephine Butler*, 1954, pp. 55–6.

destitution, desertion, death, or misfortune acquired its last drop of bitterness when women had to turn to the poor law.[41] Sometimes a woman guardian was there to help.

Nursing the Sick

Lady visitors over many years had found the infirmary wards among the worst in the workhouse. The sick were to be found in attics, basements, and sheds, and alongside dustbins and privies. When Miss Spottiswoode got Guildford to build new infirmary wards, all they could do with the old ones was store potatoes in them. The sick wards were frequently overcrowded, dirty, smelly, and noisy. They were also dangerous; smallpox cases, for example, were admitted to the general wards. They often lacked even the most primitive washing and sanitary arrangements, let alone WCs, any screens for the dying, or cupboards for stores. Chamber-pots did double duty as washbasins. A not untypical infirmary was in the Marland Workhouse, near Rochdale, newly built in 1864, for two hundred and ten inmates at two a bed.[42] In 1870 there were one hundred and ninety-four paupers, of whom forty-seven were imbeciles, and not one able-bodied among them. Medicines were administered by paupers who could not read. Diets ordered by doctors were not given. Infectious cases roamed at will. Medicines were kept in unlabelled bottles, together with blacking and firewood, in one box. Bodies, bottles, and linen went equally unwashed. All inmates urinated into a tub in the corner of ward—and the urine was sold by the guardians for scouring cotton when it was rank enough. The inmates had sores, vermin, and diarrhoea. The smell, the unrelenting noise (the clanging of keys, the wailing of infants, the dying of the sick, the muttering of the senile, and the head-banging of the imbeciles), the dirt, and the congestion defy description.

Unless the workhouse enjoyed a conscientious medical officer, wards mixed epileptics, infectious children, lying-in cases, and the senile elderly together. Doctors were overworked and underpaid, and expected to provide

---

[41] 'Women under the Poor Law', *The Englishwoman*, 1910, pp. 23–32.
[42] This description comes from F. B. Smith, *The People's Health, 1830–1910*, 1979, pp. 392 ff.

medicines from their own purse. Worst of all was the lack of nursing care, undertaken by paupers who had to be bribed with drink to do the dirty work. Inevitably they were somewhat short of the virtues of sobriety, reliability, cleanliness, and good health—to say nothing of literacy—that nursing required. As Miss Twining reported, '*all* helpers are more or less incapable through age, infirmity, ignorance, imbecility or immoral character'.[43] The better class of pauper nurses were elderly women, grubby perhaps and frail, but often kindly and willing. In the worst wards, the wounds of surgical patients were infested with maggots, and rotted with gangrene. Sheets were seldom changed even after smallpox patients had died in them; and surgical dressings not at all.

Slowly during the 1860s general medical and nursing standards rose, as advances were made in anaesthetics and antisepsis, as doctors became somewhat more professional in their training, and as Florence Nightingale's work in the Crimea showed what trained nursing could achieve. But workhouse sick wards remained out of sight until reports by the Ladies Workhouse Visiting Society and revelations of the *Lancet* in 1865 led to pressure from the influential and newly formed Association for the Improvement of the Sick in Workhouses, which mustered Shaftesbury and Dickens among its members. The central Poor Law Board was forced into action. Its Metropolitan Poor Law Act of 1867 permitted London workhouses to build separate infirmaries; and encouraged them to employ professional nurses.

During the late 1860s, the Nightingale nurses school and the great voluntary hospitals began for the first time to produce a trickle of trained nurses for poor law work (though because it was thought socially inferior nursing, the religious sisterhoods would take no part in it). Louisa Twining knew what the problem was. In 1879 she founded the Workhouse Nursing Association to promote the training of nurses for poor law infirmaries. By 1890 the Association had trained over a hundred probationers, and had appointed almost three hundred staff. By 1898 they had placed eight hundred nurses in workhouse service, but applications always outpaced supply, for the turnover was rapid as many young nurses

43 L. Twining, *COR*, 1 Dec. 1890.

could not handle the demands made on them by workhouse life. The LGB's General Nursing Order of 1897 finally banned the employment of pauper inmates as nurses; but promptly allowed them to be re-employed as assistants provided they were under the supervision of trained nursing staff.[44]

Women guardians were always the first to demand trained nurses and better medical care. Medical Officers like Dr Rogers of the Strand paid generous tribute to women's work at the board in extracting supplies and staff. Miss Winkworth was elected with Miss Clifford in Bristol in 1882, and within a year had obtained trained nurses for the infirmary. Miss Hannah Cadbury was persuaded to become a guardian by her younger sister who was working under Florence Nightingale. Miss Muller of Lambeth, Miss Glossop of Richmond, and Mrs Lawrie of St George's Hanover took medical training to prepare for this work. The middle-aged Mary Royce trained as a doctor in Leicester, stoically resitting her examinations until she passed. Elected as a guardian, she became voluntary doctor to the Infirmary until she died of a patient's infection in 1892.[45]

In London the rate-funded separate infirmaries permitted under the 1867 Act were soon attracting three times more patients than the private hospitals of Guys, St Thomas, and Barts put together. They were no longer a cluster of sick wards for pauper inmates, but, said the COS, 'the infirmaries are assuming the position of hospitals', by training their own nurses and serving their neighbourhood. By 1899, of Kensington's three thousand eight hundred patients, 70 per cent came from outside the workhouse. Poor law hospitals were providing a free general medical service for the urban poor. Kensington's Tory Miss Alexander typically deplored the trend. The socialist Mrs Despard typically thought it all rather splendid—for the first time the poor below the poverty line had access to free medical care.[46]

[44] Workhouse Nursing Association, 10th Annual Report, 1890. They received requests for 200 nurses a year, but could supply only 70 or 80 p.a.

[45] Cadbury, *The Making of a Woman Guardian*; F. Wardley, *The Founder: the Life of Dr Mary Royce*, 1970, p. 5. More generally, see Dr J. Rogers, *Reminiscences of a Workhouse Medical Officer*, 1889.

[46] Miss Alexander, Mrs Despard, SE Poor Law Conference, 1901, p. 454; *COR* 1 June 1887, p. 246.

London's problems were those of success: poor law medical care was losing its stigma, its recipients were after 1885 no longer disenfranchised, and consequently care was acceptable. Outside London, the situation was very different. Only a few cities had separate infirmaries by the time women became guardians. Almost all the nurses were still pauper inmates. Miss Fullagar, a Leicester guardian, issued notes of guidance to newly elected women in 1895; among other tips she suggested that they learn some First Aid and elementary nursing, so that they could teach workhouse 'nurses' how to change sheets, make fomentations, and apply bandages. Miss Gibson, an experienced matron, told the central poor law conference in 1895 that there were still large provincial workhouses with two hundred and thirty medical beds in sixteen wards under one trained nurse, eight untrained assistants, and no night staff. At night, patients were locked in and left. Plymouth (without women guardians) was perhaps the worst, with no sanitation, and the sick sleeping in the corridors. In the better London infirmaries, one trained nurse would be caring for twenty beds; outside London one untrained nurse could be struggling with up to eighty beds. Not surprisingly, wrote Louisa Twining, 'The old and infirm were put to bed and kept there, for there was no one to dress them, and passive cruelty was general; the bed sores were frequent, though called "eczema"; and yet what could one nurse, much less an untrained girl, do with eighty to ninety cases under her care?' Another young Nightingale probationer with one sixty-year-old woman to help her, struggled, she said to keep up the traditions of her training as she cared for one hundred and twenty beds; but she was defeated by her own fatigue.[47]

The new women guardians after 1894 gave all the help they could. They brought in more trained nurses, hired night staff to ensure the nurses got some sleep, and tried to provide them with decent food, fresh air, and a bedroom of their own. Accounts of their achievements came back from Southport, Bradford, where Mrs Moser tried also to get a woman doctor appointed, and Yarmouth (although male guardians thought

[47] Miss Fullagar, *Parish Councillor*, 18 Oct. 1895; Miss Gibson, Central Poor Law Conference, 1895; Twining, *Workhouses and Pauperism*, p. 202.

trained nurses were wasted on the 'walking man traps' in the female wards).[48] Another woman guardian reported that her infirmary was run by a 'drunken untrained woman and two very inferior assistants'. Within a year she had got three trained nurses from the Workhouse Nursing Association. Louisa Twining joined Tunbridge board in 1893. Though she introduced screens and thermometers, even she with her national reputation could not persuade them to employ trained nurses. A Norwich male guardian commented in 1901 that the elderly infirm who filled the workhouse needed kindly motherly women, not qualified nurses, to care for them. One lady guardian in Shropshire succeeded in building an infirmary, but 'as she had to do the fighting' and got blamed for the cost, she lost her seat.[49]

Fortunately, the voluntary hospitals set standards of good practice against which women guardians would measure their workhouse infirmaries. By 1909, voluntary hospitals were employing one nurse to every two or three patients, immeasurably better than twenty years before. As private hospitals entered the era of the X-ray, so they were followed at a distance by the workhouse hospital in the larger towns. But unfortunately, as the voluntary hospitals would take only the curable, the clean, and the comfortably off, workhouse infirmaries remained a warehouse of the morally as well as physically terminal. Indeed the situation in rural unions may have worsened. Workhouses designed for several hundreds were only a quarter full. The able-bodied did not enter, the children had left. Miss Chapman, a Tisbury guardian, estimated in 1900 that nine of Dorset's twelve workhouses had less than one hundred inmates in them; the only able-bodied women were the retarded who had to be entrusted with all the housework, nursing, and baby-care, scalding some of the babies in the process. The rural workhouse was too small to employ more than one nurse or any night staff, yet had to care for elderly or infirm inmates, suffering from epilepsy, rheumatism, lameness, severe heart conditions, incontinence,

[48] Southport, NUWW Conference, Liverpool, 1891, p. 131; *Bradford Observer*, 28 Dec. 1899; *Bradford Weekly Telegraph*, 27 Oct. 1911; *Yarmouth Times*, 27 Jan. 1896.
[49] Central Poor Law Conference, 1901, p. 767; for Shropshire, RC on Poor Laws, Q. 70561.

and senility. The work was hard, unpleasant, and monotonous. The nurse was isolated and lonely, often at war with the
master's wife. One nurse found her patients unable to eat the
official fare, so tried to introduce sago and rice puddings. She
was refused by the master's wife for 'fear the patients will fare
too well'. Unable to eat their food, many of the patients were
seriously malnourished. Another country nurse wrote to
Louisa Twining, 'surely in no other branch of the nursing
profession can the work be so very unsatisfactory and
discouraging as in a lone country union. How one longs for
lady Guardians knowing how helpful they can be.[50]

Guarding the Feeble-minded

Another group of inmates that had reason to bless the arrival
of women guardians were the feeble-minded. Dangerous
lunatics were detained in the county asylum; but the feeble-
minded were within the workhouse though they were in law
free to leave if they chose. They were first 'seen by women
guardians', claimed Miss Lidgett. As the 1909 Royal Commission noted, if the feeble-minded were in mixed wards, visitors
found 'imbeciles annoying the sane and the sane tormenting
the imbeciles'; the sick, young, elderly, and respectable were
disturbed and upset both by 'the revolting habits and hideous
appearance' as well as by the constant noisiness of the half-
witted. But segregated, they were confined to one room, living
out a minimalist existence, and 'never went out from one
year's end to the next'.[51] Out of mind, out of sight.

Guardians like Mary Clifford of Bristol had particular affection for their 'simple susans', taking them home for teas of
sausages and shrimps, cake and gooseberry pie, and out to
treats. They taught them modest tasks like making bandages.
'Poor puzzle headed folks' said Mary Dendy, who as a member of Manchester's school board, was to found the National
Association for the Care of the Feeble-minded in 1895.

Obedient to any influence, either from without or within, rarely able

---

[50] See also Cuttle, *The Legacy of the Rural Guardians*, pp. 29, 33, 45, for similar
descriptions of the elderly sick going without food; Twining, op. cit., pp. 194, 197; for
rural workhouses, see ch. 7.

[51] Miss Lidgett, NUWW Conference, Croydon, 1894, evidence of the WLGS to the
RC on Poor Laws, appendix lxxxv; *Minority Report*, pp. 238–9.

to give any reasons for their actions, they pass through life, in ordinary circumstances, as through a bad dream. Everything happens to them, they never rule their own existence. Birth happens, death happens, and in between parenthood happens, and too often crime and degradation happen. All are equally beyond their own control.[52]

Miss Agatha Stacey of Birmingham had founded a training home for the feeble-minded where her girls made rugs out of carpet waste that glowed with colour and pride. She, Mary Clifford, Mary Dendy and later Mrs Ellen Hume Pinsent of the Birmingham LEA, went from conference to conference, teaching guardians to 'see' their feeble-minded, describing their special needs, and urging that they be withdrawn from the workhouse into small sheltered homes where they might do simple jobs and be protected from the street life. The 1895 Royal Commission on the handicapped had drawn attention to the special needs of the retarded. But even where guardians were willing to pay for them to attend small homes, there were so few places it was considered useless to write for one.[53]

By the turn of the century, compassion for their plight was overlaid with eugenic concerns. Mary Dendy claimed that 75 per cent of imbecility was inherited; up to two-thirds of the offspring of an imbecile would also be severely retarded. At Bridgwater, one feeble-minded workhouse inmate had twenty-nine descendants, of whom twelve at least, down to great-grandchildren, it was alleged were feeble-minded. All in the workhouse, all on the rates. As family sizes were reduced, the retarded continued to have large families; and there was much talk of national degeneracy and racial deterioration. Arguments of humanity, economy, and national efficiency came together in the call for a Royal Commission on the Care and Control of the Feeble-minded (1904–8); and their recommendations were embodied in the Mental Deficiency Act of 1913, which authorized guardians to segregate and detain them.[54] Women guardians had pioneered a better way

[52] Miss Dendy, S. Midlands Poor Law Conference, 1905–6, p. 158.
[53] *Parish Councillor*, 15 Nov. 1895; RC on Poor Laws, Q. 70543. The guardians had no legal powers to detain feeble-minded adults against their will, but counted on them not knowing their rights.
[54] More generally, see G. R. Searle, *Eugenics and Politics in Britain 1900–1914*, 1976, p. 32.

forward, the sheltered home, but in practice most feeble-minded remained in workhouses until the workhouses themselves often became the mental handicap hospitals of the 1930s. Only in recent years have small-scale community homes and community care come back into favour.

Another group, often classed with the feeble-minded, were the epileptic. In an era without modern medication and when fits were still sometimes thought to indicate possession, women guardians fought hard to bring them outside workhouse walls. Mrs Moser of Bradford said that with fresh air, good food, and a tranquil environment, epileptics could live full and stable lives in colonies of their own. Many boards sent epileptics to the Lingfield colony in Surrey. Others started their own homes. Over in Bolton Mrs Haslam, Sarah Reddish's colleague on the board, tried to get TB cases out of the workhouse, by offering the board the use of a large hilltop cottage she owned, which she would adapt with open shelters. The guardians agreed, but the proposal was vetoed by the town council who said they feared that soakaways from the cottage would pollute the town's water-supply, despite evidence from the surveyor and the medical officer of health that the cottage drained in the opposite direction.[55] Again, both groups remained largely within the workhouse, despite the efforts of women guardians.

By the later 1890s, the workhouse was being 'deconstructed'. Most pauper children were no longer in the workhouse. Few able-bodied entered it, as prosperity improved, and attitudes to outdoor relief changed. The urban sick were cared for by nurses in separate infirmaries, and some moves had been made to bring the mentally and physically handicapped out of the workhouse into small certified homes. Outdoor relief apart, most of these developments had been led by women. As Mr Bousefield, himself a member of the London school board, said, 'Men had helped in these reforms, but the necessary pressure came from the ladies'.[56]

The vagrant tramp remained, often alcoholic, usually diseased, sometimes feeble-minded, turning up at unsavoury casual wards to sleep in leaking lean-to sheds. There were

[55] *Bolton Daily Chronicle*, 9 Dec. 1908.
[56] *EWR* 15 Feb. 1886; *WS* 18 July 1895.

thought to be around nine thousand of them on the road with perhaps eight hundred women, and two to three hundred children, but their numbers were falling. Women guardians often worked hard to get them better sheds, cleaner clothing, beds rather than boards, somewhat more palatable food, and help in finding casual work. Some women guardians, like Mrs Moser and Mrs Priestman of Bradford, called for a more humane regime, as the tougher the guardians were, the more tramps voted with their feet and entered common lodging-houses to spread infection—'Would it not be better for the community if vagrants could be helped to live a better life?'— but women could make little headway against prevailing attitudes of deterrence. Women on the tramp were particularly vulnerable; their conditions were 'very bad, medieval and brutal', said a woman guardian to the Royal Commission in 1908. 'Everything possible is done to lower the women lower still . . . and one lady guardian is impotent to fight it or to get things altered'.[57] Pamphlets by Mrs Higgs of Oldham and Julia Varley of Bradford describing their experiences on the tramp were widely circulated by the NUWW and may have helped to mitigate their treatment.

Comforting the Elderly

One group remained within the workhouse, those elderly unable to live independently outside, mostly men, presumably because they were less able to look after themselves and less welcome in their children's homes. Miss Thorburn found the old women 'simply waiting for the end. Their chief desires are warmth and hot tea, and their chief dislike being bathed. They doze a good deal and sit or lie hour after hour, and year after year, doing nothing, and seem to me more like human vegetables than human beings'.[58] Mrs Fuller, the daughter of Sir Michael Hicks-Beach, became a guardian for Chippenham in 1894. Looking back seven years later she was struck by the 'extraordinary difference in public opinion' that had occurred since then. When she joined, even the smallest

[57] *Bradford Observer*, 30 Nov. 1899; evidence of WLGS to RC on Poor Law. See also Preston Thomas's comment—'the police treat him as a criminal but do not punish him, while the poor law authorities treat him as a pauper but do not relieve him', op. cit., p. 351.

[58] Miss Thorburn, RC on Poor Laws. Q. 35693.

and most timid suggestion for improving the comfort of the
elderly was brushed aside. So they sat in shabby ill-fitting
workhouse dress on hard benches around the walls, locked in.
The building with its bare stone floor, curtainless windows,
and harsh clanging locks seemed 'more like a prison for criminals than a lasting home for aged men and women'. At her
first few conferences, every suggestion that the elderly be
treated more leniently was met by 'interruption, contradiction
and disorder, and we were told on all sides, that any relaxa-
tion would inevitably cause great abuse and great increase of
pauperism, and the Poor Law Act must be carried out in its
entirety'. She argued that respectable, elderly women, in
particular, who had managed their own homes for years,
found it especially demeaning to enter a workhouse and
overnight 'become a nobody'. When to that was added the
monotony of workhouse life, the abrasiveness of its discipline,
the intimidating authority of its staff, and the unwelcome
association with the rough and the retarded, she considered
that the problem was not so much deterring the undeserving
as persuading the deserving to enter it. Slowly, women and
working men, 'even now only grudgingly admitted to seats on
the board', were, she said, humanizing the workhouse.[59]

Women guardians and lady visitors continued to bring in
comforts to soften the austerity of workhouse life: books and
back-rests, cushions and snuff. Diets were slowly improved,
as the LGB in 1895 signalled to guardians that they might
relax the notion of deterrence for the elderly. Eggs, bloaters,
cocoa found their way on to the menu; tea, sugar and bread
for toasting were provided so that the elderly could share tea
and memories with their friends. In workhouse after
workhouse the official dress was abandoned, rooms were
made more comfortable, and furnished with curtained reces-
ses, lockers, and bedside tables. 'Inmates' became 'residents',
able to enter and leave the workhouse at will of an afternoon.
Many of the less mobile responded to the Brabazon schemes
of Lady Meath.

Back in 1882 Lady Meath had offered Miss Donkin of
Kensington £30 to spend on fancy materials to give the elderly

[59] Mrs Fuller, SW Poor Law Conference, 1901–2, p. 396.

and infirm something to do, and stop their incessant grumbling. The matron objected, Miss Donkin insisted, and soon lady visitors were teaching inmates wood-carving, lace, quilt, and rug making, wrought-iron work, and painting on crockery. By 1900 some one hundred and eighty workhouses had adopted the Brabazon scheme, although even in late Victorian England there were not enough skilled and leisured ladies to meet the demand for what we would now call occupational therapy. Some of the craftsmanship of Brabazon work was superb. One old man was taught wood-carving professionally and left to become self-supporting. Another of eighty was taken round town to study the fashions and produced 'fascinating straw hats'. A disabled landscape gardener designed flowered carpets. At one conference, Miss Baines of Leeds described an improbable linen chair-back. The linen was spun by a one-legged epileptic labourer; woven by a former butler; and embroidered by a young paralysed plasterer. It was exquisite, and won an award at the National Home Arts and Industries Exhibition. The goods were sold and the money spent on treats and comforts. Pianos were popular purchases.[60]

However, the more women guardians domesticated the workhouse with their cats and teapots, the more male guardians feared that 'the whole parish' would flock into the workhouse. The women knew differently. Mrs Fordham of Royston explained that many would starve rather than leave their home. She described one old man of eighty who had been persuaded to enter the workhouse. 'He was well looked after there, but the dreariness and routine of the place together with immense homesickness drove him almost crazy'. In despair, he discharged himself, and though lame and fragile tried to walk the six miles home in the wind and snow, 'wishing for nothing better than his old life to flicker away there, in sight of the trees and surroundings which were so much part of his life that it was impossible for him to live away from them'. Another woman guardian from Shipley considered it 'my

---

[60] *WS* 13 Dec. 1894, 8 July 1897; Annual Reports of the WGS; N. Wales Poor Law Conference, 1905–6, p. 311. The energetic Southport ladies encouraged inmates to make 41 rugs, 5 footmuffs, 6 iron flower stands, 14 bead curtains, and an unspecified number of dishcloths and envelopes. 17th Annual Report, WGS, 1899.

special duty to persuade some of our poor wretched old men and women, lost in condition, to go to the Palace (as I have named it) and it is truly a comfort to me when I see them clean and happy . . .'. With surer instincts, some women guardians went further and tried to bring the elderly out of the workhouse into almshouses. Miss Mosely, a Hastings guardian, had her board rearrange the old people's accommodation. At Bradford, Mrs Moser got her board to build separate cottages in the grounds, an early form of sheltered housing that was widely advocated and frequently copied by the new ILP members.[61] Slowly the workhouse became more comfortable. Concepts of deterrence ebbed as its rooms were filled almost entirely with the elderly and infirm. But the image of those vast gaunt buildings, threatening and intimidating, still continue to terrorize the last years of the elderly, as even in the 1950s they begged not to be sent to the workhouse, not to be given a pauper burial.

Out-relief, Pauperism, and Poverty

Much of the change in poor law policy towards the young and old, the sick, the handicapped, and the fallen, was woman-led. Is this to claim too much for their contribution? Would the innovations associated with them have occurred without their presence? It is never easy to untangle pressure and influence, the more so as many women, Joanna and Florence Davenport Hill, Mary Clifford, Agatha Stacey, Louisa Twining, Sophia Lonsdale, Henrietta Barnett, Elizabeth Lidgett, Miss Brodie Hall, Margaret Baines, and Mary Dendy, to name just some of the better known, were prominent in official poor law circles, had access to inner LGB officials, gave papers to conferences and evidence to Select Committees, wrote for the reviews and to the press. Even the most insular all-male board would not have been impervious to their indirect influence. In addition, many such boards had ladies visiting committees, softening in womanly ways the workhouse regime.

One possible modest test is the effect new women guardians had on boards that were already rightly regarded as well run,

---

[61] Mrs Fordham, *Parish Councillor*, 31 July 1896; *WS* 18 July 1895.

such as Norwich or Bradford. Take Norwich.[62] Its all-male board attended conferences, had decent staff, recruited trained nurses from Miss Twining's Nursing Association, had a separate infirmary and boys home, and adopted kindly and humane policies towards its indoor elderly. Its Ladies Visiting Committee brought in kettles and company. Though Moderate-controlled, it was in practice apolitical and forward-looking.

With the removal of the property qualifications, the 1894 elections in Norwich were fought on sharp party lines. The Progressives took control with three women guardians on the slate. As the party approached the April 1898 elections, they looked back on what they had achieved in the past three years; more nurses, better food and clothing, more comforts for the inmates; a separate cottage home for girls; the end of workhouse schooling—all children now went to local schools in ordinary clothes; and the abolition of workhouse dress and discipline for the elderly. Virtually all of these improvements came from the board's workhouse committee and subcommittees, where the women members and Mr Crotch instigated and implemented these changes, unreported by the press, but with the tacit approval of the full board and the credit claimed by the entire party. The men as ever remained busy with out-relief, admissions, supplies, staff, and buildings.

The impact of women on more backward boards—Miss Spottiswoode at Guildford, Ethel Leach's trinity in Yarmouth, Kate Ryley's quintet at Southport—was correspondingly more dramatic. Those urban boards, like Plymouth, with virtually no women guardians at all, were also held to be among the worst in the country, their workhouses overcrowded, understaffed, poorly resourced, casual, and dirty. Without women, there was often no one to fight for the inmates; for unlike school boards, the property qualification debarred working men from making much of a contribution to humanizing the workhouse before 1894. The Fabian, J. F. Oakeshott, thought that women had worked 'a complete revolution in many workhouses'. The COS agreed. Indeed,

[62] This account is drawn from the Norwich guardian records, Progressive leaflets, and the local press. For Bradford, see below, pp. 293–4.

said the Webbs more ambiguously, without women guardians the workhouse would not have survived for as long as it did.[63]

Like school board women, women guardians shared a woman's view towards those in the workhouse. They faced a large forbidding institution, designed to frighten people, and they worked to subvert that intent. They tried to domesticate the system, make it knowable, reduce it to human scale, and see its inmates as an extended household. They sought to deconstruct its bricks and mortar into smaller, more homely, and appropriate units: the infirmary, the nursery, the cottage home, the grouped home, the certified home, the reception centre. For those left behind, they struggled to rework the institutional arrangements so that they met the needs of inmates rather than stripped away their identities and individuality (Grampsie 1, Grampsie 2, Grampsie 3) to fit the system. Eva Muller in the 1880s insisted that children be known by their names, not by number as was customary. Miss Florence Hill and Mrs Despard led the way in replacing the word 'pauper' with 'person' on official documents. When women guardians accompanied a seduced girl and held her hand in the magistrates' court, or protected the feeble-minded from bullying, or mothered abandoned children, or sat with the elderly over endless cups of tea reliving their lives, they were asserting that the State's social and economic outcasts were God's poor, had immortal souls, and were entitled to citizen rights within the moral community.[64]

The twin concepts informing their work were of course the family and the home. Women guardians cast themselves as surrogate mothers to pauper children, elder sisters to servant girls and unmarried mothers, and family friends to the sick and elderly. They brought with them a culture of homeliness. They cared about draughts and dustbins, they understood the hunger for colour, decoration, and variety. Above all they taught their male colleagues that the workhouse as a quasi-

[63] *COR* Feb. 1900, pp. 102 ff.; S. Webb, quoted in *Catholic Citizen*, 15 Dec. 1919; J. F. Oakeshott, *The Humanizing of the Poor Law*, Fabian Tract 54, 1894.

[64] Marion Phillips, later Kensington borough councillor and Labour party women's organizer, visited one workhouse, spotlessly clean, full of extra comforts, but where the ladies were known as Grampsie 1, etc. *WLL*, 9th Annual Report, 1913–14, p. 62; *London* 14 Mar. 1895, 17 Mar. 1898. The word 'pauper' was finally and officially discontinued only in 1931.

penal institution damaged rather than sheltered the women and children within its walls. Time and again they explained at board meetings that people came out worse than they went in. Miss Augusta Brown of Camberwell insisted, 'The Poor law system is bad from beginning to end. I believe it perverts most of the poor people who come into contact with it'.[65]

Where the women led, the inspectors and the Local Government Board followed. The LGB's Boarding Out Order of 1870, the recommendations of the Royal Commission on the Elderly in 1895, the Workhouse Nursing Order of 1897, as well as dozens of LGB codes and circulars, were largely authorizing best practice already urged by women campaigners and adopted by leading women guardians. LGB inspectors were highly appreciative of women's work. They had welcomed the arrival of lady visitors back in the 1860s; inspectors in 1888 could 'hardly measure' the vast improvement in schools and infirmaries made by lady guardians and the 25th Annual Report of the LGB in 1895–6 paid tribute to their work. 'The lady guardians have been most useful'. Another noted, 'the very much larger number of lady guardians, whose welcome presence was seen on the new boards' . . . 'doing excellent work' . . . they were 'brightening' the workhouse . . . working with the old, sick, children, and girls (who else were there?) 'they are especially valuable'. Their value 'can scarcely be overestimated'. And so on.[66] Much of what individual women guardians pioneered, was carried by the inspectors to other workhouses, and other unions.

Women remade the workhouse. They did it through an 'earnest and unremitting attention to all the details of routine', as was said of Miss Whitehead after her election defeat in London. In the words of Miss Hope, a Bath guardian, 'the work is essentially one of detail. Cases must be

[65] Miss Brown, *London*, 22 Aug. 1895.
[66] *EWR* 15 Apr. 1888. For inspectors' comments on workhouse visiting, see SC on Poor Relief, 1861; on women guardians, 25th Annual Report of the LGB, 1895–6, p. 169 (Mr Lockwood and Mr Davey), p. 175 (Mr Murray Browne), pp. 180–1 (Mr Preston Thomas), p. 187 (Mr Baldwin Fleming). See also 26th Annual Report, LGB, 1896–7 (Mr Davey), 28th Annual Report, LGB, 1898–9 (Mr Fleming). More generally, Preston Thomas, op. cit., for comments on East Anglia and the South-west. For their part, women guardians could and did call in the inspector over the head of their boards, and count on his support.

treated individually if any permanent good is to be effected'. Or as Daisy, Countess of Warwick, exclaimed, 'It is no good dealing with humanity wholesale: you need to grapple with each case, taking individual by individual.'[67]

That was precisely the strength and the weakness of women's work within the poor law. By 'grappling with each case', 'individual by individual', women humanized the workhouses. They refused to collude with its mechanisms of institutionalization. But their belief that this was the only way to effect 'permanent good' stopped them from standing back and seeing the system as a whole, or challenging prevailing concepts of pauperism and poverty. Indeed as so many women began their poor law work by workhouse visiting, they were slow even to interest themselves in matters of outdoor relief and wider questions of poverty. When thirteen women guardians described their duties and responsibilities for the Women Guardians Society in 1885, only one of them even thought to mention outdoor relief. This was left to the men. Their concern was exclusively with the workhouse, its domestic detail, its extended household. Yet, in 1880, some 500,000 of the 600,000 people receiving outdoor relief were women and children; most of them either elderly women, struggling on a pittance and a bag of flour with help from family and friends; or widows with children, some sick, others taking in washing perhaps, their income topped up by outdoor relief. Only 4 per cent were able-bodied men, and if Norwich was anything to go by, half of those were over sixty.[68]

[67] Miss Whitehead, *EWR* 15 May 1886; Miss Hope, 19th Annual Report, WGS, 1901; Countess of Warwick, *Life's Ebb and Flow*, 1929, p. 109. With five other woman guardians, Lady Warwick obtained trained night nurses, took children out for fostering, introduced the Brabazon scheme, opened cottages homes for crippled children, and toured the country in her red Wolseley campaigning for school meals. After nine tempestuous years, her resignation was 'received with mingled feelings by the guardians', *Daily Express*, 22 May 1905; and *Life's Ebb and Flow, passim*.

[68] Of outdoor paupers in 1880, some 25,000 were male able-bodied (many of them elderly), 75,000 female able-bodied, and just under 200,000 dependent children; there were also 78,000 non-able-bodied men, 193,000 non-able-bodied women, and 35,000 dependent children. A further 45,000 were classed as insane and outdoors. Altogether a half were elderly and infirm, a third were receiving medical treatment, and most of the rest were widows. Local evidence confirms this—of Norwich's 63 able-bodied men claiming outdoor relief in 1887, 26 were under 50, five under 60, 19 under 70, 13 under 80. Minutes, Norwich Board of Guardians, 21 Dec. 1887. See 26th Annual Report, LGB, 1896–7 appendix E. More generally, see P. Thane, 'Women and the Poor Law in Victorian and Edwardian England', *History Workshop Journal*, 1978.

Despite all the evidence suggesting that most outdoor paupers were destitute because they were sick, elderly, or widowed, rather than idle and unemployed, women guardians none the less retained tough, simplistic, and moralistic views towards outdoor relief, drawn from their COS training, their temperance work, and their education in orthodox political economy. They were still quoting Thomas Chalmer's aphorism of the 1820s, 'Character is the cause, comfort is the result', with apparent satisfaction in the 1890s. Taken together, this meant that women guardians adopted with unusual firmness the conventional hostility to outdoor relief that was the wisdom of the time, the view of their social circle, and the policy of the LGB.[69] The best-run union had simultaneously the least outdoor relief, combined with the most considerate, personal, and careful aid to indoor paupers. The fact that reducing outdoor relief put almost intolerable pressures on voluntary hospitals, private charity, and common lodging-houses from those who would not enter the workhouse at any cost was thought to be an acceptable price to pay. Women guardians were willing enough to bind the wounds of the army of the vanquished, to paraphrase Mary Clifford, but they suspected that the army was vanquished by its lack of valour and that the wounds were to that degree self-inflicted. Miss Sarah Ward Andrews put it forcibly at a speech she gave to Nottingham's women's movement in 1882:

One reason why she wished to become a guardian was because she desired to know what pauperism was. So many children were brought up paupers, and the desire to remain so seemed to pass from family to family. The great cause of pauperism had been the want of thrift and this she believed had arisen in great measure by the thought on the part of the people that they had out-relief or the workhouse to go to in case of necessity. They [the ladies of England], ought at once to attack this evil . . . They had allowed children to grow up at somebody else's expense rather than their own. There were people in this country who were willing to be paupers . . .

[69] The LGB had begun its crusade against outdoor relief in the late 1860s, as the Lancashire cotton famine began to bite. Goschen's Minute of 1869, followed by the new LGB circular of 1871, reaffirmed the principles of the workhouse test; LGB inspectors urged boards to cut out relief; Henry Fawcett, who represented advanced liberal thinking, urged boards to abolish it (*Pauperism*, 1871). In Whitechapel, Canon Barnett would give no outdoor relief.

Every issue of the *Englishwoman's Review*, edited by Caroline Biggs, urged ladies to 'stem the increasing evil of pauperism', to favour 'the judicious repression of pauperism', and the like.[70]

There is little evidence that many women guardians had digested the findings of Charles Booth's massive survey of the London poor which showed in 1889 that a third of London's poor were below a subsistence-level poverty line; and that for the vast majority (85 per cent was Booth's figure) this was clearly not their own fault—it had nothing to do with idleness and intemperance and everything to do with very low wages, chronic unemployment, sickness, large families, and old age. Women joined poor law boards in the 1880s and 1890s, when attitudes to poverty and pauperism were in ferment; when to the research of Booth and Rowntree was added the writings of the Fabians, the experience of the settlements, and the revelations of the early Labour party.[71] Yet women guardians evinced little sense of the precariousness of the working-class family economy, which was never more than one week's wage from destitution; and little sense either that there was simply no surplus for saving against the catastrophe of sickness or the inevitability of old age. Mrs Rowlandson, a Birmingham guardian, was confident in 1892 that Birmingham's administration was most humane but as '99 per cent of pauperism was caused by drunkenness' it had to be dealt with on repressive lines. All their philanthropic experience notwithstanding, the class, cultural, socio-geographical distance of women guardians from the working poor was never more clearly marked than on matters of out-relief.[72]

Mrs Amelia Charles, elected in Paddington in 1881, was one of the few women guardians before 1894 to challenge such rigid and narrow orthodoxies. She was the daughter of a Somerset landowner who had come to favour repeal of the corn laws in the 1840s when he saw his labourers collecting

[70] Miss Andrews, *EWR* 15 May 1882, *EWR* 15 Sept. 1881, 15 Mar. 1883. Dr Summers argues that women 'consistently worked to reverse the most fundamental tenet of the 1834 poor law, the abolition of relief outside the walls of the workhouse', in 'Women's Philanthropic Work in the Nineteenth Century', in S. Burman, ed., *Fit Work for Women*, 1979, p. 49. On the contrary: women were the most devoted opponents of outdoor relief.

[71] E. P. Hennock, 'Poverty and Social Theory in England: The Experience of the 1880s' in *Social History*, 1, Jan. 1976, pp. 67–91.

[72] *WH* 4 June 1892.

nettles to eat. She stood as a WGS candidate—in a careless moment they failed to ask her what she stood for, and her tory-radical views were to prove, in her phrase, somewhat unexpected. She was utterly and entirely opposed, she said, to offering the workhouse or nothing. To insist that an ordinary working man be self-supporting 'under all the circumstances, seeing that he is constantly liable to industrial pauses, illness and accidents', was doctrinaire and dangerous. To expect him to leave his home, and bring family and furniture into a semi-penal workhouse, was tyrannical. She favoured out-relief. 'By being assisted at home in times of emergency, the poor are saved from becoming permanent paupers'. Once the poor entered the workhouse and lost their home and possessions, they had little hope of returning to respectable life. The workhouse turned a temporary emergency into permanent destitution. Even more provocatively, she cast doubts on the virtue of thrift. The French peasant was proverbially thrifty, and also grasping and avaricious. The one followed from the other. 'The qualities of the [English] poor are not often considered; they are however sympathetic and unselfish; their experience of hardship [means] . . . they are almost universally charitable to each other . . . They have pre-eminently the art of thrift. They spend their money not only to supply but to conceal their wants'. The editor of the *Review* carefully, and I think uniquely, disassociated herself from Mrs Charles's views. But at every poor law conference she attended, at every meeting of women guardians, there was always one good-humoured, but courageously dissenting voice to the conventional wisdom that pauperism was a moral leprosy compounded and communicated by out-relief.[73]

At first, hers was a lone voice on the Paddington board, but over the years she built a majority, enlarged out-relief, and cut both pauperism and the rates. From the 1890s other women, and Progressive politicians more generally, came to share such views and placed adequate outdoor relief, pensions, and public works for the unemployed on the agenda.

[73] *WPP* 13 Apr. 1889; *EWR* 15 Dec. 1883; *London*, 22 Apr. 1897. Giving widows outdoor relief was cheaper than taking their children into the workhouse. Mrs Charles later became hon. sec. to her local COS branch, though one wonders on what terms. See *COR* lists for 1895.

Chamberlain's espousal of pensions in 1891, voluntary and contributory though they were, had at least recognized the common structural, rather than personal, problems of old age. Women guardians in London, the South, and the resort towns continued to speak COS language; but many of the men and women now representing the textile towns of the North came from a very different background. Take Keighley for example. In 1892 the Liberals ran a slate of nine, offering two seats to the recently formed women's Liberal association. Mrs Greenwood, one of their candidates, had at the last moment to rent a warehouse and an empty house to qualify. Their Liberal programme demanded reform of the poor law electoral system; better care for children, the elderly, and the casuals; a labour programme of direct employment, no sweating or subcontracting, and a programme jointly with the town council to promote 'temporary and honourable employment' during recession.[74] A similar programme was advanced by London Progressives in 1893; boarding out, out-relief for the aged, better medical care, public works for the unemployed— but reformation of the idle residuum by a rigid refusal of outdoor relief except in emergencies. Mrs Reynolds had been recently elected in Dorset, 'returned by the working women of her town', she told the NUWW conference in 1892. 'People said they had paid the rates, and now it was their turn to have help; they looked at it very much as they would putting into a club. She felt that if they looked at it that way it was not so pauperising as the effect of their going into the workhouse'. Which was precisely what those opposed to outdoor relief feared.[75]

As the economic depression deepened, heightening the plight of the unemployed; as town councils themselves moved into the field of public works, removing the stigma attached to relief; as attitudes to the elderly pauper softened; above all, as working men, Progressive, and Labour candidates joined poor law boards after 1894, so attitudes to outdoor relief underwent a sea change. Advanced Liberals were by now calling for state

[74] *WH* 2 Apr. 1892. For progressive thought on pensions, see P. Thane, 'Noncontributory versus Insurance Pensions' in P. Thane, ed., *The Origins of British Social Policy*, 1978, M. Freeden, *The New Liberalism: an Ideology of Social Reform*, 1978.

[75] *London*, 2 Mar. 1893 (the pen was presumably the Webbs). For Dorset, *WH* 4 June 1892.

pensions, recognizing that the pauperism of old age was 'the normal fate of the poorer classes' independent of character. They welcomed the increased economic consumption pensions would generate; and insisted on the ethical obligation of all members of society each to the other. Older women guardians with older COS attitudes found themselves beached. The NUWW conference of 1894 heard a series of papers on pensions and the elderly by Miss Pauline Townsend and Miss Tabor, both quoting Booth and Canon Barnett, to show that working men could not possibly provide for their old age on thrift alone, but needed a pension of five shillings a week. As Miss Tabor pungently said, for the elderly to enter the workhouse was 'about the most dreary instance of economic inutility that our social system has to offer. Even the ethical value of old age is absent there'.[76]

The following year, the NUWW turned its attention to out-relief and the unemployed. Mrs Calvery, a Brixworth guardian, offered the conventional COS view. Destitution was a refusal to be self-supporting; outdoor relief rewarded that refusal. She criticized those new guardians calling for outdoor relief as a 'safe road to popularity'. She was supported by Mrs Shaen of Kensington: 'We must learn to realize the fact that whoever gives a penny to a beggar, gives an additional blow to a man who is already down'. Miss Sophia Lonsdale, a Lichfield guardian and Bishop's granddaughter, had never yet come across any deaths by starvation as she rode her horse to hounds. Mrs Amie Hicks, a trade-unionist and friend of Margaret Macdonald, turned angrily on the well-heeled conference, quoting case after case of malnutrition and starvation that she knew of and said bitterly that as the workers 'built the wealth of this country', they were entitled to poor relief as of right. 'We workers are not going to receive private charity when there is a fund, the rates, to which we have contributed'. Mary Clifford wound up the conference soothingly by reminding her audience that guardians must do nothing to undermine independence, thrift, and 'the desire to rise'.[77]

But that view was on the retreat even within NUWW

[76] NUWW Conference, Glasgow, 1894, p. 39.
[77] NUWW Conference, Nottingham, 1895, pp. 153–62.

circles. Increasingly, new guardians were coming to favour outdoor relief for those elderly and widowed able to look after themselves, as a quasi-pension; and public works as an alternative to the labour test for men unemployed during economic recession. Over at Yarmouth the newly elected Mrs Leach was pleading the cause of the unemployed to the indifferent ears of her fellow guardians. In Lambeth, the socialist Mrs Despard had joined a board which in a district noted for its severe underemployment was notorious for its harshness to the outdoor poor. She urged both public works and general relief. Ada Nield, a factory girl and member of the ILP, had become a Crewe guardian in 1894 and within a few months was pressing her colleagues to acquire fifty acres of land for the unemployed to farm. She was no more successful than the other women. However, Mrs Pankhurst had been elected in Chorlton on a Labour ticket. With her husband she organized the Unemployed Committee as well as their soup-kitchens; and forced her fellow guardians to ask the town council to start public works. At the north-west poor law conference the following year she suggested that other boards might also support farm colonies, road mending, cottage maintenance. Inevitably she was told that if the work was profitable, it would be done privately, and if it was not, it should not be done at public expense.[78]

However, even Mary Clifford was by 1896 agreeing that outdoor relief—for the respectable (but not the intemperate) elderly, and for widows (but not deserted wives)—had its place within the poor law. Pensions she could not support. Mrs Fuller, a Chippenham guardian sharing Mrs Charles's views, told the NUWW that guardians were faced not with 'won't workers' but with 'past workers and can't workers'. Refusing outdoor relief was disastrous for small rural communities. It stripped the villages of the elderly, and depressed living standards and labourers even lower as they were expected to contribute to their parents' keep. Whatever the system of relief, indoor or outdoor, it made little difference,

---

[78] Mrs Leach, *Yarmouth Times*, 26 Jan. 1895; Mrs Despard, Minutes, Lambeth Board of Guardians, *passim*; Ada Nield Chew, *The Life and Writings of a Working Woman*, 1982 edn., p. 19; Mrs Pankhurst, *The Making of a Militant*, *passim*; NW Poor Law Conference, 1895.

she said, as 'the ratio of pauperism rises and falls with poverty'. If unions were strict enough, they could reduce pauperism, 'but the poverty and human misery is increased'.[79]

At Bradford, such changes in view can be clearly traced. Having had Edith Lupton on the school board and Eva Muller on the poor law board, Bradford liberals counted themselves unlucky and tried to keep women out of public office. But on to the 1894 board and coinciding with the onset of severe distress came a wave of talented women drawn from Bradford's leading families, such as the kindly Jewish Mrs Florence Moser, wife of an alderman, who was at the centre of Bradford's Guild of Help, its NUWW, its crèches and clubs; Mrs John Priestman, friend of Mrs Byles, followed shortly by her Quaker daughter-in-law, Mrs Edith Priestman, an early member of Bradford's ILP; Mrs Ellen Stott, sharp-minded and Progressive wife of a local ironmonger; and later on, Julia Varley, a trade-union organizer. A further fourteen women joined the board before 1914, a remarkable number: many of them seeking quite explicitly to build bridges between the town's employers and its workforce in the aftermath of its troubled labour disputes.[80]

Bradford's board, under its chairman, Bentham (subsequently a member of the 1909 Royal Commission on the Poor Laws), was like Norwich widely regarded as one of the most advanced in the country. Even so, it was the newly elected women members who urged that children should be brought out of the workhouse into small scattered homes; that the elderly should be offered grouped cottages of their own; that they should try the Brabazon scheme; that pauper nurses should be replaced by trained staff; that the casual wards should be made less deterrent and brought into better order. Their contribution was gracefully recognized when Mrs Moser was made vice-chairman of the board.[81]

[79] Mary Clifford, 'Outrelief', NUWW Conference, Manchester, 1896; Mrs Fuller, SW Poor Law Conference, 1897–8, pp. 291–2. Mary Clifford remained indomitably opposed to outdoor relief for able-bodied men.

[80] J. C. Scott, 'Bradford Women in Organisation 1867–1914', Bradford MA, 1970; T. Wright, 'Poor Law Administration in Bradford 1900–1914', Huddersfield MA, n.d.

[81] *Bradford Observer*, 19 July 1895, 20 Feb. 1896, 30 Nov. 1899; *Bradford Weekly Telegraph*, 27 Oct. 1911.

Significantly, however, it was also the women, led by Mrs Stott, who became the voice of the Bradford Unemployed Emergency Committee and those men who were walking the town 'so long without food that they had music in their insides'. Unemployment, said Mrs Stott, could not be solved by 'commercialism' [capitalism], because capitalism needed the unemployed to keep wages down. Public bodies must intervene. She suggested that the board should buy land to start a farm colony. Responding in part to the women's pressure, the board agreed to establish a new industrial committee, on which all its women members sat, for 'the finding and supervision of all work for the employment of recipients of relief, whether indoor or outdoor'. Mrs Stott and Mrs Priestman argued that families should not be required to exhaust their savings before being offered help. All three ladies insisted that contractors should be forced to observe fair wage clauses. They encouraged their board to call a national conference on the unemployed; and meanwhile asked that the guardians and the town council should together set up a local labour exchange. The movements in Bradford are to be found elsewhere in the major towns of the North, though they were not so clearly women-led as here, and did not have the same background of industrial strife. But this suggests that even on the most progressive boards, with able male guardians, much of the policy initiative still came from its women members.[82]

Meanwhile, when at the central poor law conference of 1903 Miss Sophia Lonsdale asked, 'Why should these allowances be drawn forcibly from the reluctant pockets of the ratepayers, instead of being given by friends and relations and those who are genuinely interested in the poor old people?', many guardians clearly thought she was living in another social and economic world. Miss Vandrey, a Derby guardian, explained that it was the independent-minded respectable elderly who were most reluctant to enter the workhouse. 'Living out the end of their days, as all present would wish to do, in their own way, afforded greater happiness to the majority than the best classified wards in the very best managed Unions, and it was much less expensive to the long-suffering ratepayers

[82] *Bradford Observer*, 7, 14 Mar. 1895; 7 Feb. 1897; 2 May 1895; 18 Apr. 1895; 17 Oct. 1899.

(applause)'. What is noticeable in her remarks, apart from the applause, is the closing of the distance, as 'we would wish to do', and the test of happiness, not deterrence. One sour local government inspector commented that this showed not a willingness to be independent but only a willingness to receive relief in the form the elderly found acceptable.[83] Meanwhile, Miss Fanny Fullagar, for fifteen years a Leicester COS guardian, was, like Miss Baines of Leeds, 'turned off by the Labour party—according to them I had no sympathy with the poor, because I advocated stricter enquiries and said the small struggling ratepayer should be considered'. Increasingly COS guardians found themselves on the defensive and in the minority; older-style LGB inspectors were ignored. Mrs Florence Ramsay, a Fareham guardian, argued in 1906 that widows should not be given full out-relief, but, as the LGB had said in 1871, should put some of their children into the workhouse. She clearly irritated the conference, though this may have been because she tried to give them a lengthy reading-list on the subject. When Mrs Florence Keynes not yet a guardian, turned up at a conference in 1904 to tell guardians all about the COS, she was sharply told that the COS already had too much influence on poor law policy. When she suggested that statutory relief increased the evil it set out to relieve, she was greeted with not-so-polite scorn.[84]

The COS itself sadly and perceptively commented:

The general advocacy of Socialism, as it is popularly understood, had altered the standpoint of many people in regard to social obligation. The State and the Municipality are now often regarded as instruments for equalizing the economic condition and adjusting the interests of different classes of the population . . . The elections have become political . . . and the distribution of relief is often controlled by members of a class who on individual and social grounds desire to increase the amount of it and the number of its recipients . . . With this change has been associated yet another, entirely consistent with the general trend of thought. There is a contraction of interest in personal and voluntary work. Relief, not charity, dominates the

[83] Central Poor Law Conference, 1903, pp. 714–18; Twining, *Workhouses and Pauperism*, pp. 248–9.

[84] Miss Fullagar, RC on Poor Laws, 1909, vol. xli, appendix cxxx; Mrs Ramsay, N. Midlands Poor Law Conference, 1906–7, p. 317; Mrs Keynes, S. Midlands Poor Law Conference, 1904, p. 102.

position. Societies for the care of young women and servants find it
more and more difficult to obtain the help of volunteers. There is a
decrease in candidates for the ministry. There is not the same
demand for admission to the Settlements for men. The thought of
the younger generation of the middle class is turned to new
problems. In part, but only to a small extent, are its members active
in poor law or municipal work.[85]

Most boards had in the past come to an informal agreement to
keep out-relief off the political agenda. It required only a
handful of determined working men, advanced Liberals, or
ILP guardians to reopen the question, and force the parties on
the board to start an auction for popular support, by offering
generous outdoor relief.

George Lansbury's explosion is well known. Elected a
guardian in 1892, 'From the first moment I determined to
fight for one policy only, and that was decent treatment for the
poor outside the workhouse, and hang the rates! . . . I never
could see any difference between outdoor relief and a state
pension, or between the pension of a widowed queen and
outdoor relief for the wife or mother of a worker. The nonsense
about the disgrace of the poor law I fought against till at least
in London we killed it off for good and all'. Many of the newer
women guardians elected in London, in the North, and on a
Labour or Progressive ticket, shared his views, and followed
his lead.

Yet experienced women guardians were right to be worried.
Boards which relied on outdoor relief seldom made it
adequate. Farmers in particular would dole out 2/6d. a week,
plus flour, to casual labourers, knowing this would keep their
wage bills down. Investigators for the 1909 Royal Commission
on the Poor Laws found that many families on outdoor relief
were suffering malnutrition. 'The families possessing them
[such incomes] do not live on them; they beg or borrow; they
pawn their furniture or clothes; they don't pay rent but move
from place to place as landlords push them out. Such a life can
be continued for a long time.' The COS had at least an
honourable record of insisting that if outdoor relief was to be
given, it must be sufficient. Nor were guardians who favoured

---

[85] COS Annual Report, 1899–1900, quoted in H. Bosanquet, *Social Work in London
1869–1912*, 1914, p. 87 (1973 edn.).

out-relief necessarily, or even usually, among the more progressive or humane when it came to their indoor paupers. Miss Flora Joseph, co-opted to a Somerset board that gave out-relief, tried to take the children out of the workhouse. She was refused because it cost too much. She tried to improve the children's playground, which also served as the drying yards. She was refused because it cost too much. She rented a field herself for the children to play in.[86] She, like Margaret Baines of Leeds, believed that guardians were buying cheap popularity with doles of outdoor relief. More typical of newer women guardians, however, was Mrs Clara Rackham, elected in 1904 as a Cambridge guildswoman, who told the NUWW in 1909 that guardians faced problems not of idleness and intemperance, but of unemployment, underemployment, and casual labour, which together engrained casual habits which were hard to eradicate.[87]

Meanwhile, the Royal Commission on the Poor Laws was from 1906 taking evidence. Until it reported, few boards were willing to experiment further. The Commission's members included six COS activists on one wing, and four Labour/Fabian voices on the other; after three years of detailed inquiry the result was a Majority and a Minority report which echoed the divisions that by now existed on almost every board in the country.[88]

Both reports endorsed what women guardians had long been saying, that the general mixed workhouse was a misconceived and misbegotten institution; that children, the sick, the handicapped, and the elderly should be aided outside its walls; and that the able-bodied should be covered by insurance against unemployment. Where the two reports parted was over the respective roles of charity and the poor law, whether insurance should be voluntary or statutory, whether old-age pensions should be non-contributory, whether a

---

[86] RC on Poor Laws, vol. xxi, appendix, report by Mrs Harlock, p. 72; Miss Flora Joseph, ibid., Qq. 69293, 69410.

[87] Mrs Rackham, NUWW Conference, Portsmouth, 1909, p. 33.

[88] The RC visited 200 unions, 400 institutions, heard from 450 witnesses, and received 900 statements of written evidence. M. Rose, *The Relief of Poverty, 1834–1914*, 1972, p. 44. For the inquiry, see B. Webb, *Our Partnership*, 1948. Its right wing included Mrs Bosanquet and C. S. Loch, its left wing Mrs Webb, Lansbury, Francis Chandler, a Manchester trade-unionist, and the Revd Wakefield, Dean of Norwich.

minimum wage and public works programme were necessary;
and whether poor law boards as such should be disbanded
and their residual functions pass to town councils.
Controversy raged, opinions polarized, and the Liberal
Government went its own way, drawing instead on the very
different (and Continental) tradition of social insurance.
Lloyd George's Minute of 1911 said it all: 'Insurance a necessary temporary expedient. At no distant date hope State will
acknowledge a full responsibility in the matter of making
provision for sickness, breakdown and unemployment. It
really does so now, through the Poor Laws, but conditions
under which this system had hitherto worked have been so
harsh and humiliating that working class pride revolts against
accepting so degrading and doubtful a boon'.[89] The Liberal
Government's legislation of 1905–11 began the slow road
away from concepts of pauperism, doles, and workhouse, to
the notion of insurance and rights of benefit in return for
contribution.

Accordingly, local authorities were given powers under the
1905 Unemployed Workmen's Act to set up local joint
committees to promote public works. Women everywhere
were co-opted on to Distress Committees. In 1908 modest
non-contributory pensions were awarded to the very old (over
seventy), very respectable, and very poor: 5s. a week for a
single person, 7s. 6d. a week for a couple, where their income
was less than 10s. a week and where they had not hitherto
received poor relief. (This last clause was abandoned in 1911.)
Flora Thompson's *Lark Rise to Candleford* describes the relief
and gratitude felt by Oxfordshire villagers free from fear of the
workhouse for the first time in their lives. 'At first when they
went to the Post Office to draw it, tears of gratitude would run
down the cheeks of some, and they would say as they picked
up their money, "God bless that Lord George . . . and God
bless you, Miss", and there were flowers from their gardens
and apples from their trees for the girl who merely handed
them the money'.[90]

Half a million elderly received their pensions in January

[89] Lloyd George Minute, quoted in W. Court, *British Economic History 1870–1914*,
1965, p. 413. Poor Law powers passed to the local authorities in 1929.
[90] F. Thompson, *Lark Rise to Candleford*, ch. 5, final paragraph.

1909. The number of elderly on outdoor relief fell by a quarter. Meanwhile the Trades Board Act had established wages councils in a handful of trades to eradicate sweating of those too powerless to form unions. The 1911 National Insurance Act provided some cover for those unable to insure themselves. Above all, the First World War was to end easy assumptions about who lived off whom, who were the productive, and who were the parasites; and was to transform the notion of citizenship for women as well as for working men. Meanwhile in the preceding forty years, women had helped to improve beyond recognition the quality of life for some of the most marginal members of society. Elizabeth Lidgett, one of London's most able guardians, could well say, 'I wish to speak of the pride I have felt in being associated with my sister Guardians from whom I have received fresh ideas of work'.[91]

[91] NUWW Conference, Liverpool, 1891, p. 133.

WOMEN IN COUNCIL

6

Women and the Government
of London 1888–1900

London Government

By the mid 1880s, women were firmly placed on school and poor law boards and helping to shape education and poor relief policy. A third cluster[1] of responsibility had so far eluded them—that for the 'built environment' of their towns, the streets and houses, public health and policing. These were functions of town councils whose statutory authority came from the 1835 Municipal Corporations Act. This Act established a common ratepayer franchise in place of fancy franchises and the old historic vote. It had however referred to 'males' throughout. The few women who were freemen lost their votes. The many women who were ratepayers were denied theirs. But the Act had not applied to London, and when in the 1880s local government reform seemed imminent in London, the women's movement was determined that this time round their rights should not be linguistically extinguished, and that they should be free to take part in any new structures that might be devised.

In the first half of the nineteenth century, London was still run by its ninety parishes and vestries, cross-cut by several hundred commissions, grand juries, and assizes, all with separate rating powers, overlapping territory, and conflicting jurisdiction. At its centre was the wealthy City of London, buttressed by medieval privileges, as one MP put it, 'like a strange animal pickled in spirits of wine'. The problems of draining and sewage, however, were no respectors of persons, property, or parish, as cholera testified in 1848. So in 1855 the LGB devised the administrative expedient of the

---

[1] A fourth responsibility, the magistracy, had to await 1918 when women became JPs.

Metropolitan Board of Works (MBW), essentially to take the daily quota of fifty-two million gallons of sewage out of the Thames. The MBW was indirectly elected, its members drawn from the vestries, to drain and sewer London.[2] Vestries remained, the smaller ones grouped together in district boards. In the absence of any other common body, the functions of the MBW expanded. It embanked the river, drove thoroughfares through the worst slums, created and protected open spaces, such as Hampstead Heath, and sought to regulate utilities and nuisances. It was acting as a city council (and rather a good one) on the fragile and insecure constitutional basis of an indirectly elected joint board. It therefore lacked the cash and coercive powers to handle slum landlords and greedy utilities, or the moral authority to prod vestries into sanitary action. When in the 1880s its architects' department was found to be corrupt, there was no one to defend it. The MBW was scrapped. The LCC was to inherit its work, its problems, and its boundaries.[3]

For behind the rise and the fall of the MBW, the LCC and the GLC was a small difficulty: people could not agree on how London should be governed. On the one hand its vast sprawling population of four and a half million could in no sense be regarded as a knowable community, like Birmingham or Manchester. How then could London have a directly elected and properly accountable city council when the essence of local government was that it defined local communities? The Trafalgar Square riots suggested that the East End and the West were separate nations and not one city. Nor was Westminster comfortable with the thought of an all-embracing London authority, over the water so to speak, wrapped around its deliberations. (Hence police powers ultimately remained with the Home Office, and did not, as in other cities, go to the council.) On the other hand, London's vital services, roads, sewers, fire, and river, were too interdependent to be 'balkanized' down to a multiplicity of vestries. In any case

---

[2] For the MBW, see R. Macleod ed., D. Owen, *The Government of Victorian London 1855–1889*, 1982; and in summary, A. Briggs, *Victorian Cities*, 1963, ch. 8; G. Gibbon & R. Bell, *The History of the LCC 1889–1939*, 1939. The 22 larger vestries selected one or two members to sit on the MBW; 56 smaller parishes were clustered into 15 district boards with one member apiece; the City had three representatives.

[3] See Macleod, Owen, op. cit.

should London, capital of the world, be denied what even the smallest borough possessed, a single voice with which to press its claims on government, a single forum in which to resolve the meta-problems of city life?

Most people came to favour a two-tier structure for London, though whether the top tier should be directly or indirectly elected, and whether it should rape the City of London was less clear. Nor was there any consensus about the lower tier— existing vestries, new parliamentary constituencies, or not-yet-invented boroughs? Liberal plans for reforming London local government got lost in the politics of 1884–5. The Tories on coming to power put the matter aside until the scandals around the MBW, along with the growing need to restructure rural local government, forced Salisbury's cabinet to reopen the question.

Rural England had acquired the parliamentary house-holder vote in the Third Reform Act of 1884, but at local level was still run by an oligarchy of JPs., alongside poor law unions and a plethora of parishes. As the agricultural depression deepened, rural landowners called for financial help with their rate bills. Ritchie, the Conservative president of the LGB, introduced a bill to solve both problems, by establishing directly elected county councils in the countryside and, as that would carve London up between Kent, Surrey and Middlesex, making London a county council in its own right. Ritchie would have liked to complete his reforms by settling the nature of the lower tier as well, but his bill to that effect fell with the change of government in 1892. So the LCC was ushered into life over the top of the existing vestries. The Liberals were to complete the restructuring of rural local government when in 1894 they introduced district councils, based on the old poor law unions, and secular parish councils, in the process scrapping the poor law property qualification which had debarred so many women and working men from office. Without quite realizing it, they also admitted women to London vestries.[4] The LCC immediately elected a Progressive majority so, well satisfied, the Liberals left its lower tier alone. Only when the Tories returned in 1899 was London local government reform completed, and old

---

[4] See below, ch. 7.

vestries became the new boroughs. This time round, the Lords overruled the House of Commons and deliberately excluded women from office.[5]

The LCC continued the work of the MBW in matters of sanitation, slum clearance, river works, and open space, as well as its duties to regulate baby farming under the Infant Life Protection Act of 1872, and license places of entertainment. The LCC also took over the functions of the former county justices for asylums and reformatories, and acquired new responsibilities of its own, for technical education, for housing and planning, and for transport. The LCC had extensive powers, budget (£1½ million), rateable value (£31 million), and staff (over 2,500). Though it lacked the powers over police and utilities that other great cities had, the new LCC could expect to attract councillors of high standing. It also attracted two women, Jane Cobden and Lady Sandhurst; and its Progressive majority appointed a third, Emma Cons, to be an alderman.[6]

The Battle for the LCC

The County Council bill had received its first reading in the Commons in March 1888. A month later, Ritchie was asked whether women electors would be able to stand. He replied that 'Women, as regards their eligibility to be elected as members of County Councils, will be in the same position as they are with respect to election as councillors under the Municipal Corporations Act 1882. Though they may vote at municipal elections, they are not qualified for election as councillors'.[7]

That seemed definitive, but in law the position was less clear than Ritchie claimed. The 1882 Municipal Corporations Act had confirmed the right of women ratepayers to vote, won in 1869, but had excluded them from election. However, the 1888 County Councils Act said that 'every person shall be qualified to be elected and be a councillor who is qualified to elect to the office of a councillor'. Women were 'qualified to elect'. Did this formulary overrule, or was it merely to be

---

[5] See above, ch. 1.

[6] The LCC franchise of householders, freeholders, and £10 occupiers included some 80,000 women among its 600,000 voters.

[7] Mr. Ritchie, Hansard, House of Commons, 20 Apr. 1888, col. 27.

understood in the light of, the 1882 Act? After all Lord Brougham's Act had ensured that words of the masculine gender were deemed to include women unless they were expressly excluded, which the 1888 Act did not do; and in any case the previous Acts of 1835, 1869, and 1882 had not applied to London as it had no municipal corporations. There was obvious room for argument. With their work on school and poor law boards to their credit, women thought that this time they might get the benefit of the doubt.

So, early in November 1888, a dozen prominent women Liberals and guardians met with friendly barristers and Liberal election agents at Mrs Sheldon Amos's house to plan their strategy. They formed the Society for Promoting Women as County Councillors (which became in 1893 the Society for Promoting Women to all Local Government Bodies, and shortly thereafter the Women's Local Government Society). Annie Leigh Browne was its secretary, Eva McClaren its treasurer, Louisa Twining and James Stansfeld its god-parents. They quickly raised enough money for overheads and, significantly, 'for defending, if necessary, their right to sit upon the County Council': and turned their minds to finding candidates, seats, and all-party support.[8]

Their first thought was Mrs Fawcett but, absorbed by the Irish question, she refused. Richard Cobden's younger daughter, Jane, in her mid thirties and active in advanced Liberal circles, agreed to stand, as did Margaret Sandhurst. Lady Sandhurst was the sixty-two year old widow of a distinguished Indian administrator, and was well known in women's Liberalism and alternative philanthropy.[9] The WLGS scoured the lists for other public women of note—Octavia Hill, Mrs Charles of Paddington, Mrs Evans of the Strand, Elizabeth Lidgett of St Pancras, Miss Varley of Islington, Mrs Henry Kingsley of Kingston, the Hon. Maude Stanley (all but

[8] WLGS Minutes, 9, 17 Nov. 1888.

[9] Jane Cobden-Unwin (1851–1949), married to a publisher in the mid-1890s, remained active in Liberal, Irish reform, and suffragist circles. Lady Sandhurst after her husband's death became a spiritualist; she would do nothing without consulting him and he had the convenient habit of forbidding her to go to events like Birmingham's poultry show. (V. Martineau, *Recollections of Sophia Lonsdale* 1936, p. 86). She cured physically handicapped children with powers of 'magnetic mesmerism', *WPP* 5 Jan. 1889.

Octavia Hill with experience as guardians), pressing them to
stand for the LCC—but they refused. However, they heard
that Mrs Massingberd, a wealthy young widow of Lincolnshire, and Mrs M'Ilquham, a married woman guardian of
Tewkesbury, both proposed to fight county seats. Mrs
Massingberd, with land in three counties and temperance
activities in almost as many, was outraged to find that the
Third Reform Act had given her labourers a vote but not
herself. Mrs M'Ilquham was determined to test the issue for
married women, though she had no great expectation of
victory.[10]

Meanwhile, in London, the WLGS had begun the delicate
task of finding seats. The Primrose League thwarted Lady
Sandhurst's chances in Marylebone, a place where, since
Elizabeth Garrett, women always did well; she was adopted
instead by Brixton Liberals for one of their two seats. Jane
Cobden was to stand for Bow and Bromley. It was a courageous decision on the part of the local Liberal and Radical
associations. Women might prove unpopular with the electorate as the LCC was not obviously women's work (sewers?
drains? roads? utilities?) in the way that the LSB was. More
problematically, their nomination papers might be rejected by
the returning officer. Worst of all, a legal challenge to their
seats might be mounted if they won, and then the beaten Tory
would take over Liberal seats by default. However, when the
WLGS offered money, workers, speakers, literature, and lots
of legal opinion to the local organizations, their nervousness
was somewhat allayed, and both associations were to remain
faithful to their women in the rough times ahead.[11]

Outside London Mrs Massingberd and Mrs M'Ilquham
both surmounted the first obstacle when the returning officer
accepted their nomination papers. Mrs M'Ilquham was
standing for the Cheltenham division of Gloucestershire as an
independent, and without much in the way of help or cash
fought a lonely battle against strong party candidates for
whom 'feeling ran very high'. Mrs Elmy urged her 'to fight it

<hr/>

[10] Mrs Massingberd, *WPP* 5 Jan. 1889; Mrs M'Ilquham to Mr. L. Courtney, MP
24 Oct. 1888, Fawcett collection.
[11] WLGS Minutes, 6, 10, 13, 20 Dec. 1888; W. Saunders, *History of the First L.C.C.*,
Minutes, 27 Jan. 1891.

out. It is one of the biggest services you can render the cause of women and particularly of wives . . .'. Though she only collected thirty votes, she had, as she wrote to the WLGS, at least established 'a precedent for other married women to stand'.[12] Mrs Massingberd's candidature, on the other hand, electrified the division of Lindsey. 'In the railway carriages, in the markets, and in all the countryside, nothing else is talked of. Some gentlemen, including some of the clergy, were rash enough to go to a meeting . . . with the idea that they would move a resolution against this popular lady candidate. They were not even allowed a chance of opening their mouths'. Instead they circulated unpleasant leaflets warning voters that they would be wasting their votes as her election would be invalid. The combined forces of parsons, publicans, and Primrose League proved too strong even for the lady squire and she was narrowly defeated by just twenty votes.[13]

In London the returning officer instructed his staff to accept the women's nomination papers, despite objections lodged by the Tories. Miss Browne promptly moved into Brixton to secure the return of Lady Sandhurst and her running mate, Captain Verney; in Bow and Bromley, Jane Cobden's campaign was organized by George Lansbury and a team of working men, immensely proud of their lady candidate. *The Pall Mall Gazette* helpfully reminded readers that

most of the work is simply keeping house for the county. Like a good housewife, the council has to keep the house and the staff of servants in order: to see that the floors are swept (or rather the roads mended), that the menage is without waste or disorder, that the children don't run wild (reformatories etc). That the charities are administered without cheating or pinching, that the plant, the furniture, the shire halls, the lock-ups, police stations etc. are kept in repair, and that the various servants are well chosen, well managed, and rightly paid, and the accounts well looked at. What are all these but the very things in which women have always had their fingers? Why should men, who would shudder at the idea of keeping their own home, rush in, disdaining women's assistance and cooperation, to keep house for the county?[14]

[12] Mrs M'Ilquham, *WPP* 26 Jan. 1889; Mrs Elmy to Mrs M'Ilquham, BM Add. MS 47449, 20 Nov. 1888; WLGS Minutes, 21 Jan. 1889.
[13] *WPP* 19 Jan. 1889; WLGS Minutes, 21 Jan. 1889.
[14] *Pall Mall Gazette*, 22 Oct. 1888. George Lansbury, still straddling Lib-Labism

The WLGS issued their Progressive manifestos which committed them to rate equalization, fair labour policies, and public control of the utilities, charities, and open spaces of London: organized the committee rooms, summoned meetings of women electors, and canvassed door to door.

On 17 January 1889, both women were triumphantly returned, Lady Sandhurst and Captain Verney in Brixton, Jane Cobden (and a Tory) in Bow. Their divisions celebrated by dancing the night away in quadrilles. When the LCC Progressives met for their first group meeting and decided to take eighteen of the nineteen aldermanic seats, they added to the women's triumph by nominating Miss Emma Cons as an alderman for her housing and social work. Margaret Sandhurst had defeated Mr Beresford Hope in Brixton, a viciously anti-suffragist MP, and ominously he promptly lodged an election petition against her, contending that her election was invalid. Jane Cobden was more fortunate. Her runner-up was the defeated Liberal who immediately wrote to say that he 'would not take the initiative in any such proceedings'; and gracefully offered her his help.[15]

The LCC held its first meeting in the Guildhall. Jane Cobden and Margaret Sandhurst took their seats to the right and back of the chamber amid its arches and draperies. Lord Rosebery was elected chairman. Jane Cobden cast her first vote at that meeting. Margaret Sandhurst, with a legal action ahead, thought it prudent to abstain. The women moved smoothly into their committee work, Lady Sandhurst devoting herself to baby farms, Jane Cobden to children in reformatories, Emma Cons to women patients in asylums.

Beresford Hope v. *Sandhurst* came to court on 18 March 1889.

was invited to take over her LCC seat in 1892 but threw his lot in with the SDF instead. As MP for Bow and Bromley, he fought a by-election on women's suffrage in 1912. G. Lansbury, *My Life*, 1928, p. 74, *WPP* 5 Jan. 1889.

[15] Emma Cons, 'keen, quick, very clever and very good, most philanthropic, with a passion for helping anyone in difficulties, absolutely careless of herself and her comfort and devoted to the people among whom she worked'. A housing manager, she started the Old Vic because 'she was determined to give her tenants in S. London and their friends a decent place of entertainment' with bowdlerized music hall fare. Martineau, op. cit., pp. 160–2. See her own comments to 12th AGM of WLGS, 1905. Jane Cobden received her assurance in a letter from E. Rider Cook, 21 Jan. 1889 (Cobden papers); see also G. Lansbury to J. Cobden 21 Jan. 1889, Cobden papers, *East London Observer*, 16 Mar. 1889.

Mr Costelloe, a fellow LCC councillor and feminist, acted as the WLGS barrister. The judges ruled that, given the common law disabilities of women, if parliament had not expressly included them its silence must be construed as excluding them from office. Lady Sandhurst went to appeal in May but the judges thought the whole business was a nonsense and dismissed her appeal as absurd. Lady Sandhurst was now unseated and Beresford Hope took over her seat without further election. Jane Cobden, though vulnerable to a similar action, knew she would not be challenged by a defeated candidate. Emma Cons as an alderman was in a different position again. However, both women recognized that they held their seats only by grace and favour. They would not win a court case if they were taken to law. But could they change the law?

Given the successful track record of women on school and poor law boards and the substantial number of suffragist MPs in the House, the WLGS thought it might get enough support to extract parliamentary time even out of a Conservative Government. Within the LCC the first of many motions supporting the right of women to be county councillors was passed (forty-eight votes to twenty-two), Jane Cobden asking members, to laughter, whether any of them wished to do what Lady Sandhurst had done, inspect twenty-three baby farms? Friendly MPs Mr. Channing, Professor Stuart, Walter McClaren, Sir John Lubbock, and Mr Firth (the latter two both LCC councillors) introduced a bill in the Commons qualifying women to become county councillors. The Earl of Meath, himself an LCC alderman whose wife had started the Brabazon workhouse schemes, introduced a similar bill into the Lords. He urged their Lordships not to fear that this was the first step to the parliamentary vote; it was their womanly work that the LCC and the voters valued. Lady Sandhurst's son and the Archbishop of Canterbury joined him in the lobby but the vote was lost by twenty-three votes to one hundred and eight. A few days later the Commons bill was tabled, but the government made it clear that they had never intended to make women eligible for county council office and that they never would, so it fell.[16]

[16] *WPP* 18, 25 May 1889; *WSJ* 1 June 1889; *EWR* 15 June 1889.

Questions were now being asked in the LCC about the position of Jane Cobden and Emma Cons. Lord Rosebery told members he had taken legal advice. 'The seats of Miss Cobden and Alderman Miss Cons on the Council were not affected by the [Sandhurst] decision', he said, 'although the decision went to show that at the time of their election they were disqualified . . . He might properly invite the ladies to resign their seats, but if they declined to do so he was not bound to do anything further in the matter.' He did and they didn't, and he, Lord Rosebery, therefore asked councillors to take no further action which would prejudice their position. This was agreed.[17]

The WLGS had also been taking legal advice. Margaret Sandhurst had been unseated by an election petition which asked the courts to declare the result void. Such a petition in electoral law had to be filed within twenty-eight days of the election. From that threat Jane Cobden was now safe, as of course was Emma Cons. Electoral law also held that even if someone had been improperly elected, after twelve months that election was deemed to be valid. Mr Costelloe therefore suggested that the women adopt a low profile, not attend meetings, and wait out the rest of the twelve months. After that, their position for the rest of their term of office should be unassailable, though they would not of course be able to stand for re-election in the light of the Sandhurst judgement. Emma Cons accordingly resigned her seat on the parks and open spaces committee, and on the housing committee. Her colleagues, proud of her work, promptly invited her back on the committee as a visitor, and, as she ruefully noted, made her pay for the privilege by giving her an even heavier burden of work.[18]

By February 1890 the twelve months were up. The women consulted Lord Rosebery and formally resumed their seats. Both took the occasion to go public. Jane Cobden told members, to applause, 'I am now entitled not only to be, but to act as a councillor; and I feel strong in the knowledge that my present action is heartily approved of by my constituents (to whom I am primarily responsible) and I believe by the

17 *EWR* 15 June 1889; W. Saunders, op. cit., Minutes, 21 May, 1889.
18 WLGS Minutes, 4 Apr., 29 May 1889.

majority of this Council'. Emma Cons added, that

> in spite of the continuous good will of my fellow workers, I have felt the drawbacks of my position, the serious lack of a vote, and the corresponding diminution of influence, and much valuable time has been wasted by the fact that I could not be counted as one of a quorum. My selection . . . laid upon me a call to aid in the work of this Council, which I ought not to disobey. I believe that for such work as ours women are not only qualified but needful, and I desire to defend their franchise as far as lawfully I may . . . those duties are to me a trust, not only for women but for the whole community, (cheers)[19]

Emma Cons rejoined six committees (asylums, housing, industrial schools, parks, sanitation, theatres) and eleven subcommittees. Jane Cobden rejoined housing, asylums, and industrial schools. The Progressive majority on the LCC continued loyally to support them.

While the women had voluntarily withdrawn from official council work, the Tories had been quiet. Now that they resumed their work publicly, matters speedily came to a head. The parks committee proposed early in May 1890 to appoint Jane Cobden to a vacancy. Two Tory members, Mr Boulnois and Colonel Rotten, said that in that case 'An application would immediately be laid at one of the courts of law to enforce a penalty against one of the ladies for having voted in the Council'. The Progressives greeted this statement with cries of shame. Undeterred, Sir Walter De Souza, a Tory LCC member, on behalf of his group filed writs upon both women, claiming penalties of £250 from each woman for the five votes they had cast at council when, he said, they were not legally qualified to vote. The Tories could no longer unseat them; but they could seek to stop them acting and voting as councillors. What made it more unpleasant was that De Souza was acting as a common informer and therefore entitled to half of any penalties imposed. Miss Emma Cons let it be known that she would go to prison rather than pay any fines, and De Souza, sensibly not wanting a martyr languishing in jail, persuaded the courts to hear the case of Jane Cobden first.[20] The case was

[19] Saunders, op. cit. Minutes, 11 Feb. 1890; *WPP* 15 Feb. 1890; notes of speech, Cobden papers, u.d.

[20] Costelloe to Miss Browne, WLGS Ms 30 Apr. 1890.

not heard until November, and the women worked steadily on, Emma Cons putting in thirty-two out of a possible forty-one attendances, as well as dozens of visits of inspection, Jane Cobden making twenty-six out of thirty-nine possible meetings.[21]

The WLGS fearing with some justification that the courts would be no friend of the women, again mounted pressure on parliament to change the law. Lord Meath reintroduced his bill into the Lords; the WLGS and friendly LCC councillors went on deputation to Lord Granville and Lord Salisbury; seventy-five members signed a petition of support. Granville and Salisbury were sympathetic. As Granville said, if the electors did not want women, they need not elect them. Meath argued in the Lords that the electors had shown that they preferred devoted and educated women to represent them rather than publicans, jerry-builders, and slum landlords. But Earl Cowper produced the usual anti-feminist line: women needed to be protected 'from the struggles, enmity and bitterness of public life'; women did not want the public spotlight; and if they did, that merely went to show that they were emotionally unsuited for it. The bill was lost by one hundred and nineteen votes to forty-five.[22]

Attention now moved to Jane Cobden's court case. The judge agreed that though the women were not qualified to be elected, having been elected and not challenged, their election was now valid. However, having been disqualified from being elected when they were elected, they could not act as councillors without incurring penalty. If, however, having been elected as councillors they were to resign, they would incur penalty for that. It was an Alice-in-Wonderland position. Emma Cons wrote a savage letter to *The Times*. The judge's decision, she said, denied ratepayers their free choice of representatives, denied women the right to be representatives, and denied the LCC the benefit of their skills and experience, in her case thirty years of work in the field of housing. She would continue to fight 'for equal municipal rights for women as for men. It is a bitter experience when one for the first time

---

[21] Mr Beresford Hope attended 29 of 98 meetings 1889–90, 29 of 132 1890–1.
[22] *WPP* 14 June 1890, *WSJ* 1 July 1890.

realizes that even a long life spent in the service of one's fellow citizens is powerless to blot out the disgrace and crime (in the eyes of the law) for having been born a woman'. If ratepayers only knew the 'crying need' for women on the LCC, they would be sick at heart.[23]

The WLGS reviewed their position. So far their legal costs had come to over £700 even though Mr Costelloe and their QC had waived fees. Members raised funds by selling pamphlets, giving lectures, and soliciting donations. The Countess of Aberdeen took a deputation of eighty women's Liberal associations to Gladstone, who thereupon promised his moral support. The women also took revenge by holding rallies in De Souza's Westminster division. But to no avail. Jane Cobden's appeal was heard and dismissed in April 1891, although the penalties were reduced from £250 a time to £5. Emma Cons was in yet a different position. She had not been improperly elected, but properly or improperly selected as an alderman by the ruling majority. She wanted not to pay the fine, 'to fight her case and possibly go to prison'. She was persuaded otherwise, since this would have disqualified her from public office for five years.[24]

Where did that leave the women? They were still validly elected to the LCC, but they could not act as members of it. There was nothing to stop them attending meetings, and this Jane Cobden proposed to do 'believing that the presence of the women councillors could help to keep their cause before the minds of the other members' as the last few months of the present LCC ticked away. Miss Cons thought this was somewhat '*infra dig.*, and a waste of time' but decided in the best theatrical tradition that the show should go on. Both women attended the next LCC meeting, ostentatiously but silently taking their seats.

De Souza would not leave the situation alone. He insisted that if they attended they must sign the attendance book; if they signed the book, they were acting as councillors and in contempt of court. The chairman refused to give a ruling, Progressive members taunted De Souza with the money he was collecting from the women as a very common informer,

---

[23] Quoted *EWR* 15 Jan. 1891.
[24] WLGS Minutes, 28, 29 Nov., 3 Dec. 1890; *WH* 14 Feb., 28 Mar. 1890.

and proceedings dissolved into uproar. The women thought they had done rather well.[25]

Meanwhile, one last attempt was being made in parliament to regularize their position before the next round of county council elections was due and women finally lost their seats. In the Commons, the Tory Sir Richard Temple, a leading member of the LSB, gave them a character reference. Women in public work possessed 'an assiduity, a constancy, and a sympathy not to be surpassed, and seldom equalled by men'. This was not enough to save the WLGS bill from defeat when Labouchere argued that women on the LSB were all very well, as it was a specialist body, but the LCC was 'a local parliament', and if the House agreed to this, they could not resist the parliamentary vote. In the Lords, Meath wanted to introduce his bill for the third time but was beaten by his own poor health. However, as Ritchie, the Conservative president of the LGB was tabling a bill for the further reform of London local government, to replace vestries with district councils that might take on poor law work, the WLGS switched their attention to that. The WLGS secretary, Mrs Stanbury, stood patiently in drenching rain outside the National Liberal Federation Conference in December 1891, handing out leaflets to hurrying delegates, asking them to support Ritchie's bill and watch over women's rights. If women could not sit on the LCC, they might at least sit on the district councils. Before Ritchie's bill could progress very far, the government fell.[26]

Within a few weeks, the first LCC would dissolve, and out would go the women. Even so, they would make a fight of it to the end. They threw all their formidable canvassing skills and speakers into the Westminster and Brixton divisions in the hope of taking revenge on De Souza and Beresford Hope. Beresford Hope was defeated, to their satisfaction. De Souza held a safe seat. The Progressives were again returned with a handsome majority and asked the women to continue their work for the LCC informally as visitors. The WLGS was tempted by the offer—women continued to need women—but decided that they must hold out for elected office or nothing.

[25] *WH* 9 May 1891; correspondence of solicitors for Miss Cobden and De Souza, 7, 9 May 1891, Cobden papers.
[26] *EWR*, 15 July 1891.

For the time being, there could be no further initiatives on the county council front. The WLGS turned their attention to the incoming Liberal government which was proposing to introduce parish and district councils into rural England. As these district councils would absorb poor law work, it was essential that women's position on these authorities be protected and their legal rights be unassailable.

In February 1892 Margaret Sandhurst died. Unlike Emma Cons and Jane Cobden, she was never to see the second generation of women—Nettie Adler, Susan Lawrence—become London county councillors. Nor was she there to hear perhaps the biggest tribute the LCC could have paid its former women members. In one of its first motions, the new, all-male LCC agreed to petition parliament to permit women to be county councillors as this was 'highly advantageous' for the community of London.[27]

The Women's Local Government Society

The battle for London had been fought by the Women's Local Government Society. The WLGS is one of the least known, yet was one of the most effective women's organizations in the late nineteenth century. It brought leisured upper-middle-class Liberal ladies together in London drawing-rooms for tea and talk, in order to run in the process a highly professional and formidable lobbying machine. At its centre was one woman, Miss Annie Leigh Browne (1851–1936), who founded, funded, and sustained the Society until, after the War, it became the Women's Citizens Association.

Miss Browne's family moved in Bristol's wealthy suffragist circles, which included the Davenport Hills and the Carpenters. On her father's death in 1870 she moved her family and modest fortune to London where she became close friends with Mary Kilgour, a doctor's daughter and maths lecturer. Together they helped the Shirreff sisters launch the Girls Public Day School Trust, supported Emma Paterson's fledgeling women's trade unions, and raised funds for Elizabeth Garrett's London Medical School for women. In 1886 they brought together some women already active in local government (Mrs Charles, Mrs Evans, Caroline Biggs of the

[27] LCC Minutes, 3 May 1892.

Englishwomen's Review, Lucy Wilson of the Vigilance Association), into a modest Local Electors Association. They wanted
to prise open vestry politics. Mrs Charles of Paddington was
willing to test the law but her nomination papers for Paddington vestry were rejected by the Returning Office in 1887. While
they were still deciding what to do next, the Metropolitan
Board of Works collapsed amid charges of corruption. Bills for
London and county government were introduced by the
government; and the group of women turned their efforts
there. They were joined by social purity workers who wanted
the new LCC to clean up London, by Liberal party workers,
and by suffragists. As we have seen, they elected and defended
Jane Cobden and Margaret Sandhurst. During the winter of
1892, with further local government reform in the air, the
WLGS decided not to disband but to promote women to all
local government bodies.[28]

The WLGS was an upper-middle-class, Liberal, and
London group, which functioned through family, social,
philanthropic, and Liberal connections. It was also stoutly
feminist, supporting the claims of all women to elected local
office, whatever their politics and their views on suffrage,
while refusing to support men, however sympathetic, to
anything at all.[29] Its inner circle of Annie Leigh Browne, her
sister Lady Lockyer, and close friend Mary Kilgour, also
included Mrs Sheldon Amos, active in Liberal and social
purity work, Mrs Maitland of the London school board, Mrs
Louisa Mallett, a candidate for the London school board,
Elizabeth Lidgett, a guardian and National Vigilance Association worker, as well as Eva McClaren and Mrs Heberden
(formerly Sarah Ward Andrews) who helped found the Women
Guardians Society a few years before. Their longest serving
chairman was Jane Strachey (a suffragist Conservative), who,
like Margaret Sandhurst, was the wife of a retired Anglo-
Indian administrator, and known for her writing, smoking,
and billiards, as well as her eleven children, of whom Pippa

[28] *EWR*, 15 Aug. 1887; *WH* 4 Feb. 1893. 1st Annual Report, 1893. It seems likely
that women were eligible to serve on vestries before 1894.

[29] The WLGS refused to join the NUWW and WGS in promoting machinery for
the selection of suitable persons, both men and women, as poor law guardians in
1901, because they would only aid women. Minutes, 15 Jan. 1901, 11 Dec. 1907.

acted as secretary to the women's Suffrage Society, and
Lytton as one of its occasional stewards. Its President, the
Countess of Aberdeen, was also the titular head of women's
Liberalism. Indeed, the executive of the Women's Liberal
Association in 1893 included Lady Aberdeen, Emma Cons,
Jane Cobden, and Mrs Massingberd, as well as Mrs Maitland
and Mrs Mallett.[30] Eva McClaren, also a high-profile
Liberal, was with Mrs Massingberd a key figure in women's
temperance whose network of six hundred branches across the
country was anxious for women in public to extinguish the
local drink trade. Miss Browne kept up their connections with
women's trade-unionism: all WLGS printing went to women-
only printing offices. They also became close friends of Miss
Emily Janes, the secretary-organizer of the NUWW which
brought women in active public, philanthropic, and religious
work into conference.[31] Like the British Women's Temperance
Association, the NUWW offered a countrywide network of
contacts that were ostensibly free from party or political
colour; Mrs Sheldon Amos was on their executive; and by the
late 1890s their interdependence was so strong in some towns
that the local government committees of the NUWW were
also the local branch of the WLGS. They had access to Mrs
Elmy's and Mrs M'Ilquham's Women's Emancipation
Union, whenever Mrs Elmy could bring herself to speak to
wealthy Liberal ladies without feeling patronized; and cordial
if somewhat formal relations with women's suffrage societies.
COS secretaries, the women's settlements, and the Women's
Co-operative Guild attended their AGMs. Annie Leigh
Browne herself served on London Progressive committees and
attended meetings and worked for the London Reform Union.
The inner ring of the WLGS, in other words, was at the centre

[30] For Annie Leigh Browne, see *The Times*, 14 Mar. 1936, *The Vaccination Inquirer*,
1 Apr. 1936; For Mary Kilgour, see *The Times*, 24 Sept. 1951, *The Manchester Guardian*,
1 Jan. 1955; Mrs Sheldon Amos, widow of a famous lawyer, granddaughter of Jabez
Bunting, sister of the editor of the *Contemporary Review*, (whose wife was a sister of
Elizabeth Lidgett), and at various times on the committees of the WLF, the NUWW,
and the LNA. She founded the London Morality Council with Bishop Creighton,
whose wife was president of the NUWW. See *EWR*, 15 Apr. 1908. For Lady Strachey,
see C. R. Sanders, *The Strachey Family 1588–1932*, 1953, as well as biographies of Lytton
Strachey. For the Countess of Aberdeen, see *WH* 14 May 1892. For Mrs Maitland
and Mrs Mallet, see ch. 2.
[31] For the NUWW, see ch. 1.

of overlapping circles of women's progressive, philanthropic, and purity causes. At drawing-room meetings, their elegant, cherished, and well-thumbed address books were passed from hand to hand, as they pooled their membership lists, exchanged vice-presidents, and sought guest speakers for their public meetings.

At least as important as their women's network, however, were their entries into male clubland, through family and social ties. WLGS members had staunch support from husbands, some of whom were parliamentarians or lawyers and had often the good sense to be both. Within the Commons, they could count on a cohort of suffrage MPs of whom James Stansfeld, former president of the LGB, Leonard Courtney, Walter McClaren,[32] Dr Shipman, and Mr Channing were the most energetic. Other MPs were members of the LCC (Sir John Lubbock, Sir Joshua Firth, Baron Dimsdale) or of the LSB (Sir Richard Temple, Mr Bousefield) men who, while not necessarily suffragist, were persuaded of the value of women's public work. In the Lords they could rely on Meath and Aberdeen, and two former LSB chairmen, Londonderry and Reay, to introduce their bills, grace memorials, and preside over public meetings. There were a wealth of other connections, through Lady Frances Balfour (and the Tory Cecils and Balfours), Mrs Fordham, whose father was a junior LGB minister, and Lady Frederick Cavendish.

At least as valuable was the supply of instant, cheap, and expert legal advice. At various times the WLGS had on their committee Channing and Dr Shipman, both Northamptonshire MPs and distinguished barristers; the Fabian Mr Costelloe who chaired the LCC's parliamentary committee and defended Margaret Sandhurst and Jane Cobden for the WLGS;[33] Mr Torr, an expert in election law and later a revising barrister; and Mr Cyril Dodds, later KC. These men

[32] Walter McClaren, son of Priscilla Bright MacClaren, married Eva Muller, and became MP for Crewe. As a 16-year old, he acted as escort to Sophia Jex Blake and other medical students at Edinburgh who were being stoned for their temerity. Mrs Cobden Unwin's speech, WLGS, *AGM* Mar. 1913.

[33] For Costelloe, see interview with wife, *WPP* 13 Dec. 1890; B. Webb, *Diaries*, 20 Mar. 1895. His daughter, Ray, married a Strachey, and was biographer of Mrs Fawcett.

drafted parliamentary bills for the Society, prepared appeals, argued cases in court and before revising barristers, offered Counsel's opinion on a host of matters, and turned their Opinions into pamphlets for the women to sell and circulate. In its life, the WLGS took up dozens of legal cases on behalf of women voters and women candidates. With the exception only of the battle for seats on the LCC, it won them all. Country solicitors, parish clerks, and hostile election agents were quickly cowed by the opinions of eminent Counsel that came winging down from London. Appeals by backwoodsmen to the LGB got them nowhere, as the LGB was invariably sympathetic to women's public work.

Yet, despite their worldly and sophisticated connections, there was nothing pretentious or notably self-seeking about the WLGS women. Inevitably, the constituency for whom they worked was propertied just because most women voters and most women candidates had to possess a property rate-paying qualification. So did most men. Like the suffrage societies, the WLGS wanted propertied women to enjoy the same rights as propertied men. They were feminists. But they campaigned ardently for the removal of all disabilities on women, those of gender, those of marriage, and those of property. They worked equally hard to protect the votes of working women in Colne, or the right of an obscure widow to be appointed a rate collector. Their leaflets encouraging working-class women to stand for parish and district councils were straightforward, well informed, and uncondescending. The inner circle of the WLGS was undeniably upper- and upper-middle-class, and it never occurred to them to be embarrassed by it. To that extent they did not differ from most philanthropic or political organizations of the day, the WLF, the NUWW, or early suffrage societies, to say nothing of the Primrose League. Annie Leigh Browne in particular, had leisure, money (she left £55,000 in her will), education, and connections; and she laid them all out for the Society. As Mrs Elmy waspishly noted, she lived for the WLGS and was 'a bundle of nerves, poor thing', but conceded 'still they *are* working . . .'.[34] Miss Browne was quiet, fiercely determined, hard-working, demanding. She hated the limelight, dreaded

[34] Mrs Elmy, BM Add. MS 47452, 1 Aug. 1901, 23 Oct. 1901.

public speaking, and was always searching for charismatic women to 'front' for her. She bought the best for the Society—offices in Westminster (at first shared with the National Vigilance Association, then with the Women Guardians Society and finally, having rejected space at the Suffrage Society premises, offices in Tothill St.). She employed secretaries who were university educated and commercially trained, and paid them well. Mrs Stanbury, their most effective, was paid £150 a year for a thirty-eight hour week, with six weeks' holiday a year. In her absence, Miss Browne's private secretary ran the office. She bought the best legal advice going, when she could not get it for free from committee members. WLGS expenditure rose to around £400 a year, income always fell short by £100 or so, but Miss Browne and her mother would anonymously bridge the deficit, as well as propose, and pay for, better offices, furniture, more lectures, rural organizers, clerical assistance, train fares, pamphlets, and legal opinion. Understandably she quailed at the thought even of her money financing the black hole of endless rural propaganda, and so WLGS work was inevitably constrained by finance to London. Every Annual Report sought extra donations. Yet even when Miss Browne was funding the Society, she was always scrupulous to ask for and accept committee direction. Entirely sensibly, however, she used her money to buy help with the trivia, thus avoiding incessant fund-raising bazaars. She spent her time, instead, organizing Margaret Sandhurst's committee rooms, writing letters to *The Times*, standing patiently in the lobby of the House of Commons to accost MPs who would take on her bills as a private member's motion, attended endless conferences of numerous organizations to place women's local government work on their agendas, and drafted dozens of leaflets. The result was a small, painstaking, highly influential, and remarkably successful society.

The WLGS was, however, worried by the closeness of its ties to party Liberalism. Its members knew that if they were to bring out the women's vote for women candidates, or get the attention of Conservative politicians, they had to broaden their membership base and keep a public distance from official Liberalism. They assiduously cultivated the Primrose

League, rebuffed though they usually were, the League's Ladies Grand Council believing 'that they had always thought it wiser to keep clear of all questions affecting women's work as apart from men'.[35] As one WLGS member muttered when they received the letter, women's sphere was in the background. They were delighted to get Baron Dimsdale, a Conservative LCC councillor and later MP for the City, on their Executive, and headhunted the Conservative and feminist Lady Strachey as their chairman. They relied heavily, as did many women's organizations, on the racy intermediatory skills of Lady Frances Balfour—'a sort of liaison officer' she described herself between women's societies and parliament—who though daughter of the Whig Duke of Argyle, was sister-in-law to the Tory prime minister, Balfour, as well as to Princess Louise.[36] They noted that the women's suffrage societies were split by the issue of party affiliation; and women's Liberalism by the issue of women's suffrage. The WLGS had good reason to avoid lining up officially behind other feminist causes which could divide their support. So the WLGS always preferred to sponsor all-party local women's committee and independent women candidates, free of party label; and to avoid pronouncing on women's suffrage, even though their personal views were clear enough.

Despite their best efforts, the Society remained also a largely London society, and this worried them too. The WLGS had about one hundred and fifty directly enrolled members and a further twenty or so provincial correspondents who sent them details of local results, lobbied local MPs, planted motions in local councils, and helped set up local election committees for women candidates. In practice, membership figures were not that significant as the WLGS (like the Women Guardians Society before it), relied on its networking with women's liberalism, temperance, suffrage, and the NUWW to gain signatures for petitions and to put pressure on MPs.

Most towns had women's local government committees of some sort, either a branch of the WLGS or of the NUWW, or of the Liberal party. But the cost of keeping in touch and

[35] WLGS standing subcommittee Minutes, vol. xxi, 13 May 1903.
[36] Lady F. Balfour, *Ne Obliviscaris*, 1930, vol. ii. p. 172.

informed was expensive (£20 in postage to circulate every school board, £50 and vast clerical effort to write to 8,000 parish clerks). The Society did its best to extend its work beyond London, to bring in northern and Cornish women to the AGMs, to send out free literature and lecturers to Norfolk and Northumbria, but they lacked the resources to do much more. When the Women Guardians Society folded in 1904, the WLGS noted that 'a great deal of work has been given up throughout the country that the committee feels to be of imperative importance, when the right of women to work on public bodies had already been seriously diminished . . . '.[37] Unable to finance local work, they could not bring in the local finance with which to finance local work. Sensibly, they brought their parliamentary, publicity, and legal skills to bear on Westminster instead. How successful were they?

The WLGS set itself two goals: to increase the number of women holding elected office, and to increase the number of women holding the local government vote. This meant promoting two pieces of parliamentary legislation. The first would permit women to sit on town and county councils, the only local authorities still barred to them. The second would permit married women who were ratepayers to have the local government borough vote and not to lose it on marriage. Year after year for fifteen years, Miss Browne wrote to over a hundred or so sympathetic MPs, asking them to ballot for one of their two bills. She sometimes got a first choice promise, more usually half a dozen second choice promises, but none of her promises ever came up. So she would attend the Commons on private members' day to lobby MPs who were lucky in the draw, but here again she met little success. Friendly MPs tried to introduce her bills as unopposed business after midnight. Mr Spicer tried several times in 1894, other MPs in 1896, and Mr Channing and Dr Shipman thereafter, but as Dr Shipman ruefully noted, one die-hard Tory stockbroker MP, Sir Frederick Banbury, boasted that it was his mission in life to oppose women's bills.

Sometimes I dodged a bit. I thought Sir Frederick Banbury might nod occasionally so I missed a few days. I once missed a week, and then I brought it out again. I tried in all the quiet ways that occurred to me to bring the bill forward, but Sir Frederick Banbury beat me.[38]

Sir Frederick Banbury was still giving Nancy Astor problems in the 1920s: but she gave as good as she got. Supportive peers continued to introduce bills into the Lords without hope of success. It was frustrating. The women could reasonably count on a Commons majority for any bill introduced there, if only they could obtain parliamentary time; whereas in the Lords they could always get time for a debate but never votes for a majority. Increasingly it seemed that only Government would offer time in the Commons and be able to impose a bill on the Lords, and this no government was willing to do before the 1905 Liberal landslide.

The deep disappointment of the LCC years of 1889–92 was set aside, however, when it became clear that both Conservatives and Liberals proposed to introduce district and parish councils into England. Fowler, the new Liberal president of the LGB, made it clear in March 1893 that he proposed to make women capable of serving on these district councils since they were absorbing poor law unions on which women were serving with distinction. The electors would be 'all the men and women who are registered as County Council electors'.[39] The women's position seemed safe enough, but the Society prudently took legal advice rather than risk interminable court actions in future years. Their QCs told them that women's right to vote would be vulnerable to legal challenge because the electoral register was to be compiled not from a simple list of parish ratepayers, married women among them, but by revising barristers who, on past precedent, would strike off married women. The bill needed amendment. The WLGS swung into action, circulating 50,000 leaflets and petition forms to 300 provincial Liberal associations, 650 branches of the BWTA, 53 MPs, 67 working men's Liberal clubs, all the Fabian societies they knew of, their 165 provincial

---

[38] WLGS Minutes, Report of a reception at Lord Brassey's house, 14 Feb. 1907.
[39] Mr Fowler, Hansard, House of Commons, 21 Mar. 1893, cols. 698, 689. The Liberals would have liked strong parish councils but as 'poor men could not go long distances to attend district councils' then in Mr Torr's cynical view, the Tories came to favour strong district councils instead. WLGS Minutes, 16 Dec. 1891.

correspondents, and any other addresses the suffrage societies and Mrs Elmy could provide. Fowler was not disposed to argue, and their QC (Haldane) and Mr McClaren safely shepherded amendments through the House which gave all women ratepayers the vote, and allowed any woman, ratepayer or resident, to stand for election. At one point they seemed close to bringing off another major coup by tacking on to the bill a similar voting clause to apply to other local government bodies (such as borough councils which allowed only single women ratepayers to vote) but the Speaker ruled them out of order. The women considered adding an even more ambitious clause, making women eligible to serve as borough and county councillors, but were warned that this might lead to defeat for the entire bill in the Lords. They drew back.[40]

As the bill also abolished the remaining property qualifications for poor law guardians, and made explicit the right of women to be elected to London vestries, the bill could be counted a major success. The number of women poor law guardians rose from 119 to 692; 13 women joined London's vestries. 140 women were elected as rural district councillors. After the failure of women to maintain their seats on the LCC, it was even more important to gain seats on the new districts, since it seemed likely that the specialist poor law and education boards on which women served would, like health, in time lose their functions to multi-purpose local authorities. If women were not already serving on these generalist councils, they would find their local government work disappearing with the end of local government boards.

The WLGS was right to be worried. Already some of the London vestries were petitioning to be incorporated as municipal boroughs. In 1899 the Tory government finally tackled what it had failed to do in the early 1890s, the lower tier of London local government. The old vestries, district boards, and miscellaneous authorities were to be recast with changed boundaries though little change of function as the new London boroughs. At first women thought their position as vestrywomen remained secure. When the London boroughs

[40] WLGS Minutes, 19 Oct. 1893; Memorial to Shaw Lefevre, president of LGB, reported *WS* 7 Feb. 1895; WLGS Annual Report, 1894. See above, pp. 45–7.

bill was discussed in the Commons in February 1899, Balfour was asked whether women could become borough councillors. His reply was inaudible, because, it turned out, he had failed to realize that women were already serving on London vestries. However, he proved not unsympathetic. Coached by Lady Frances Balfour, he declared to the House that women could become councillors, though not mayors or aldermen, and Clause 24 was safely taken through the House to that effect. But the women's clause faced defeat in the Lords, where it was argued that the new London boroughs were not reformed vestries, or urban district councils, on which women already sat, but new municipal councils on which women did not. As Lady Frances noted,

What the Cecils call 'Frances' lobby on Monday will contain Salisbury, Selbourne, all the bishops and liberals peers and such independent sit-on-the-fence peers as will be affected by the speech Salisbury will make for us: into the other lobby, three to one against the women, will go the Government and the mass of Tories . . . No sense of constituents, or pleasing anyone but their own prejudices and consciences . . . Of course we mean to try and insert 'Councillors' when it comes back from the Peers, and abandon (too late in my opinion) Aldermen, but though Salisbury says we should do this, he thinks that fear of wrecking the bill will prevent a clear women's vote.[41]

Her predictions were accurate. The Lords filled up with peers to vote Clause 24 down and as quickly emptied again. The Annual Report from the usually decorous NUWW attributed its defeat to the 'Jockey Club which, as one of the peers said, had resolved that women should not serve on the London Borough Councils'. Too close to clubland. The women worked desperately hard to get MPs to reverse the Lords' decision; but found Chamberlain and Jesse Collings against them. Though John Burns made a powerful speech on their behalf, the government was unwilling to delay the bill any further, acquiesced in the Lords verdict, and whipped its back bench.[42]

The WLGS tried gallantly to mend matters by introducing

---

[41] Lady F. Balfour to Prof. G. Saintsbury, quoted in Balfour, op. cit., vol. ii. p. 156.
[42] WLGS 7th Annual Report, 1900, 8th Annual Report, 1901; NUWW Annual Report, 1903, p. 129. See also ch. 1.

a bill of their own permitting women to be borough and county councillors, but before this had got very far they were faced with an even greater threat, the abolition of school boards and with it their women members.

The return of the Tory government in 1895 had, as we have seen, coincided with the return in 1897 of a Progressive and expansionist majority on the London school board. Tory ministers proved willing to use the notorious Cockerton judgement limiting school boards to elementary school work as a reason for transferring all education from school boards to county and borough councils. The government's first bill, which had so offended Dissent, had been amended. Advanced Liberal opinion now agreed that multi-purpose councils rather than specialist *ad hoc* boards were the most appropriate authority to administer education.[43] Women of course were not eligible to serve on those multi-purpose councils.

WLGS women were in despair. At a stroke, women would be wiped off the educational map. The WLGS had few friends still in the House (many had lost their seats in the 1900 general election), they were low on funds, and they could not rally Progressive opinion. The LCC, long their staunchest ally, had everything to gain from the bill as they would become London's education authority. Women's Liberalism was split down the middle with most members in favour of the bill. Lady Aberdeen and Lady Frederick Cavendish resigned from the WLGS; Mrs Fawcett publicly reprimanded the WLGS for putting women's claims above children's needs; drawing-rooms were closed, donations fell off, letters criticizing their hostility to the bill flooded the Executive. The Committee doggedly and wretchedly reaffirmed their view, that until women could serve as borough and county councillors, the WLGS was opposed to those bodies becoming educational authorities. The claims of women transcended mere party considerations. The WLGS had reached the nadir of their influence. Only the money and rock like determination of Miss Browne, her mother, sister, and friend, prevented the WLGS from dissolving itself.[44]

When it became clear that the women's movement was

---

[43] See ch. 2.
[44] WLGS Minutes, 18 June 1901.

unwilling to defend school boards and unable to have women elected as councillors to the new local education authorities (LEAs), the WLGS scratched round for some crumbs to salvage. They rallied the outgoing school boards and persuaded the incoming county councils to testify to the value of women's educational work. As a result Balfour offered women the right to be co-opted direct on to the education committees.[45] The WLGS sadly decided that this was better than nothing, had the right made a duty on LEAs and then worked hard to raise women's minimal presence on the new bodies. Such changes did mean that women acquired seats for the first time on the more rural and reactionary authorities, but as co–opted members they had lost the authority of elected members and lost the ability to fight for increased resources at finance committees or to shepherd their proposals through full council.

For the WLGS was well aware that co-option let even friendly men off the hook. All the experience of the last thirty years showed that if men could benefit from women's aptitude, expertise, and hard work without actually giving them power and authority, they would do so. At every point, the WLGS had a proud record of resisting co-option in place of election. They had refused to accept the LCC suggestion that women should serve as 'visitors' to LCC committees. They had warned local women's groups not to promote ladies visiting committees as they would pre-empt women poor law guardians. They had forcibly attacked their Fabian friends when, during the London borough debates, the Fabians had suggested ladies sanitary committees to aid borough work—they deplored this 'advocacy of unofficial work as a substitute for that of women members of vestries'. The Fabians, who should have known better, were suitably chastened.[46]

To make matters worse, it looked distinctly possible that poor law work would go the same way, and that women would be wiped off that map too in the urban areas. The WLGS started to mobilize its scant resources to protest against that possibility, but in the event, for reasons unconnected with the women's movement and much connected with urban politics,

<hr />

[45] Strachey papers, WLGS notes 12 May 1902.
[46] WLGS Minutes, 23 Mar. 1900.

that threat receded, and poor law boards remained in the towns an independent authority until 1929.

From 1899, women seemed to have suffered a series of defeats. They had lost their London vestry work. They had lost their rights, though not their labour, in the field of education. One further anomaly was proving more and more distressing. The 1894 Parishes and District Councils Act had permitted women to be councillors on all three authorities, parishes, urban, and rural districts. UDCs larger than 20,000 population in size had the right to be educational districts, acting with delegated powers on behalf of the county council, as well as possessing the more usual sanitary and highways powers. But as urban districts grew in population, they petitioned to become borough councils, with further statutory responsibilities and the dignity of a mayor. Often they would seek to take in an adjacent parish or RDC as well. Whenever they were successful, women found themselves disadvantaged twice over. Married women lost their ratepayer vote; all women lost their seats. When Colne and Todmorden were incorporated in 1894, thirty married women voters promptly lost their votes. When Ealing UDC became a borough in 1902, women who had been standing as candidates, had to step down. Miss Margaret Ashton, a leading figure in Manchester's suffrage circles, was a Withington urban district councillor and chaired its education committee. When Withington was incorporated into Manchester, she lost her seat and her work, though in her case the WLGS was able to persuade Manchester to co-opt her on to its education committee. There were abundant constitutional precedents for protecting existing voting rights for life (in 1832 for example), but the LGB saw no reason to protect women's rights whenever a UDC became a town council.[47] At the 1904 NUWW conference, Mrs Booth pointed out that it was now 'more difficult for women to obtain responsible positions on public bodies than was the case five or six years ago, and that it was probable that in the near future many women might suddenly find themselves outside their work'. Miss Kilgour agreed. 'The possibilities of women in public life . . . are

[47] WLGS Minutes, 15 May, 8 Oct. 1897; WLGS 5th Annual Report, 1898. For Withington, WLGS Minutes, 4 Oct. 1904.

narrowing, narrowing almost every year and latterly every year. It behoves us clearly to realize what is left to us, in order that we may use it'.[48]

But in 1904, in the *post mortem* of the education debates, the tide started to turn. Mr. Channing, who had chaired the very first WLGS meeting fifteen years before, had in September 1903, tried to introduce a bill to permit women to become borough and county councillors before the old LSB was finally dissolved. In July 1904 the Lords debated a similar bill; and the women's network of temperance, moral purity, NUWW, Liberal, suffrage, and local government branches picked themselves up and stirred themselves into activity again. Then came the first genuine piece of luck they had had for many years. Dr Shipman won the private members' ballot and in March 1905 he introduced the WLGS bill permitting women to become borough and county councillors into the House. The WLGS were exuberant. At last they would get their Commons debate and, with work, their Commons majority. The WLGS Executive speedily met friendly MPs to plan their strategy. Petitions of support came in; twenty-five borough councils and the LCC asked that they might be granted women members; and Shipman's bill won an overwhelming majority in a Tory House of 171 votes to 21. The measure failed to make it through committee stage as the general election approached, but the straw poll had made its point.[49] Everything now hung on the general election.

The Liberals won their landslide, and the WLGS sent off delighted notes of congratulation to friends who, like Burns, might find themselves in high places. Dr Shipman now sought to introduce his bill again but it was talked out when Tory MPs suddenly found extremely contentious matter in the preceding bill which controlled the importation of Canadian cattle. The WLGS turned formally to the new government for help, to Campbell-Bannerman, a supporter of their cause, and to John Burns, president of the LGB. Questions were asked in the House, deputations mounted, annual conferences of Liberal, trade-union, and women's organizations passed resolutions of support. The WLGS was making one last

[48] NUWW Annual Conference, York, 1904.
[49] WLGS 13th Annual Report, 1906.

supreme effort. Susan Lawrence, a co-opted LCC education member, organized letters signed by almost every woman on urban education committees. Miss Browne wrote to Ramsay Macdonald asking for Labour support, asserting, quite accurately, that London vestrywomen, like London school board women, 'were for the most part the choice of the electors in the very poorest wards'.[50] By January 1907 they had the Labour Party's support, as well as that of a hundred trades councils. Fifty town councils now petitioned the government, ninety public meetings were held within a few weeks. The WLGS had their reward. Their bill was indeed included in the King's Speech and by March 1907 was on the floor of the House.

John Burns reminded MPs that over two thousand women now served on parish and district councils and on poor law boards; 615 had been co-opted to education committees, many were serving on Distress Committees, some acted as governors of polytechnics. His bill would allow women to join the 320 town councils, 29 London boroughs, and 62 county councils from which they were presently excluded. For good measure, and with a defiant gesture at the Lords, he would have women become mayors as well.

Despite mutterings about the suffragettes, the bill went safely through the Commons. The Lords toyed with the thought that women should only be co-opted as aldermen rather than be elected as councillors. They thought about sending the whole measure to a special inquiry. Finally they passed the bill, adding only that women should not be chairmen or mayors of councils. The Commons took pleasure in overturning their amendment, and the bill became law in August 1907. The question of married women, their right both to vote and to sit, was still left undefined, but Burns had warned the WLGS not to insist on additional amendments as this might lose the whole bill. In practice, and with the help of friendly returning officers, married women like Mrs Pinsent of Birmingham were to stand and succeed.[51]

The WLGS took a deep breath and then swiftly moved into

---

[50] Labour party /GC/1/346/ 7 Feb. 1906, Labour party papers, Walworth Rd.
[51] WLGS 14th Annual Report, 1907, 15th Annual Report 1908. Married women ratepayers did not get full rights until 1914.

action again, seeking candidates for the forthcoming November local elections. Seventeen women stood, six were elected.[52] Women had breached the last citadel of local government. The Society presented an illuminated Address in gratitude to Miss Leigh Browne, and held a splendid, and rather expensive, eight-course dinner at the Trocadero, to celebrate. Emily Davies, who had sat on the first London school board in 1870, and Emma Cons, who had served on the first London county council were there to hear the newly elected women councillors talk of their work and to toast each other.

Without any doubt, the 1907 Act was a triumph for the WLGS. It was their bill; and with it, all elected local government office was now open to women. Though there were still a few loose ends, in particular the rights of propertied married women, none the less the parliamentary work of the WLGS had largely come to an end. The *Englishwomen's Review* noted in 1909, 'In looking back over the past twenty years, one cannot refrain from a word of admiration of the pioneers who formed and carried on the Women's Local Government Society, and who had undauntedly kept before the public the desideratum of a two-sexed local government, until, mainly through their efforts, legislation in this direction is being achieved'.[53] From now on, WLGS work was to move out into the localities, to find women willing to stand for office, to place them in winnable seats, and to provide them with support, workers, and voters.

The annual round of setting up private members' bills, deputations to ministers, the lobbying of individual MPs, the rallying of the women's network to build backbench support, all this had represented a heavy work load carried by a handful of women. But the WLGS had always known it needed to fight on two fronts—the parliamentary, to seek a change in the law, the legal in order to protect and if possible advance the existing rights of women as voters, as candidates, and also as employees. In an unwritten constitution, parliamentary statute would always be understood in the light of common law; and common law was a coral reef of high-

[52] See below.
[53] *EWR* 15 Jan. 1909.

court judgments, invariably conservative and therefore hostile
to women, and especially to married women, whose legal
rights were held to be subsumed in those of their husbands.
Annie Leigh Browne had a healthy respect for the power of
the courts and never shirked buying the best possible legal
advice for even the most marginal of cases. Every step
forward, every achievement gained, every statute amended,
had to be defended against the crass judgements of hostile
local government lawyers and returning officers responsible
for drawing up the local government register. In Exeter, for
example, women were illegally prevented from becoming poor
law guardians on a mistaken reading of the LCC Sandhurst
judgement. The WLGS had them reinstated. Nine married
women in Derby were struck off the local government register
in 1895 by the local revising barrister. The WLGS solicitor
worked with the local Liberal agent to have them put back on
again. In the Salford school board elections of 1897, the local
clergy put out leaflets asserting that women were not allowed
to stand—this at a time when women were becoming chairmen of other boards. The WLGS disabused them of their
notions. Miss Eve in 1895 was appointed by her colleagues on
the LSB to serve as an almoner (governor) of Christ's
Hospital School. Christ's resisted her appointment, and it was
left to the WLGS to take the case through to the Charity
Commissioners and the law courts before Christ's would
accept her.[54]

Legal decisions also touched the rights of women to work for
local government, and here the LGB was much less helpful.
When the government was appointing further inspectors of
infant life, and of industrial schools in 1896, the WLGS had
the words 'male or female' added to the job description. In
Oswestry, Mrs Price had been the *de facto* relieving officer on
behalf of her husband during his long illness. On his death,
and after a competitive interview, she was appointed to his
post. The LGB demurred on the grounds that she would have
to escort lunatics to the asylum, for which she was obviously
unsuitable. The guardians, local clergy, doctors, and magistrates insisted that she was eminently suitable, that she did it

[54] WLGS Minutes, 27 Apr. 1893 (Exeter), 14 Oct. 1895 (Derby), 13 Dec. 1897
(Salford).

all the time, and confirmed her appointment. The LGB threatened them with surcharge, Miss Leigh Browne was brought in, she turned to Mr Haldane QC, and the LGB finally acquiesced.[55] Mrs Sampson, a lady residing with her husband in Hampstead and paying the rates, lost her vestry vote when the rating officer would only accept rate payments from her husband. The WLGS successfully argued her case before the revising barrister and she got her vote back. Similarly, the revising barrister in Cheltenham took one hundred and ten married women off the electoral roll, but together the WLGS and Gloucestershire county council had them reinstated.[56]

Every executive committee of the WLGS spent time discussing their legal Opinions and drafting leaflets to give legal advice about election law. The Society became a centre for expert, speedy, and detailed legal advice on all issues affecting women's local government franchise and office, a role crucially important when so much was left to the discretion and deliberations of local revising barristers and local returning officers as to whose names would be accepted on to the voting register and whose rejected, whose nomination papers for office were valid and whose were not. Every case that the WLGS took up in England after 1889 they won, and each time the result was circulated around England, both to clarify the law and to consolidate their gains.

The WLGS had given itself two key tasks; to put pressure on parliament to remove women's local government disabilities; and to monitor the legal system to ensure that no new disabilities were surreptitiously imposed. A third task developed out of these two, that of arousing and mobilizing public opinion so that it would recognize and respect women's work in local government. In their twenty-five years of work before the First World War, the Society published and distributed several hundreds of thousands of leaflets, endless letters to the press and articles in the *Reviews*, placed friendly puffs of meetings, achievements, and incidents, in progressive papers,

[55] WLGS Minutes, 15 Mar. 1896, WLGS 5th Annual Report, 1898.
[56] Mrs Sampson, WLGS Minutes, 29 Sept. 1899, 7th Annual Report, 1900; Cheltenham, 24 Nov. 1899. They drew the line at taking on the Church of England, as some members wished, to gain access to church parish councils, 6th Annual Report, 1899. The next issue might then have been women priests.

wrote columns in the local government journals recording the work of elected ladies on school boards, poor law boards, and parish and district councils. Miss Browne, for example, circularized all women parish councillors in 1895 for their experiences, questions, comment and criticisms, to form the basis of articles in the *Parish Councillor*, as that will 'bring under the eyes of their male colleagues and the public the work women are doing and their point of view as to many questions, and should facilitate the future election of women as councillors . . .'.[57]

Their executive travelled up and down the country. In a typical three-month period the handful of WLGS committee members lectured to nurses, to the Women's Co-operative Guild, to women journalists, to women's institutes, to teachers, and to local Liberal branches. At every conference of a friendly organization, they mounted their stall and moved their motions. They cultivated the goodwill of town and county councillors as the petitions of support they received, showed; gave evidence of impressive quality to Royal Commissions and Select Committees; and in their own home patches intervened in local elections, often becoming candidates themselves rather than let the occasion go by without a woman in evidence, and in public. They even, belatedly, won the support of the Primrose League. A wry comment coming from Mrs Fawcett in May 1914 on a letter from Mrs Humphrey Ward throws light on both ladies:

> Mrs Humphrey Ward's proposal was that we women's suffrage associations should abandon our main object and purpose in return for her giving her personal support for women for local parliaments. It is as if she proposed that the Church of England should abandon Christianity in exchange for her withdrawing *Robert Elsmere* from circulation.[58]

London Vestrywomen

When in 1899 London's vestries were being reformed as London boroughs and losing the services of women in the process, Leonard Courtney told the House that the year before, it was suggested that 'a certain lady on the south side

---

[57] Letter to Miss Browne, 22 May 1895, WLGS Papers, vol. xxx.
[58] Quoted in R. Strachey, *Millicent Garrett Fawcett*, 1931, p. 271.

VESTRIES AND DISTRICT BOARDS: Boundries in 1894.

1. Chelsea detached
2. Westminster
3. St George Hanover Square
4. St Martin in the Fields
5. St James Westminster
6. Strand Board (six districts)
7. St Giles' Board (two districts)
8. Holborn Board (five districts)
9. Clerkenwell
10. Stoke Newington
11. St Luke
12. Whitechapel Board (nine districts)
13. St George's in the East
14. Limehouse Board (Limehouse, Ratcliffe Wapping and Shadwell)
15. Bow
16. Bromley
17. Poplar
18. Wandsworth detached
19. Putney
20. Wandsworth
21. Tooting
22. Streatham
23. Clapham
24. Lewisham
25. Penge (Lewisham Board)
26. Southwark St George
27. St Saviours Board (two districts)
28. St Olave's Board (three districts)
29. Deptford St Paul
30. Deptford St Nicholas
31. Greenwich
32. Charlton
33. Kidbrook
34. Lee
35. Eltham

2. London, its vestries and district boards

of London should stand for the Board of Guardians, but she felt that she could not discharge the duties of both offices. Thereupon a meeting of workingmen was held, and it was decided that the work of the vestry was more important than that of the guardians, and they could not allow this lady to give up her place on the vestry'.[59] Guardians dealt only with the indigent poor, school boards with the children of the poor, but vestry responsibilities for local sanitation, for the built environment of the city, touched the daily life of every one.

By the 1890s some of London's worst slums had been cleared and its main streets sewered. But the large-scale railway developments of the 1860s, followed by the slum clearance empowered by the 1875 Cross Act, together with demolitions for new thoroughfares, board schools, warehouses, and the like, had seen the housing stock shrink, rents soar, and overcrowding intensify. The 1880s had witnessed a housing crisis in London, and a crisis of social conscience, a collective class guilt, (thought Beatrice Webb) to accompany it. The 1891 census showed that a fifth of the population were overcrowded; in the poorer East End, the figure rose to 40 per cent; in Whitechapel in 1901 it was nearer 50 per cent. And it was a tolerant definition of overcrowding: more than two adults to a room. As children under ten counted for half, and babies not at all, a three-roomed tenement could contain four adults, four children and any number of babies without being statutorily overcrowded. The result, Victorians feared, was moral pollution. As Sir John Simon, the government's chief medical officer said, 'It almost necessarily involves such negation of all delicacy, such unclean confusion of bodies and bodily functions, such mutual exposure of animal and sexual nakedness, as is rather bestial than human'.[60] Physical pollution was equally obvious and offensive. The City was still not removing its refuse efficiently. Cesspools still festered in subterranean swamps. Privies still overflowed into courts where barefoot children paddled. Manure heaps grew until worth removing. Horse dung

[59] L. Courtney, Hansard, House of Commons, 6 July 1899, col. 48.
[60] As quoted by J. Burnett, *A Social History of Housing*, 1978, ch. 6; G. Stedman Jones, *Outcast London*, 1971, pt. 2 'Housing and the Casual Poor'; B. Webb, *My Apprenticeship*, ch. 4, 'The Field of Controversy'.

smeared the streets with sludge and hosted the horse-flies that carried disease. Pigsties, cowsheds, and corner abattoirs slaughtering 13,000 animals a week in London in 1892, still discharged their blood and waste into neighbouring yards and alleys. Water remained polluted with sewage, corpses, and industrial waste; food, apart from that of the Co-op., was contaminated, adulterated, or both. In Finsbury in 1903 a third of the milk tested contained pus, nearly half contained dirt. The LCC's Medical Office of Health sampled ice-cream in 1898 which had in it 'cotton fibre, lice, bugs, fleas, straw, and human, cat and dog hair'. Domestic fires and chimney fumes wreathed London in smog. The stench, noise, dirt, and dung overpowered the senses; disease and dysentery overtook the young. One infant in two still died before the age of five in St. George's Southwark in the 1890s, figures as bad as any town at any time for which records were kept. Morbidity and mortality rates stubbornly refused to fall.[61]

These were all matters for the vestry. Sewage and refuse removal, cemeteries, street works, cleansing, lighting, and paving, the removal of nuisances, the abatement of overcrowding, and the inspection of workshops, were the minimal statutory duties imposed on vestries. Whether they did more than that, or even did that efficiently, was up to them.

By 1894 some of London's vestries were larger than most county boroughs. Paddington had a population of 118,000, St Pancras 236,000, Lambeth 253,000, and Islington a population over 320,000. The more progressive and ambitious vestries such as Chelsea, Lambeth, Islington, Kensington, and Camberwell, which prided themselves on their civic administration, or Battersea and Bermondsey which had a strong working-class presence, had by 1894 extended their functions and adopted permissive legislation well beyond the statutory minimum. They were building baths and washhouses; providing public libraries; finding winter unemployment relief works; enforcing the Acts against adulteration, encouraging the LCC to build cottages to replace those cleared as slums. They were employing considerable

---

[61] This paragraph draws heavily on A. Wohl, *Endangered Lives*, 1983, pp. 22, 60; see also F. B. Smith, *The People's Health*, 1979.

staff, meeting weekly or fortnightly, had full committee structures and often their own direct labour organization. With up to 120 members each, such vestries were indeed surrogate town councils.

Other vestries were more sleepy. St Lukes was controlled by the Moderates, St George's in the East by its rector, and both had to be forced by the LCC into cleaning and lighting their streets. Rotherhythe, which employed no full-time staff at all, was found by the LCC to have jerry-built its sewers so that they all drained the wrong way. Ninety per cent of its property was known to be in need of major repair. Westminster, despite or because of its wealth, appeared largely indifferent to its slums. The tone of such vestries was summed up by the comment of a vestryman to a newly appointed MOH in 1872, 'Now doctor, I wish you to understand that the less you do, the better we shall like you'.[62]

More offensive still were those vestries known to be corrupt; whose membership was largely drawn from shopkeepers, slum landlords, coal merchants, undertakers, grocers, publicans, and small builders, who placed contracts with each other and their friends. Mile End was known for its civic feasting, Hackney, Fulham, and St Saviours for their secrecy and jobbery. In Clerkenwell the worst rack-landlords not only sat on the vestry, but also on the sanitary committee where they felt free to traffic in unsavoury property. In St George's Southwark, when Progressive members argued for a direct labour organization to perform the vestry's building work, local contractors packed the public gallery, heckled, issued writs, and came to blows.[63]

Political will, parsimony, or local pride, meant that some vestries were models of local administration with affinities to Chamberlain's Birmingham, while others were almost as stagnant as the old unreformed municipal corporations of pre-1835 days. Polls were low—often under 10 per cent—and vestrymen attendances were poor, with under a third attending regularly. The Parishes and District Councils Act of 1894 had left the structure and functions of London's vestries

[62] Wohl, op. cit., p. 187; *London*, 22 Nov., 6 Dec. 1894; J. Roebuck, *Urban Development in 19th Century London*, 1979, p. 155.

[63] *London*, 22 Nov., 6 Dec. 1894.

untouched; but the Act had introduced the secret ballot and the Corrupt Practices Act into London local government; had abolished the property qualification and weighted property vote for guardians and vestries alike; and permitted women residents to stand for office. The *London and Municipal Journal* thought that of all these developments, 'the most hopeful sign for the regeneration of London' was the arrival of women. 'Women will prove a new municipal type: pacifying, stimulating, and elevating. The mere presence of a woman on the London vestries will purge them of those elements of rowdyism, of vulgarity, and of personalities which too often characterize their meetings.' Party bickering would diminish, the moral tone would improve, and member involvement in administration would rise.[64]

Would women be willing to stand? As *London* noted, 'In some aspects the work is perhaps less agreeable, and appeals less to their sympathies, than poor law work'. One of the problems, said Alice Busk, a Southwark vestrywoman, was that people widely believed 'that Guardians dealt with persons, and that vestries and Councils dealt only with things, and that the details of official work was wearisome, because the results were not commensurate with the labour involved'.[65]

Most women candidates were to stand as Progressives. The Progressives, brought together in the London Reform Union, offered a common manifesto across London with four main planks. The first was to end jobbery and secrecy. Vestries should be open and accountable, meet of an evening so that working men could attend, their accounts should be audited, their reports printed. Next, Progressives demanded stronger sanitary action in London. Refuse needed to be removed more efficiently and not left to mount up until men could expect a generous tip for removing it. The streets needed watering, cleansing, and paving with macadam and not wood so they could be sluiced. Vestries should provide baths and washhouses, so that families living in one or two rooms without running water, paying 3*d.* for nine buckets of it, might

[64] *London*, 29 Nov. 1894.
[65] *London*, 15 Nov. 1894; Miss A. Busk, 'Women's Work on Vestries and Councils' in Revd J. Hand, ed., *Good Citizenship*, 1899.

wash themselves and their clothing. Public lavatories were needed for both men and women. At present, men relieved themselves in doorways, staircases, and the streets, and women away from home not at all. The third plank of the Progressive manifesto, shared with the LSB and the LCC, were fair labour policies, trade-union rates (24*s.* for a 48-hour week), an in-house direct labour organization to end corrupt contracting, and winter public relief works for the unemployed. Finally, Progressives argued that London's problems interlocked: that therefore the LCC should run all utilities, trams, water, markets. Empty properties should be rated, and all rate burdens equalized, so that richer parishes helped poorer parishes finance the streets and sewers that were London's arteries and the green spaces that were London's lungs.[66] (Victorians were rather keen on the body politic image.) The Moderates tried less hard with their manifesto, asking voters simply to support business men and economy.

Special literature was put out by the Progressives targeted at the women's vote. Beatrice Webb headed hers, 'From Westminster Women to the Women of Westminster'. Dirt and disease was every woman's business, it said. For many years, local ladies sanitary associations had sent out leaflets and itinerant lady lecturers to teach child care and the virtues of carbolic. Women's sphere was customarily to make healthy homes, men's to build healthy streets. But, Progressives argued, reports from district nurses, medical officers, and sanitary inspectors showed that family homes were no fortress against water borne disease or fly-infested food. Private health still depended on public health, healthy homes on healthy cities. Women could not 'stand idly by and see the slaughter of the infants', as small babies fell victim every summer to the dirt disease of gastro-enteritis. Women must insist on soundly built homes, free of damp; fresh air, efficient sewers, pure water, and clean streets. Only then would the moral and physical health of the cities improve. Only then would the race grow strong, walk tall, and become fighting fit.

Men alone, they argued, could not decently supervise

---

[66] London Reform Union, *Women's Work in London under the Local Government Act,* 1894; *London,* 13 Dec. 1894, 7 May 1896.

women's lavatories, inspect the sanitary arrangements of sweated workshops, control the common lodging-houses that were all too commonly brothels, supervise women's laundries and girls' swimming-baths, and ensure that parochial charities distributing coal and cheer to the deserving poor did not think that only men were cold. 'Men and women are different, and every variety of power ought to be placed at the service of the poor'.[67] Henrietta Barnett of Whitechapel urged the affluent members of the NUWW to use their local government vote, if not for themselves, then for 'the ugly powerless women in mean back streets whose lives were all harder because she had not used her vote. Let the ladies who are now indifferent to their municipal privileges confer with half a dozen decent working mothers, and they will learn how large a bearing a well-lit street has on the upbringing of a family in a large town.'[68]

Responding to her call, some thirty women, mostly under the Progressive banner, stood for the vestry elections at the end of 1894. Some were still digesting Booth's findings, others were moved by the *Bitter Cry of Outcast London*, others had had their consciousness heightened by the settlement movement. Kensington selected five names from among the twenty-five submitted by the women's associations, though here in a Tory vestry they prudently stood as Independents. Under the experienced chairmanship of Jane Cobden-Unwin, and with the support of the McClaren clan, all polled well but only Mrs Willis was elected.[69] In Paddington stood Annie Leigh Browne herself, Mrs Charles, the tory-radical poor law guardian who had tried to stand for the vestry in 1887, and Miss Hodges, secretary and accountant to an educational institute. Mary Kilgour acted as their election agent. Mrs Charles alone won a seat. The St Pancras list produced by the London Reform Union included two women, of whom Mrs Philimore, the Fabian researcher and friend of Beatrice Webb, was successful. In Camberwell two women guardians, Miss Augusta

[67] WLGS pamphlet, *Why Women are Wanted on Vestries*, 1894. For women's earlier sanitary work, see *Englishwomen's Journal*, 1 Apr. 1859, *Fourth Report of the Ladies Sanitary Association*, 1861.

[68] Mrs Barnett, NUWW Conference, Manchester, 1896.

[69] For the Kensington list see *WS* 29 No. 1894; it included Chadwick's daughter, and women known for philanthropic and school management work.

Brown (she of the inedible soup) and Miss Evans, both stood for the vestry. As they favoured generous outdoor relief, state pensions, and winter public works, they were unusually popular for women guardians, and were swept to success. Marylebone returned two women, the WLGS Mrs Sheldon Amos, and the Tory Miss Beatrice Willoughby who once on the vestry insisted on voting with the Progressives to the dismay of her party. Down in Southwark, Miss Alice Busk, a barrister's daughter and housing worker, was elected along with Miss Elizabeth Kenney, a trained nurse. Three women won minor vestry seats [70] in Lewisham, two in Bromley which was part of the Poplar board, and one each in Islington and Fulham: other women stood without success in Hammersmith, Hampstead, and Plumstead. One of the liveliest campaigns took place in Westminster where Beatrice Webb was a candidate.

Dec. 1st, 1894. Galton's little study turned into the central office of the Progressive candidates for the Westminster vestry elections. Certainly we have created our organization and selected our candidates with singularly little trouble. The first stage was to create a branch of the London Reform Union—Sidney chairman, Galton secretary; the second to call in the name of the LRU branch, a conference of all the temperance, trade union and political organizations and to form a Progressive Council—Sidney chairman, Galton secretary; the final step to select our candidates and to form these into one organization; Sidney chairman, Galton secretary. These three organizations, under their respective chairmen and secretaries, have worked with wonderful harmony; and have between them, at a cost of £30, initiated a really vigorous campaign. The Westminster radicals, a poor down-trodden lot, perpetually licked at all elections, hardly know themselves with their ninety candidates . . .

Dec. 1894. Crushing defeat at Westminster vestry elections; only five Progressives out of 96 . . . The slums of Westminster are as completely Tory as the palaces. I do not think there has been any lack of energy or even of skill in engineering such forces as we had. But it is obvious that our attempt to collar a constituency with three weeks' work—mostly amateur—was a fiasco which we ought to have expected. Against us we had a perfect organization with a permanent staff, a local paper and unlimited money, we had all the

---

[70] The minor vestries selected members to serve on the joint district board as well as on other bodies.

wealthy residents, nearly all the employers of labour, and the whole liquor interest . . . We had a register from which every known Liberal had been knocked off, year by year, without a protest from the feeble, flickering little Liberal association. Behind all this we had Burdett-Coutts' charities and churches. And to fight these potent powers Galton stood single handed with a mob of working-men and small tradesmen candidates . . .

and she might have added, three radical women.[71] Altogether, some thirteen women were elected on the London vestries in 1894, mostly under the Progressive label and as part of a swing to the left which brought many of the large vestries under Progressive control. London's women voters, however, were thought to have polled lightly, only five thousand of London's one hundred thousand enfranchised women coming to the polls. The women's press, foreseeing further reform of London government, was jubilant at the results:

The London vestry will be the bridge by which women will pass into the Town Council, and from thence into the County Council. If a good contingent of competent women are well established in London vestries, it will be impossible to turn them out when those vestries are converted into District councils or Local municipalities. And when women sit on London municipal bodies, it will be impossible to refuse the claim of their sisters to similar positions of privilege elsewhere.[72]

London's vestrywomen settled into their work. Mrs Evans, elected for St Martin-in-the-Fields, was one of the best known. Proprietor of a photographic shop and a poor law guardian, she had for ten years patiently (or more accurately, impatiently) attended vestry meetings, often the only observer, following a rating wrangle over her property when she found the vestry assessment committee 'very unsympathetic'. She had decided she would 'sit in that Board room among the men who refused to listen to my appeal', and make a visual statement about her rights. By 1894 she had become so much part of the furnishings, that the vestrymen simply assumed that she would now cross over and join them, and offered her the choice of any ward she wanted. Having over

---

[71] B. Webb, *Our Partnership*, 1948, pp. 66–8.
[72] *WS* 29 Nov. 1894. By 1899 there were 17 vestrywomen.

ten years acquired a somewhat better grasp of vestry affairs than most vestry members, she soon found herself on all vestry committees bar one, as well as serving on charitable trusts and the sick asylum as St Martin's nominee. 'For each of her new posts', *London* commented approvingly, 'she puts herself through an exacting apprenticeship. The elections, one after another, are simply testimonials to her proved capacity'.[73]

St Martin's was a small vestry with 13,000 population, it was central, and its problems were less those of public health (its infant mortality rate of 145 p.a. was better than the London average of 161 in 1894), than those of roads, traffic, and works associated with the Trafalgar Square/Charing Cross area. In 1895 it was employing forty-nine staff and spending £100,000 a year.

At a Women's Emancipation Union conference in 1896, Mrs Evans told other women what her work entailed (and one suspects may have frightened them off rather than brought them forward).[74] She served on the public health committee, inspecting factories, workshops, private homes, stables, and slaughter yards. The lighting committee repaired and lit the parishes six hundred gas lamps and, as they cost £10 each a year to run, decisions about their siting were an expensive responsibility. Her third and most important committee was works and general purposes, which determined the planning implications of road works, gas and water pipes, new sewers, electric and telephone wires. As the London Telegraph Company was seeking to construct a London-wide telephone network at the time, this meant digging up every single yard of pavement. Each new utility meant that the streets were dug up yet again. This committee also erected public seats and lavatories—Mrs Evans had persuaded them to build women's WCs at Charing Cross, a memorial of convenience, she thought. When she joined the finance committee she found them meeting only quarterly. 'What could they do but sign checks prepared by the officials? Now they meet once a fortnight. I said we were there to represent the ratepayers, and we had a right to know how the money was spent. The

---

[73] 'Women in Local Administration', *Westminster Review*, July 1898, p. 387, *London*, 22 Apr. 1894.

[74] 'Women in Local Administration', pp. 387–8; *London*, 22 Apr. 1897.

accounts now all pass through our hands before each vestry meeting.' In much the same way she had reported the Metropolitan Asylums Board to the LGB for its faulty financial procedures; the LGB made them change their ways. For a short time she sat on the Housing of the Working Classes Committee, but as 'our vestry does not house any working classes', this was 'a very easy task, the working classes having for the most part been removed, to make way for flats and what is called a superior class of property'. The committee did once prepare a housing scheme but the vestry turned it down. 'This didn't suit me, so I resigned. I didn't go on the vestry to play about.' She instead joined the improvement committee which widened streets and pavements. She also sat on the parliamentary committee which scrutinized parliamentary bills which had implications for the vestry, such as underground railway bills, the sale of food and drugs, officers' superannuation, and the like. She drew the line at sitting on the town hall committee which determined lettings; and instead used her spare time on the neighbouring Strand board of works to establish a local labour exchange which she thought 'has done a great deal of good'. In 1896 she got the vestry to give her a baths and washhouses committee. This was no sinecure. Within a couple of years 53,000 people were attending the baths, and 28,000 the washhouses.[75]

The annual vestry reports show that in 1898–9 Mrs Evans put in 155 of 162 possible attendances—double that of the chairman and vice-chairman—on top of her poor law work. She was sufficiently senior to accompany the chairman on a deputation to Balfour to protest about the boundaries of the new Westminster borough into which the vestry would be incorporated in 1899. As she somewhat drily told the women's conference, 'So far as I can see, there is nothing whatever in the work and duties of a metropolitan vestry that a woman cannot master if she chooses to give her mind to it. Of this I am perfectly certain, that the presence of a woman tends to make the men attend more regularly than they otherwise would'.[76] To universal dismay she was unseated in 1899 when

[75] St Martin-in-the-Fields vestry, Annual Reports, 1894–9.
[76] 'Women in Local Administration'; *London*, 22 Apr. 1897; Balfour delegation, St Martin's vestry Annual Report, 1898–9, p. 18.

she foolishly tipped the driver of a carriage lent her for election day. She was therefore found guilty of bribery under the Corrupt Practices Act on a petition brought by her opponent. In Marylebone, the Radical Mrs Amos and the Moderate Miss Willoughby worked closely together. Marylebone itself was a mixed parish, part of Elizabeth Garrett's old school board division; and included the well-to-do areas around Regent's Park as well as the slums around Oxford Street. Moderate in politics, it was held to be well administered, having spent money on drains and sewers to reduce its mortality figures. Miss Willoughby, while carefully insisting that she was no advocate of women's rights, believed that a woman 'is as much a citizen as a man'. She was particularly proud of her ability with accounts. By 1897 she was an effective member of the vestry's finance committee, as well as doing good works on baths and washhouses. 'At first, I am sure the opinion of the vestry was against women', she reflected. 'But they have changed all that on better acquaintance.' The only difficulty came with the sanitary committee. 'Mrs Sheldon Amos was put on that, but we agreed we had better not serve on that committee unless there were two of us,' as the work was distasteful and indelicate. Another Moderate, Miss Nesta Carew, subsequently joined them, and went on the works committee, as 'her special hobby is clean roads'. She fought against salting them—cruel to the horses, and maybe even to children—and waged war on the great railway companies whose extensive terminals would 'damage the whole neighbourhood'.[77]

Mrs Willis was the only Kensington woman to be elected, and wondered whether she should not resign rather than fight as a Progressive on a Tory vestry known for its parsimonious policies and harsh attitudes towards the poor. Inexperienced in public work, and steadily outvoted, she valued her vestry seat for the edge it gave her when it came to other public work. The books she recommended to the library were bought, the sanitary problems she mentioned to the MOH were seen to, the deserving poor she recommended to local charities were helped. As she confessed, somewhat naïvely, she 'got a

[77] *London*, 1 Apr. 1897.

hearing, as a member of the vestry, which I am sure they would not have given to a private person'.[78] She was probably right.

Mrs Philimore had joined the St Pancras vestry, which was dominated in equal measure by its populist radical traditions stretching back to the 1820s and by its three railway stations, gas works, and warehouses. The result was a high rate of both pauperism and mortality, until the LGB insisted they bring down the first and the LCC coerced them into doing something about the second. Vestry politics, though not its administration, were strongly Progressive, and Mrs Philimore found herself on the health and parliamentary committees, while her husband made his way up the LCC. She was soon joined by Mrs Miall Smith who devoted her energies to baths and washhouses, and earned the compliments of male colleagues for her efforts.[79]

Down in Camberwell, Miss Augusta Brown and Miss Evans had been swept on to its Progressive vestry in 1894. Both were convinced that women had much to give vestry work: they knew more about domestic sanitation and housing than men, were free from jobbery, and had more time to spend on visits and committee work.[80] Mrs Bracey Wright later joined them and specialized in the state of the sewers, persuading the vestry to experiment with ventilation schemes to reduce the dampness and nasty smells that pervaded working-class homes.

Two of the most effective vestrywomen were Alice Busk and Elizabeth Kenny in St George's Southwark, whose parish of 60,000 was thought by Charles Booth to be the poorest in London. In Alice Busk's words, 'at the request of the Working Men Candidates, I stood with them for election, their idea being that, as I had been engaged for twelve years in Whitechapel and Spitalfields in the management of tenement property inhabited by the casually employed, I might be able to help them get the poor parish of St. George the Martyr into a more sanitary condition'. The vestry's damp and crumbling property lay below the Thames high-water mark. At high tide

---

[78] *London*, 8 Apr. 1897.
[79] Ibid.
[80] *London*, 22 Aug. 1895.

cellar dwellings and basement workshops floated in sewage. The drainage was 'fearsome'. Into such noisome property pressed the worst overcrowding in London—with 211 people an acre, four times the London average, and four hundred times the national average. The life of its infants ebbed with the tide. Yet the overcrowding was so severe that its devoted MOH dare not close unfit housing as 'increasing overcrowding would then result'. He was desperate to create open space and to get the LCC to build new and decent property. In the mean time, he tried to save children's lives by removing them into day crèches for the first year.[81]

The women members studied the sanitary inspectors' diaries, the Nuisance and the Mortuary Registers, the Notification and Disinfection books, and the drainage and sewer maps (the sewers were laid on the 'switchback railway' system, Alice Busk noted, and not always joined).

Together, the two women encouraged their vestry's health committee to redrain parts of the parish. Mortality rates improved. They built women's lavatories, and supervised the common lodging-houses where women were so often battered and brutalized. They collected evidence about landlords who neglected to repair their property and pressed charges against them. They sought for new powers to light and inspect the open stairways and common tenement passages where down-and-outs dossed for the night. Alice Busk had taken over one such court as housing manager back in 1888, scene of a lurid murder, and she was determined to make even the roughest and poorest housing safe for women and children. Both women served on the committee for infectious diseases, which like most vestries tried but failed to persuade the poor to have their houses fumigated. The women persuaded the committee to let them have a free hand. They carefully and attractively furnished a reception house with baths, cots and newly laundered pink and blue dressing-gowns; and encouraged the local poor to use it by throwing an Open Evening—'a Reception for the Reception House' . . . 'to popularize the building'. They invited 1,900 prospective residents, of whom around 400

---

[81] St George's Southwark vestry, *Annual Reports*, 1896–7, p. 11; 1895–6, p. 12. A. Busk, 'Women's Work on London Vestries', paper read to the AGM of the WLGS, 1899.

came to see it. The vestrywomen courteously dealt with every query and hesitation. It was a striking success; and soon a hundred families a year were passing through it while their homes were being cleaned.

Another moment of triumph came when they had a female sanitary inspector appointed in 1896. She worked wonders in abating overcrowding and persuading the landlords to undertake repairs. She had to bring only a few prosecutions, and won every one of them. Miss Busk learnt that the entire parish had only two small open spaces, both of them disused burial grounds—just half an acre for its 60,000 people. She got the vestry to add 'a real garden' and the LCC to plan open space as the old courts were condemned. Passmore Edwards offered the vestry £5,000 towards a library, and again it was left to the women to organize their ladies network with the aid of volunteers from the local women's settlements to poll the parish and secure a handsome majority for adopting the permissive Public Libraries Act. Within three months of the library opening, they had 1,600 members, 6,500 books, and 60,000 issues a year. And, Miss Busk could report, the rates had been cut.[82]

Against such a record, must be set the experience of Mrs Charles, one of the best-known women in London public life on Paddington's vestry (population 120,000), she had been a poor law guardian since 1881, defending the virtues of outdoor relief twenty years before such views found a hearing. But she joined a Tory vestry. Only her ward elected a full slate of Progressives; and the Tories allowed just a handful of them to serve on committees at any one time. She reported, 'When the committees are being arranged, the old members vote for each other, or admit only harmless members that they hoped to keep quiet. The result is that some members are on two or three committees, while others are on none at all.' She was never permitted a single committee. The electors' wishes, she complained, were checkmated; vestry members like herself were criticized for low attendance at official work. She disliked

---

[82] The reception house, St George's Annual Report, 1897–8, p. 24; Busk, 'Women's Work', *passim*.

the deference paid to officers and the close ties of friendship
that determined the placing of contracts. Whenever she asked
a question of detail about the accounts, such as the hire of
horses, she found the item subsequently and mysteriously
disappeared, to resurface at the end of the year in Miscellaneous. She could not get the vestry to clear its refuse regularly.
Garbage piled up for two months at a time, until the stench
was so severe that no one could open their windows. In
desperation, she called in the LCC who required weekly
collection. She found vestries, in comparison with poor law
boards, incompetent, grubby, and irregular. However, her
battles arose not because she was a woman but because she
was an experienced and popular Progressive whom the Tories
were determined to marginalize.[83]

At any one time there were only thirteen or fifteen women
among the three thousand or so vestrymen of London, but
every one of them had achievements to their credit, even
though they lacked the seniority, experience, solidarity, and
contacts of their male colleagues. Whatever their formal political views, they were impatient with inefficiency, distrustful of
secrecy, sensitive to their local communities, and willing to
mobilize opinion and spend the rates to add a few bricks
to building Jerusalem. Most of them took a special responsibility for women's issues, baths and washhouses, refuse collection and street lighting; and in vestries too, more than in any
previous sphere of work, they refused to respect the traditional
boundaries of what constituted women's and what men's
public work. Many of them mastered the technical matters of
sewage and sanitation, of drain-pipes and road surfaces, of
finance and law. At a time when the declaration of financial
interest and the probity of vestrymen was far less fastidious
than would be expected today, when parish property,
charities, and jobs were the freehold of the few, then as Alice
Busk said, knowledge was power. 'The solution of some
difficulty, the defeat of some unscrupulous opponent, often
depends on familiarity with the Act or byelaw dealing with the
particular case.'[84]

    [83] *London*, 1 Apr. 1897; Paddington vestry Annual Reports, 1894–1900.
    [84] Busk, 'Women's Work', p. 382.

Not only did vestrywomen care for women's interests, it was noticeable that they served the concerns of the poor. It was no accident that within vestries, as on school boards, women were returned for the poorest wards. Lady Frances Balfour wrote to *The Times* when the London boroughs bill was under discussion:

'There is no question, however, what the result of a *referendum* in all of poorer London would be. Women members of vestries standing for re-election have been returned by noticeably increased majorities, and it is solely in certain rich quarters that there is any opposition to the eligibility of women. Take Kensington . . . The Goulbourne ward, the poorest ward in all Kensington, placed two women on the vestry. The question is not devoid of a class character, with all the poor on one side and a fraction of the rich on the other.[85]

Vestrywomen had a notion of sisterhood and mothering that transcended class boundaries and had them fighting for the most down-trodden of London.

One final point. In their vestry work, more than in any previous sphere of public work, women came to appreciate the impact which the built environment, rather than any personal or moral qualities, had on family life. Comfort was about drains as well as drink, about a mended roof rather more than thrift. They learnt of the effect good local government could have on the quality and the quantity of life. Those working in 'outcast London' in particular were, like Canon and Henrietta Barnett in Whitechapel, committed to collective community action and intervention. Alice Busk quoted approvingly Joseph Chamberlain who described local government as 'offering the widest possible field for beneficient activity'. 'By it', Chamberlain had said, 'you can bring to all, those opportunities, necessaries, and luxuries which otherwise would be but for the enjoyment of the few; by good local government you can improve the condition of the people, and confer on them health, comfort, recreation and education'. In her own vestry, in a little-known part of London, Alice Busk added, where she represented poverty-stricken, uneducated, and voiceless people, she had been striving, however imperfectly, 'to reach out after this idea'. Now that

[85] *The Times*, 22 May 1900.

women, after five years of devoted service, were to be disqualified by the London Borough Act, who, she said, would
uphold this ethic of 'civic service'?[86]

[86] Busk, 'Women's Work', p. 383; cf. Revd Robert Dale's call in Birmingham, to
'the sacredness of what is called secular business'. 'The gracious words of Christ,
"Inasmuch as ye did it unto one of these my brethren, even these least, ye did it unto
me", will be addressed not only to those who with their own hands fed the hungry and
clothed the naked and cared for the sick, but to those who supported a municipal
policy which lessened the miseries of the wretched and added brightness to the lives of
the desolate. And the terrible rebuke, "inasmuch as ye did not do it unto one of these
least, ye did it not unto me", will condemn the selfishness of those who refused to
make municipal government the instrument of a policy of justice and humanity.' *The
Laws of Christ for Common Life*, 1884. See also E. P. Hennock, *Fit and Proper Persons*,
1973.

7

Rural Life and Local Government 1894–1914

The Parish and District Councils Act, 1894[1]

IN 1833 a Norfolk village would have been ruled and rated by its rector and church vestry, perhaps with the help of a resident squire; and governed at one remove by justices of the peace in petty and quarter sessions. By 1893 that same village enjoyed the attentions of poor law guardians, possibly a school board, a rural sanitary district, the county council, and in some places a highways and a burial board. Each body had different boundaries, electorates, qualifications for office, voting procedures, and rating powers. It was all parliament's fault, said the Liberal Henry Fowler, President of the Local Government Board, in the Commons, because it had created special authorities and special districts for special purposes every time it wanted something done. No one denied that the result was confusion, inefficiency, and probably extravagance.

The Third Reform Act had extended household suffrage to the counties in 1884, and made the restructuring of rural local government inevitable. When in 1888 the Tories introduced county councils, they had contemplated a tier or two of local government below the counties to tidy up the mess. There was broad agreement that parishes on the one hand, and poor law unions-cum-sanitary districts on the other would make sensible tiers. Norfolk, for example, had nearly nine hundred parishes, twenty poor law unions, and a county council for its

[1] Contemporaneous material on parish and district councils is sparse; as far as I am aware, there is no modern research at all. More background has been included in the text than would otherwise be justified because there is no secondary literature to which to refer readers. This account is drawn from the records of Norfolk's 21 RDCs (20 after 1902 when Guiltcross in the south was dissolved), its poor law boards, and the local press. Norfolk seems to have had more working men on its parish councils and more women on its district councils than elsewhere.

200,000 people.[2] But since the Swing riots of 1830 and before, a sullen surly hostility of labourer for master had simmered below the surface charm of thatch and roses, frequently flaring into cattle maiming, incendiarism, rick burning, and constant poaching, a not-so-petty lawlessness that approached at times to rural guerrilla warfare.[3] The Tories were not at all keen to revitalize parish government, give it worthwhile powers, and make it a platform for latent rural village radicalism. The tier above parishes, the poor law unions, was equally problematic. At a time of deep agricultural depression, Tories were understandably apprehensive about reforming poor law boards and making them more democratic by removing the weighted property vote and the presence of ex-officio JPs. Without these safeguards, the propertied might find their landed wealth legally confiscated as they succumbed to demands from the rural poor for more generous outdoor relief. As Balfour put it, those who hoped to get relief would be voting for it; that would mean the end of 1834 and agricultural security. They left well alone.

The Liberals naturally did not see it in quite that light. They could expect political gains as well as administrative benefits to flow from reformed rural local government. Democratic parish councils would challenge the hegemony of Tory parson and Tory squire, and would provide a robust platform from which to demand three acres, a cow, and something more. Rural protest and rural discontent might now be given constitutional outlet and Liberal form. Combined poor law–sanitary districts, would then logically sew up the rural scene.

Accordingly, Fowler introduced his Parish and District Councils bill into the Commons in March 1893. He proposed secular parish councils which would have powers to provide village amenities: allotments, footpaths and stiles, baths and washhouses, burial grounds, libraries, and street lighting. They would administer parish charities, taking them away from Anglican Church and Tory vicar, and have concurrent powers with the district council for sanitary matters. They

[2] This excludes the municipal boroughs of Norwich, Yarmouth, Thetford, and King's Lynn.
[3] See J. Dunbabbin, ed., *Rural Discontent in 19th Century Britain*, 1974; E. Hobsbawm and G. Rudé, *Captain Swing*, 1968; D. Jones, 'Rural Crime and Rural Protest' in G. Mingay, ed., *The Victorian Countryside*, 2 vols., 1981.

might levy a rate of up to 6*d.* in the pound. The council would be annually elected by show of hands, though a poll could be demanded.[4] Rural district councils would inherit the functions of the old poor law unions and the former rural sanitary districts; they would also take over from highway boards their highway functions, and from JPs their licensing powers. They became responsible for school attendance where there was no school board, although all other educational matters remained with the school managers until in 1903 education as a whole passed to the county council. The bill also established urban district councils for smaller townships which had no pretentions to borough status. These might be suburban districts around large cities who wanted to avoid paying the county highways rate and might in time become boroughs themselves; or old small market towns, such as North Walsham or East Dereham in Norfolk, subregional centres of their localities. These UDCs had responsibilities for sanitation and urban amenity similar to the London vestries, but poor law powers remained with the rural district council (RDC) in which the UDC was situated.

Fowler made it clear from the beginning that women ratepayers would be able to vote both for parish and for district councils; married women who owned property distinct from their husbands were also enfranchised. Any woman, married or single, resident or ratepayer, was eligible for election to either body, just as she was to the school boards, though she remained excluded from the county councils. As women were already serving with distinction on poor law boards, not a single voice was raised in the Commons to suggest that they should not continue to serve on their successor bodies, the new RDCs. And if parishes were electing women to represent them on the RDCs it would be an obvious nonsense to bar them from the lesser authority, the parish council itself. In any case, there were plenty of examples of women churchwardens, highway surveyors, and overseers, to suggest that parish office in the past went not with gender but with land.

At the committee stage, however, Fowler found himself

---

[4] For parishes above 300, councils were compulsory; below 300, optional, obtained by petitioning the county council.

challenged by both Left and Right. From the Left, Walter
McClaren and James Stansfeld tried to widen the local
government franchise by including single women who were
not ratepayers, but only lodgers. Tories such as Gibson
Bowles from King's Lynn wished to remove married women
from the register altogether. 'It was a bill to turn women into
men', he stated to calls of disbelief. 'Yes, absolutely', he
insisted. Only married women separated from their husbands
would have the vote,

a class who least deserved it . . . A woman that was not married was
one of the failures of her sex. If they were not going to give these
advantages to women at large, they should give them only to those
who were the successes of their sex—the married women. But the
Government proposed to give the advantages only to the failures of
the successes of their sex—the married women who had first been
married and then separated from their husbands . . . They were to
have all the feminine men and masculine women coming down to
the House in force to ask them to change the ancient institutions of
the country.

The Radical MP, Labouchere, who consistently opposed any
extension of rights to women, came up with a better
argument. Wealthy husbands would make over property to
their wives. He claimed to object to any measure which
enfranchised rich women but not poor women.[5] The government's position held, but women's rights were again raised
when the question of chairing the new councils was discussed.
Fowler proposed that women should be eligible to chair
meetings but not thereby to become ex-officio JPs as male
chairmen would. This was challenged, predictably, from both
Left and Right, but Fowler again carried the vote and after
thirty-six nights of debate in the Commons and extensive
discussion in the Lords, the bill became law in March 1894.

The WLGS began to prepare for the December elections as
best it could. It faced problems that did not exist in urban
areas. Rural parishes were isolated, rural poor law boards
were insular, seldom attending poor law conferences, making
few visits to other bodies, other boards. Only the Primrose
League had any rural network and it was indifferent to

[5] Hansard, House of Commons, 21 Nov. 1893, col. 1390; 5 June 1894, col. 979.

women's public work. The WLGS rallied women Liberals, temperance and Fabian groups, as well as the Women's Co-operative Guild, to seek out candidates and to mobilize the women's vote. Mrs Fordham, herself to become a parish councillor, told rural women reassuringly, in a well-worn metaphor,

The government of the village is but the government of the home, only on a larger scale; families make homes, homes villages, villages increase and become towns, and so on. A capable woman, who manages her home well and economically, is just as able to help in the government of her village as her husband, and quite as much needed.[6]

Mrs Abbott, a guildswoman from Tunbridge Wells, reminded WCG women that they would now be able 'to get the dark corners in our streets lit and ponds and places that are a danger to children fenced in'. They could defend village greens, protect the footpaths that their men used as short cuts to work, and save the roadside wastes that add to the beauty and breathing space around our homes'. Every parish council needed at least one woman who would know how difficult it was to keep the cottage clean when the only water was carried by buckets from a ditch or stagnant pond; how hard it was to keep the family in good health when there was no sanitation, no means of getting the refuse removed, no pure drinking-water.[7]

All of these were natural womanly concerns. Women felt that they were also called to join in a more momentous exercise, nothing less than rebuilding the villages of England. For rural society seemed to contemporary observers close to collapse. The onset of the agricultural depression in the mid-1870s had forced landlords to abate their rents and farmers to abandon their land. Many acres went out of cultivation, deso-

---

[6] Mrs E. O. Fordham, 'Why Women are Needed as Parish Councillors', *Parish Councils Journal*, 1 Mar. 1896. Mrs Fordham was the daughter of W. Long, sometime junior minister at the LGB. See also WLGS leaflet, *Women's Work Under the New Local Government Act*, 1894. During 1895–6 Mary Kilgour wrote some 50 pieces for the *Parish Councillor* on 'Women in Council', encouraging, briefing, and networking women in rural local government.

[7] Mrs Abbott, *How as a Guildswoman Shall We Most Fully Use Our Powers Under the New Local Government Act*, WCG pamphlet, p. 2. See also J. Brownlow, *Women's Work in Local Government*, 1911, p. 114.

late with thistles and weeds. Cottages crumbled into disrepair. Men were still performing hard, monotonous, back-breaking work for half the wage of the industrial worker and for much longer hours. Any protest, any dissent, whether religious or political, was hounded; poor relief remained punitive, charity remained a tool of social discipline: 'blankets, soup, coals, rabbits and the rest were all paid for . . . in subservience', said the Countess of Warwick. 'We paid in cash and took in kind'.[8] By the late 1880s, latent anti-clericalism and popular radicalism was being channelled by George Edwards of Norfolk into agricultural trade-unionism. But though some men stayed to fight, many more were voting with their feet. One hundred thousand men a decade were leaving the land, the best and most skilled men of their generation, it was thought, migrating to the nearest town, emigrating to the nearest continent. They left behind a deeply depressed rural economy, depopulated rural villages, a demoralized labour force, and a legacy of social bitterness. As the Editor of the *Eastern Weekly Leader* put it, villages were

almost emptied of their population, their natural industry nearly ruined, with farmers bankrupt, and the land going back to its primeval barrenness, this is the condition of things when the labourers are called to the rescue . . . [The parish councillor] has to make his village fit for his children to stay in, so that the cockade of the recruiting sergeant or the dignity of the railway porter or the London police constable shall no longer shine in his eyes . . . He must rescue the village from its present position as the recruiting ground for the sad army of the unemployed and the reserve of the sweater.[9]

The 1894 Parish and District Councils Act, it was hoped, would reverse rural depopulation, by giving men a stake in their village, and thereby revive rural life. The very health of

[8] Countess of Warwick, *After Thoughts*, 1931, p. 242. More generally see G. Mingay, *Rural life in Victorian England*, 1977; id., *The Victorian Countryside*, 2 vols. 1981; P. Horn, *Labouring Life in the Victorian Countryside*, 1976; R. Samuel, *Village Life and Labour*, 1975. For Norfolk see L. M. Springall's fine book, *Labouring Life in Norfolk Villages 1834–1914*, 1936; and more recently, A. Howkins, *Poor Labouring Men: Rural Radicalism in Norfolk 1870–1923*, 1985. George Edwards's autobiography, *From Crowscaring to Westminster*, 1922, is valuable.

[9] *Eastern Weekly Leader*, 1 Dec. 1894; see also R. Jeffries, *Hodge and His Masters*, 1880; H. Rider Haggard, *Rural England*, 2 vols., 1902, from an extensive literature on rural decay.

the nation depended on it. If villagers continued to leave the land, said one MP, then the nation would lose 'that country blood that was wanted to revivify the rapid degeneration that was going on in the cities'.[10] As ever, the main crop of the countryside was not its corn or its cattle, but its people. Parish councils would encourage men to remain on the land, and help to halt national decay.

Parish Councils

Most of the hopes and fears of the 1894 Act, therefore, focused not on the rural district councils, but on the seven thousand new secular parish councils. They were to be 'The People's Magna Charta', as the not especially fanciful *Norwich Mercury* described them. The *Eastern Weekly Leader* considered election day as 'Emancipation Day in Rural England . . . the democratic curfew will ring out the vestiges of feudal power and ring in the new era of equal self-government'. And it continued:

For weeks past there have been crowded meetings in the village school, and heated discussions at the bar of the inn; knots of men have lingered on the threshold of the chapel, unwilling to let go the subject they will resume the minute the service is over; strange nods and winks, with whispered hints and cautions, have passed between the labourers at their work, and an unwonted eagerness has been shown in their examination of the weekly paper. At the markets, farmers have forgotten their standing complaints about low prices to vent their wrath at the appearance of the new enemy. In the rectory, there has been keen questioning of the servants, and anxious counting up of reliable dependents, while up at the Hall Bluebooks and Handbooks have taken the place of the Times and silenced the mirth of Punch. The finger of the labourer has stopped an inch short of his cap when the parson has passed . . .

Reports came in from all over the country that 'never were the villages of the land so stirred as today'.[11]

During the autumn, advanced Liberal MPs, trade-union leaders, and the red vans of the Land Restoration League had toured the countryside, explaining the terms of the Act, setting up small committees of working men to select

---

[10] Mr Price, Hansard, House of Commons, 7 Nov. 1893, col. 373.
[11] *Norwich Mercury* 1 Dec. 1894; *Eastern Weekly Leader*, 1 Dec. 1894. See also R. Heath, 'The Rural Revolution', in *Contemporary Review*, Oct. 1895, p. 190.

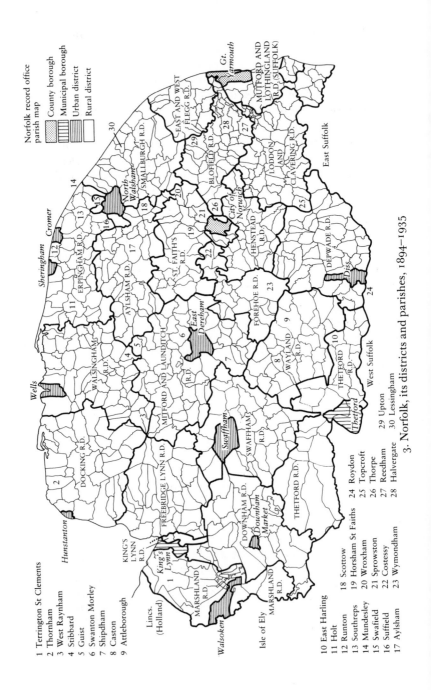

3. Norfolk, its districts and parishes, 1894–1935

candidates, and mobilizing and channelling both the deep anti-clericalism and desperate land hunger of the village labourer into parish politics. George Edwards pedalled Norfolk's lanes leaving election committees and trade-union branches behind him.

Anti-clericalism, in counties like Norfolk, was often focused on the parson, who controlled the vestry and the school, was poor law guardian and magistrate, and dispensed village charities as though they were his private property.[12] Despite the Burial Act of 1880, many continued to deny Nonconformist ministers and chapel families the right to a decent burial in the churchyard. Not surprisingly, some of the earliest demands of the new parish councils were to control the charity accounts and to provide additional burial grounds. One of the nastier responses of certain Norfolk parsons, who came out of the 1894 elections in a very unpleasant light, was to persuade labourers quite falsely, that if they became parish councillors and charity trustees, they would personally lose any rights to charitable cash, coal, or flannel.

As the depression deepened, village politics grew less concerned with the church and more concerned with the land.[13] The general election of 1892 had been fought in Norfolk, for example, largely over the issue of land reform. Norfolk had gone Liberal, Joseph Arch had regained his seat in north-west Norfolk, and a sizeable continent of agricultural labourers and Methodist local preachers (at least a dozen of them) had fought their way on to Norfolk county council to demand allotments and smallholdings.[14]

Now, two years later, labourers and craftsmen swept to power in parish after parish, unseating parson and squire. Across the

[12] For village charities see J. Arch's autobiography, *passim*; M. Ashby, *Joseph Ashby of Tysoe*, 1961, where the women ostentatiously scoured the gift of red flannel to wash the charity out of it (pp. 146–9).

[13] Springall, op. cit., p. 115.

[14] This was an extraordinary achievement since the county councils were, and have remained, a preserve of landed families. G. Phillips, *The Diehards: Aristocratic Society and Politics in Edwardian England*, 1979, has shown that a fifth of all county councillors were gentry and aristocracy; and many of the other seats were held by their stewards, agents, and staff; two-thirds of all peers held local office of some sort. In 1889 half the county seats were contested, in 1892 only a quarter, reflecting not lack of political interest but prior electoral arrangements. Its staff were part-time, generally inherited from quarter sessions. See J. M. Lee, *Social Leaders and Public Persons: A Study of County Government in Cheshire since 1888*, 1983.

country as a whole, the *Contemporary Review* calculated in 1895 that between a third and a half of the seats were won by farmers, about a quarter by craftsmen, and most of the rest by labourers, with a sprinkling of clergy, gentry, professional men, Nonconformist ministers, and women.[15] In Norfolk, the results were more dramatic. In West Raynham, Attleborough, and Swanton Morley the labourers took every seat, in Terrington St Clements they took control. The vicar got just three votes and stormed out of the meeting in pique declaring it 'a one-sided and scandalous bit of business'. In East Harling the parson managed things rather better. He received twenty-three nominations for eleven seats—twelve of them he declared were conveniently invalid. They had misspelt 'parochial', or omitted a Christian name; the three women were excluded by virtue of the chairman's ruling. Some dissatisfaction was expressed but the parson was well content. In Costessey, a parish as quirky as its name, the villagers ensured the success of their candidates by putting up both hands. The chairman noted ruefully, 'I tried to spot those who voted too often, but it was beyond me'.[16] In north Norfolk, Miss Buxton was elected to Runton parish council alongside a fisherman, a gardener, a farmer, a bailiff, a wheelwright, and a builder. At Sprowston, just outside Norwich, three labourers, a painter, a hawker, an engine fitter, a blacksmith, a spinning smith, two school masters, a baker, a miller, and a fish merchant were elected alongside Mrs John Gurney.[17]

In some parishes, the contest turned savage. At Wroxham when one of the working-class candidates made an election speech, he found himself sacked and evicted two days later, even though he was a steady workman and chapel steward. Working men turned out *en bloc* to vote but could not get him back his home or his job. Similarly, in St Faith's, where the labourers had been trying to get allotments, the elections became a battleground for lock-out farmers and labourers on strike. In Wisbech, a schoolmaster who chaired the parish meeting and ruled the vicar out of order, found himself dismissed by the vicar from his job the next day. The *Norwich Mercury* summed it up: 'The labourers have every reason to be

[15] Heath, op. cit.
[16] *Norwich Mercury*, 1, 8 Dec. 1894.   [17] *Norwich Mercury*, 1 Dec. 1894.

satisfied with the results of the parish meetings, and the parsons and squires, we fear, have just as much reason to feel dissatisfied . . . [at] the almost entire effacement of both these classes from the parochial governments of the future'.[18] It seemed indeed a 'rural revolution'.[19]

The group that had made the largest advance were the labourers; but the WLGS knew of at least eighty women returned in these first parish council elections and Mrs Elmy thought the number was probably nearer two hundred in England and Wales. Some twenty women,[20] about the same number as Nonconformist ministers, were elected to Norfolk parish councils. They were usually the wives and daughters of prominent families.[21] The more eminent and able were often also elected to the RDC.

In any parish council, much depended on the initiative of one or two people. Although women made their presence felt, it was clear that most of the energy for change came from labourers determined to extract allotments from the farmers, the charities from the parson, and clean water from the RDC.

One Norfolk woman who took the lead on many issues in her parish was Mrs Milner, a doctor's wife in Shipdham, a parish bordering East Dereham UDC. She had helped to negotiate the agreed list of candidates between the overseers and George Edwards, who represented the labourers. The newly elected council's first task was to provide allotments and they bought enough land at 25*s.* an acre to satisfy demand. They next turned to the local coal charity and Mrs Milner was, with two others, deputed to arrange its transfer from the parish. She became one of the trustees. The council tidied up the village green, which was used as a dump for

---

[18] *Norwich Mercury*, 15 Dec. 1894; cf. R. Heath, op. cit.

[19] *Eastern Weekly Leader*, 4 Dec. 1894.

[20] The 24th Annual Report of LGB lists 15 women among a sample of some 3,000 parish councillors. Mrs Elmy, BM Add. MS 47450, 7 Jan., 9 May, 1895; WLGS Annual Reports.

[21] I have traced two in Reedham, one in Sprowston, two in Topcroft, one in Thornham (Mrs Ames-Lyde), one in Rodwell, one in Guist, one in Scottow (Lady Durrant), one in Mundesley, one in Shipdham (Mrs Milner), one in Upton, one in Runton, one in Sheringham (Miss Pigott), one in Roydon, one in Caston (Mrs Partridge, rector's wife), two in Stibbard, one in Lessingham, one in Suffield, one in South Repps. There were probably a few more whose seats were uncontested and thus unreported.

timber and show vans; found an enclosure map and registered footpaths and common land; and acquired additional burial grounds. Mrs Milner, worried at the fire risk to thatched cottages, persuaded the council to invest in a fire engine and drew up the regulations for its use. She had less fortune with another motion, to consider lighting the village, when no one would second her proposal.[22]

When the re-elections came in March 1896, much of the earlier enthusiasm had withered, and in two-thirds of Norfolk's parishes councils were returned unopposed. In Shipdham, only thirty electors turned up to the parish meeting, and Mrs Milner, now inevitably unpopular as a charity trustee, tied for last place with a man. The chairman gave his casting vote against Mrs Milner. A couple of weeks later, the parish called another meeting to discuss charity funds. Villagers complained that wood was being sold off privately. Mrs Milner was not allowed to speak, the meeting grew heated, clerk and chairman walked out, and one of the labourers ran to a nearby cottage to fetch a copy of the radical *People's Weekly Journal* from which he read an article denouncing the charity trustees. A year later, her husband, the local doctor, stood for the parish but was roundly defeated.[23]

Elsewhere, the conservative Blanche Pigot, a clergyman's daughter, steered her north Norfolk parish council into attending to the crumbling breakwater, sea defences, and sewage outfall, and did her best then to push their preferred options through the Erpingham RDC. She lent them her Fisherman's Mission Hall for meetings. Over towards King's Lynn, the local squire, Mrs Ames-Lyde of Thornham, helped to regenerate her village by developing a flourishing wrought-iron industry. When Sandringham placed an order for garden gates, success was assured. She travelled the world for new designs, winning gold medals in international exhibitions for her craftsmanship, dying in Shanghai in 1914.[24]

Outside Norfolk, Mrs M'Ilquham who had fought Cheltenham for the county council in 1888, had now become

---

[22] *Eastern Weekly Leader*, 17 Nov., 1 Dec. 1894; *Norwich Mercury*, 23 Feb., 18 May, 7 Dec. 1895. For an account of village fire services, see Ashby, *Joseph Ashby*, p. 189.

[23] *Eastern Weekly Leader*, 14 Mar., 4 Apr. 1896.

[24] *EDP* 24 Jan. 1881; P. Brett and F. S. Franklin, *Thornham and its Story*, 1973.

chairman of her parish council at Staverton. She got the cottage water-supply improved, protected the roadside 'eatage' of berries and nuts for village use, and had the best of the village midwives made an overseer so that she could give medical aid on her own initiative. In Cambridgeshire, Mrs Fordham did similar work in her parish of Guilder Morden. The existing village allotments were inconvenient and expensive. She persuaded the parish council to use compulsory purchase powers to get land in the very centre of the village on which the men raised corn, potatoes, and pigs. A fellow parish councillor in the next village found the local well so polluted that children suffered skin diseases and gastro-enteritis. After two years of struggle she got a new well sunk. As she drove through the village, the women would flock out to greet her in gratitude.[25]

Two of the best-known parish councillors were Mrs Barker of Sherfield-on-Loddon, near Basingstoke, and Miss Jane Escombe of Penshurst in Kent. Mrs Barker was a clergyman's daughter, workhouse visitor, and wife of one of the larger landowners. When her husband was asked to stand for the council, he proposed his wife instead and transferred some cottages to her to give her the parochial vote. In her own words,

There were only eight nominations for the six places and the six were chosen by show of hands . . . As soon as possible after the election the council met and I was unanimously elected to the chair. The other five members were a butcher (VC), a carrier, a builder, a market gardener, and the village shopkeeper, with a most experienced and courteous assistant overseer as clerk; the majority Nonconformists and Radicals as to politics. From the very outset I earnestly deprecated any sectarian or political 'cleavage', and in my first address to them I took my stand on the broad Christian basis of doing our best for 'our neighbour' in the largest sense of the word. On these terms we got to work, and I may say they were never departed from, and from first to last I was treated with the utmost consideration, the most absolute courtesy.

The first concern was to place the charities of the parish on a new footing; they were four in number, and were, so to speak of each known variety. No. 1 was disbursed by four non-elected trustees; so,

[25] *Parish, District and Town Council Gazette*, 26 Jan. 1895; Miss A. Busk, 'Women's Work on Vestries and Councils', in Revd J. Hand, ed., *Good Citizenship*, 1899.

to aid them in their labour, we appointed four others. Unfortunately, the original trustees did not meet the requirements of the Act in a friendly spirit, and resigned *en masse*, leaving the management of an income of over £100 a year to the nominees of the parish council. The second had not been in existence forty years, and so did not come within our scope. The third was one heretofore administered by the rector and churchwards; to this I, in conjunction with another member of the council, were appointed trustees instead of the churchwardens . . . The fourth, distributed at the sole discretion of the rector, we tried to deal with . . . but were defeated on a technical point.

Those matters settled, we took the footpaths in hand. I gave the council a 25 inch Ordnance map of the parish, and one by one each landowner was approached as to the various rights of way through his property; all that the council claimed, save two, were allowed, and were by me marked on the map: brown for highways, red for bridlepaths, green for footpaths; this map is now in the parish chest . . . I purchased two 6 inch maps, marked them in the same way, had them framed, and then hung up in the two public-houses lying at either extremity of the parish . . .

It has been objected that parish council business has no opening for decidedly feminine work. Well, I grant there is none that woman can do *better* than a man, as there is in workhouse management, and school boards and committees; but for all that I must strongly urge all women who have the good of their fellows at heart not to shrink from standing for their parish council . . . Beyond the softening effect which I found so valuable, women are so much more earnest about small things than men, and parish council work deals with matters of seemingly small import. A polluted well, an overcrowded cottage, a barrier across a footpath, are too trivial for men to make a stir about, and perhaps offend the wage-giver into the bargain; but an independent woman, knowing that 'trifles make the sum of human things', and that these trifles if looked into will reveal further defects to remedy, will be earnest for frequent meetings; her cry will ever be, 'Look for work, make work, never to stand still till all the good you have the power to do for the poor of your parish is an accomplished fact'.[26]

When she retired she was presented with a handsome folding morocco screen, decorated with the photographs of her fellow

---

[26] *Parish Councillor*, 27 Dec. 1895; Mrs Elmy, 'Women in Local Administration', *Westminster Review*, Oct. 1898; *Parish, District and Town Council Gazette*, 5 Jan. 1895.

councillors. As she delicately indicated, it was the thought that counted.

Perhaps the most stubborn woman councillor was Jane Escombe of Kent. Penshurst was a pleasant village of some 1,600 people, but as in so many villages had a housing shortage which it was not profitable for speculative builders to meet. Miss Escombe pointed this out at an early council meeting; and proposed that the parish petition the RDC to use the 1890 Housing Act to build cottages of their own at public expense. The parish asked her to canvass the village, door to door. She found forty families in need of cottages but thought it sensible to start with six. They asked private landlords to do the job; they refused. They asked Sevenoaks RDC to do the job, who asked the landlords, who again refused. The RDC held an inquiry in the village; and reluctantly agreed to ask the county council. The county council held an inquiry and reluctantly agreed. The RDC showed no disposition to do anything more, so Jane Escombe on behalf of the parish petitioned the county to require the RDC to start work. A joint parish and RDC committee was set up to which Jane Escombe acted as secretary. The rector sold some land, plans were drawn up and sent off to the LGB who now called an inquiry, the third to be inflicted on the village. By November 1899 the parish had its permission, by December 1900 the cottages were built. They had cost £250 each, offered three bedrooms, were let at an unsubsidized rent of 5s. a week, and were thought to be very handsome. They were naturally named Pioneer Cottages. In time a further eight were built, and the whole incident much quoted.[27]

These first parish councils had been elected with such high hopes, and many by 1914 could point to a steady record of achievement. Take Horsham St Faiths, a parish of seven hundred just outside Norwich, its parish council made up equally of farmers and labourers. Over the years it protected footpaths, acquired eight acres for allotments, brought cottages back into repair, obtained a letter-box from the post

[27] J. Escombe, paper to NUWW Conference, Brighton, 1900; see also W. Crotch, *The Cottage Homes of England*, 1901. Crotch found only one other parish, Ixworth in Suffolk, that had successfully started cottage building, and then only with the help of their county councillor, Lord Hervey.

office and a savings bank branch by guaranteeing any losses. Wet and dirty lanes were resurfaced and two hundred acres of heathland brought under parish management. Above all, 'everyone has become independent'.[28]

If Norfolk is any guide, once labourers had their allotments, and had sorted out any other immediate grievances such as charity revenues, common land, or the state of the village pond, then interest waned: though not always as sharply as Swafield, Norfolk, from where this laconic report went to the *Parish Councils Journal* in April 1896:

Two worthy members resign; two remain. Strong appeal to public to fill vacancies, no response, meeting closed . . . Parish Council dies untimely death for lack of energy. Members hurry off; no time to pass a vote of thanks for past services. Promises at the beginning— (i) recreation ground (ii) lamps for corners of roads (iii) allotments (iv) footpaths. Results—nil.[29]

In part this was because parishes had only limited powers; most of the items the men wanted were obtained, if they were going to be obtained at all, within the first two years or so of the new council's existence. Thereafter attendances fell away, press reports are almost negligible, many of the optional smaller parish councils folded, and many of the professional journals reporting their work folded with them.[30]

There were other more sinister factors too.[31] In closed villages, where land was held by one or two families, labourers

---

[28] Fabian Tract 137, *Parish Councils and Village Life*, 1908, repr.1920. The Fabian Society calculated that 1,500 of the country's 7,000 parish councils had acquired allotments, some 52,000 holdings in all, over and beyond private arrangements. About a thousand parishes had adopted street lighting and watch powers, 700 had new burial grounds, 500 had fire protection. Many had bought land for playing-fields, but perhaps no more than 20 had provided the more costly baths, washhouses, or libraries. See also E. Bennett's assessment: 'The parish council is the only governing body in which the labourers can play a part'. (RDCs were 'farmers clubs', whose sole aim was 'to keep the rates down and do as little as possible'.) 'Here and there the work afforded by the work of these parish councils—particularly that connected with the supply of allotments—has added interest to the life of the village and contributed to self reliance and self respect.' *Problems of Village Life*, 1914 edn., pp. 117–18.

[29] *Parish Councils Journal*, 1 Apr. 1896.

[30] e.g. *Parish Councillor*; *Parish Councils Journal*; *Parish, District and Town Councils Gazette*.

[31] 'A lady canvasser in Oxford was informed, "Oh its no good your coming here, Miss, Mrs. K. (the Tory lady of the Manor) arranges our politics for us"'. Bennett, op. cit., p. 111. To be seen talking to Liberal canvassers could cost a cottage or a job.

who became parish councillors were marked men. One councillor giving information about threatened evictions was himself evicted. 'This year, no labourer dares present himself for election on the parish council.' From Lavenham in Suffolk it was reported that even where the parish provided allotments, the men were too terrified to request them as 'their masters would boycott them'. Other councillors had to inspect cottage sanitation at night, as otherwise the tenant would be evicted by day.[32]

Instead, parish councils developed in a rather different direction from that predicted in 1894. They acted less as a modest local authority in their own right, and rather more as an active lobby demanding major sanitary and highway improvements for the parish from the RDC or the county council. Commentators[33] noted that in Europe, larger bodies supervised smaller ones. In England, smaller bodies, like the parish, supervised their upper tiers of district and county. More urban and more active parishes, such as Aylsham, Attleborough, Wymondham, and Holt within Norfolk, besieged their RDCs with requests, complaints, and demands—the state of the roads, the pitfalls for cyclists, the dust and danger raised by the new motor cars. They demanded clean water, new drains, rubbish removal, and housing repairs (thus shielding the tenant from victimization), and threatened the RDC with the county council, and the county council with the LGB if they did not respond.

Women's contribution to parish council life had always been modest. Those women who took the initiative or made an impression were usually surrogate squires or connected to clerical families, though unlike male squires and male parsons they were almost invariably battling for the labourers, and supported their every demand. They had no love for the more grasping farmers on parish councils. Within three years of the 1894 Act only half a dozen or so women remained exclusively on Norfolk's parish councils, whereas between thirty and forty were district councillors or poor law guardians. Their contribution at parish council level had become minimal.

---

[32] F. E. Green, *Tyranny of the Countryside*, 1913, pp. 30, 165; for Lavenham, *Parish, District and Town Councils Gazette*, 8 Feb. 1896.

[33] T. Redlich and F. Hirst, *Local Government in England*, 1903, p. 193.

District Councils

Compared to parish councils, there was surprisingly little discussion in press or public about the new rural and urban district councils in 1894, presumably because they exhibited such strong continuity with the old poor law boards and sanitary districts. Most of the old guardians became the new councillors—generally farmers alongside craftsmen, clergy, and gentry. Compared to either parish or county councils, labourers were noticeably absent. The new RDCs met in the same workhouse, with the same officers, formed the same subcommittees under the same chairmen, and very soon had the same difficulty in finding enough candidates to contest all the seats. Even the former ex-officio members, the JPs, did the decent thing, stood for election, and were returned. At least to begin with, poor law and sanitary work continued much as before. Fresh interest was generated by the new highway duties however; and pressed by parish councils from below and occasionally prodded by the county council from above, rural district councils embraced their new responsibilities with varying degrees of enthusiasm.

One element was new, and commented on; the arrival of women in substantial numbers on to rural boards. They came to be poor law guardians. Most of them learnt to become rather effective rural district councillors.

Norfolk's pattern of rural local government may again serve as an example of the changes in the country as a whole, though its politics were probably untypical in their Liberalism. Leaving aside the boroughs of Norwich, Yarmouth, Thetford, and King's Lynn, Norfolk after 1894 had twenty-one district councils, their boundaries the same as the old poor law unions, from Docking in the north to Depwade in the south, each possessing between ten and fifteen thousand people clustered in twenty-five to fifty parishes and stretching some ten to fifteen miles across. These RDCs were the poor law, sanitary, and highway authorities for their areas. Within the RDCs were a number of smaller townships, such as North Walsham, Hunstanton, East Dereham, and Diss, which retained the sanitary powers they possessed before 1894, and which became self-sufficient urban district councils (UDCs) for highways and sanitation, while remaining within the RDC for poor law work.

Women were eligible to serve on urban district councils but even as late as 1914 there were never more than two or three in England doing so at any one time. Mrs Elmy, friend of Mrs M'Ilquham and founder of the Women's Emancipation Union, had stood for her local UDC, Congleton, in Lancashire in December 1894, hoping to get a decent water-supply for the village. As she wrote to Mrs M'Ilquham:

I have been beaten here—but I am prouder of the defeat than I could be of any victory won less honourably. The result has been to make me leader of the village party for health and sanitation as against the 'old gang' who stood *en bloc* to defend the rights of owners to keep their property in the foulest condition. The fight was a splendid one—though over 200 of the electors had to walk or drive two miles to the miserable shanty fixed as the polling place—88 per cent polled—and of the women over 90 per cent—and not one spoiled paper—I believe, but do not know, that the women supported me to a man [*sic*], but the victory was hopeless . . . We shall fight again and win, and meanwhile shame this Council into some sanitary action.

But a few years later she had ruefully to accept that she continued to be 'beaten by the wise thrift of our villages who would not *wash* or vote . . .'.[34]

Overall, there were around seven hundred and eighty UDCs in England and Wales by 1898, some of them villages with barely two hundred people, while others were large suburban centres pressing to become borough councils. Their powers were very similar to the London vestries, but as UDCs were not poor law bodies they attracted few women as candidates, and elected even fewer to power. By 1897 one woman had been elected to Romford UDC, and others to Litherland, near Liverpool, Barnes, and Dorking. The only woman councillor to gain prominence was Miss Margaret Ashton, who joined Withington UDC in 1903 and, as it had devolved educational responsibilities, chaired its education committee. Within a couple of years, Withington was incorporated into

[34] Mrs Elmy, BM Add. MS 47450, fo. 162, 20 Dec. 1894; 47453, fo. 64, 17 Oct. 1902.

Manchester, and she lost her seat. In 1907 she re-emerged to stand for Manchester City Council.

No woman seems to have been elected to the eight UDCs within Norfolk. Instead they stood for election as poor law guardians of their townships, joining the local RDC for its poor law deliberations and leaving the room when the RDC moved on to its sanitary and highway business. Neither Smallburgh RDC nor North Walsham UDC had any women councillors; but North Walsham elected three gifted women[35] as poor law guardians and they joined Smallburgh RDC for poor law work. In Erpingham, in North Norfolk, three women, Miss Blanche Piggot, Mrs George Edwards, and Miss Buxton, were elected to the RDC. For poor law work they were also joined by three women elected for adjacent UDCs, including Miss Kate Fitch, daughter of the vicar of Cromer.

As the 1894 results trickled through, returns indicated that of the 875 women elected as poor law guardians for England and Wales, some 140 were also rural district councillors. Norfolk had elected nineteen of them, most noticeably in Aylsham, Blofield, Docking, and Erpingham, three times as many as in any other county in England. In addition, another dozen women were elected as poor law guardians for UDCs. By 1900, some forty women held elected district and poor law office.[36] Just as agricultural labourers had flocked to join the new parish councils, women, no longer hindered by the property qualification, hastened to become poor law guardians.

The RDC usually met fortnightly. Its first business was poor relief. Most rural councils had four or five committees— the House or visiting committee, the finance committee, the assessment or valuation committee, sometimes a contracts committee, and a school attendance committee, where

---

[35] Mrs. Wimble, Mrs Wilkinson (a clergyman's daughter, wife of a solicitor, and a concert pianist); Mrs Petre, widowed landowner and local philanthropist.

[36] Norfolk had 24 RD councillors and 13 women guardians in 1895; most counties had no more than half a dozen RD councillors, but between 20 and 90 women guardians, e.g. Devon had 5 RD councillors, 30 women guardians, Wiltshire 7 RD councillors, 23 guardians, Suffolk 3 RD councillors, 11 guardians; Lincolnshire 9 RD councillors, 31 guardians; Yorkshire 8 RD councillors, 91 guardians; Cornwall 5 RD councillors, 18 guardians. RDC elections were by secret ballot. Each parish elected one councillor, small parishes were grouped, larger ones were divided.

farmers sat pressing children to attend school while at the same time employing them as cheap labour on their farms. Women carried more than their share of committee work, and often found themselves in a majority on the House committees, where, having served as workhouse visitors, they developed a shrewd eye for domestic detail and administration.

Norfolk was held by one LGB inspector (Preston Thomas) to be the most pauperized county of England.[37] It was economically depressed, its population was falling, work was scarce as farmers cut costs by shedding labour; and he considered that 'out relief was dispensed with the utmost freedom'. In 1895 Blofield had fifty-six inmates and three hundred and twenty outdoor poor; Mitford and Launditch had four times as many of both. More typical unions such as Aylsham, Erpingham, and Smallburgh each had some ninety indoor poor rattling around in dilapidated workhouses built for up to six hundred inmates. Half of their indoor poor were elderly men, the rest women and children. As the master of Aylsham workhouse reported to his committee, 'The name workhouse as applied to country union houses such as Aylsham is really a misnomer. They have become the infirmaries for the district and homes for the infirm and young.'[38]

So when the LGB sent out its flow of circulars, calling on boards to reduce outdoor relief, segregate and classify their indoor poor, and introduce superintendent nurses, to take three examples, their instructions were quite reasonably ignored in Norfolk. Guardians decided that the LGB was addressing itself to 'town questions', and tucked the circulars under the blotting paper. In September 1896 the LGB followed up the proposals of the Royal Commission on the Aged Poor, and suggested that guardians should be more generous in their treatment of the deserving elderly, while remaining punitive to the obstreperous. As the Aylsham master said, this hardly made sense in Norfolk.

Personally I do not believe in making any great difference between

[37] H. Preston Thomas, *The Work and Play of a Government Inspector*, 1909, p. 225.
[38] Mitford and Launditch Guardians Minutes, 16 Nov. 1896; Aylsham Guardians Minutes, 15 Sept. 1896. More generally, see A. Digby, *Pauper Palaces*, 1978, for poor law administration in Norfolk before 1894.

the old people, either in clothing, surroundings or food, as most of them are of the labouring class, not earning sufficiently in their best days to make provision for their old age. There are of course two or three more vicious than the majority but as a rule the old folks are well behaved and tolerant towards each other.

He had given his old ladies the sunniest rooms in the house, and thought they were 'as comfortable as one can make them in institutions of this kind'. He was asked by one of the women guardians whether he could at least segregate the rough from the respectable, but he pointed out that of the nine younger women in the House, three had had four or more illegitimate children, one had two, another was pregnant, two were imbeciles, and two were deaf and dumb. Who should be segregated from whom?[39] From Smallburgh came reports showing that the elderly already enjoyed better food than the new improved LGB diets and had freedom to 'stroll around'.[40] The impression from Minute-books and press reports is of kindly, reasonably competent, but unimaginative staff, who welcomed visitors into the House and were uncensorious of their indoor poor. However critical the LGB inspectors were of particular arrangements, they were always careful to credit the staff with humane intentions and the guardians were always careful to enquire.[41]

And criticism there was in abundance, both for the refusal of local guardians to follow LGB guidelines on the one hand, and their reluctance to repair their crumbling workhouses on the other. Workhouses in rural Norfolk were very large, rather old, impossible to heat and clean. Windows no longer opened, privies were noisome, bathrooms and hot water absent even on the sick wards; and unlike the towns, country guardians (if Norfolk is any test—and similar reports came from Dorset and Devon) had neither the inmate numbers, the finance, nor the political will to rebuild and replace the House. One LGB inspector asked guardians to spend some money on washing facilities. Their chairman, the Earl of Kimberley, a former Foreign Secretary, thought baths were dangerous (they had led to typhoid fever in Italy, he stated) so Wymondham's indoor

---

[39] Aylsham Guardians Minutes, 15 Sept. 1896.
[40] Smallburgh Guardians Minutes, 11 Jan. 1898.
[41] *Norwich Mercury*, 3 July 1909; see also Preston Thomas, op. cit., pp. 244–5.

poor remained dirty.[42] Later LGB inspectors tried to persuade guardians to close their worst Houses and pool their inmates and resources with neighbouring unions. To oblige, the guardians occasionally discussed the matter but would neither close their House nor send inmates to foreign bits of the county.

Women guardians, here as elsewhere, did their best to domesticate the inside and repair the outside of buildings. They brought in stoves to heat the wards, cots for the babies, new mattresses and waterbeds for the infirm, bath chairs for the elderly, washing and sewing machines to help the staff. Sometimes they got their board to permit them to introduce the Girls Friendly Society for younger women, and the Brabazon scheme for elderly women. They came with strawberries and musical evenings, and returned with complaints that their male colleagues were never to be seen within workhouse walls. They arrived on Norfolk boards when most deserted and orphaned children were already boarded out. Indeed before they became guardians, women members had often helped the board to find suitable foster homes. The children of inmates remained confined however—twenty-three in Aylsham workhouse in 1895, twenty-five at St Faith's in 1909. Usually the women members managed to persuade their boards to allow such children to attend the local school in ordinary clothes.[43] But no woman rural guardian is recorded in the Minutes as ever suggesting that they go to scattered or cottage homes, on the Sheffield or Norwich model, so that no child need remain within the workhouse. It was left to Captain Hervey, the LGB inspector, in 1909 to go the rounds of eastern rural boards pleading with them to free children from workhouse taint. Forehoe and Thetford unions asked women members to explore his proposals, but even after the First World War there were still Norfolk and Essex children living out their lives in the workhouse.[44] Individual women guardians did their best to mother particular girls: Mrs Hoare of Aylsham placed one child in a training home, Miss Buxton

---

[42] Preston Thomas, op. cit., p. 237.
[43] Aylsham Guardians Minutes, 8 Jan. 1895; *Parish Councillor*, 6 Sept. 1895.
[44] *Norwich Mercury*, 17 Apr., 12 June, 3 July, 16 Sept. 1909; see also Cuttle, *The Legacy of the Rural Guardians*, 1934, p. 118.

of Erpingham helped another; Mrs Edwards made maternity cases her especial care and stopped 'naughty' girls within days of their confinement being put to heavy labour in the laundries. But as most Norfolk unions resolutely refused to affiliate to the central poor law conferences, they were unable to learn much from the experience of elsewhere.[45] Child care remained in something of a time warp.

Also hard to help were the infirm elderly. As few workhouses now had any able-bodied women within them to do the nursing, women members took the opportunity to introduce trained staff and night staff. Given dirty and chilly buildings, and the nature of the nursing—'chronic ulcers, chronic bronchitis, chronic gout, chronic rheumatoid arthritis, chronic senile decay'—trained nurses did not stay long.[46] In the words of the LGB Inspector, 'their chief occupation is that of feeding and keeping clean a number of old people who are slowly dying. Much of their work is, in point of disagreeableness, akin to scavenging'.[47] Isolated, overworked, underpaid, and often at odds with an untrained matron or negligent doctor, many nurses stayed the bare six months of their opening contract, just long enough for them to retain their uniform.[48] On most Norfolk boards, women members were a standing subcommittee for finding nurses, a task made harder when reputable training homes refused to place their staff with rural unions whose infirmaries were unacceptable. Women guardians did what they could, offering gratuities when the nursing was especially offensive, arranging extra holidays and dainty comforts, but their scope for change was limited. However, women and inspectors together did persuade male guardians to increase the totality of trained staff. Within Norfolk there were twenty trained nurses in 1891 and a hundred in 1896, a better figure than for much of rural England. The ratio of nurses to patients also improved, from one nurse to twenty-four patients in 1898, to one nurse to seventeen patients in 1900.[49] Again, the Minutes and the press

[45] Edwards, *From Crowscaring to Westminster*, *passim*.
[46] *Lancet*, 19 Feb. 1889, quoted F. B. Smith, *The People's Health 1830–1910*, 1979, p.385.   [47] Preston Thomas, op. cit., p. 232.
[48] A uniform of two serge dresses and a dozen aprons. See Erpingham Guardians Minutes, 15 Feb. 1897, for Nurse Thornton's complaints.
[49] Aylsham Guardians Minutes, 15 Oct. 1895; Digby, op. cit., p. 171.

reports do not suggest that women members pressed hard for separate cottage infirmaries, or quasi-sheltered housing, as women guardians elsewhere were doing. Yet what then would they have done with the dozen or two people that would have been left behind in the House?

With greater imagination, political courage, and financial resources, women might perhaps have encouraged their boards to abandon the workhouse for a district infirmary, and small separate scattered homes for the less infirm, the retarded, and children. But this was never on Norfolk's agenda, which remained a county of low rates and poor services.

The policy of outdoor relief was by now relatively un-contentious in Norfolk; and bestowed without too much heart-searching on the elderly, widows, families with a sick member or in temporary distress due to the seasonal nature of much arable farming.[50] The doles were parsimonious—1/6*d.* a single person, 2/6*d.* a couple plus flour in 1894. This was pushed up in small steps to 2/6*d.* a single person, 3/6*d.* a couple plus flour on the efforts of Mr and Mrs George Edwards in Erpingham. Other unions slowly fell into line, but such sums were never enough to live on.[51] George Edwards led a team of six labour representatives on Erpingham RDC. They fought to replace the flour quota with cash. Mrs Edwards tried baking bread with it but found it too poor in quality. She made the guardians taste her bread, but she could not persuade them that money was better.[52] One item of poor law policy much hated was the docking of 2*s.* a week from labourers' wages to pay for the support of a parent in the House; but though working men and radicals in Erpingham, Aylsham, and Docking all tried to stop the deductions, women guardians abstained from the vote, and it was always lost. On the other hand, the guardians were equally impervious to pressures from the LGB to reduce out-relief. When an LGB inspector grumbled that Smallburgh was spending twice as much as adjacent unions, the guardians blithely ignored his remarks and did not condescend to discuss them.[53] Economy,

---

[50] *Norwich Mercury*, 16 Feb. 1895; Cuttle, op. cit., p. 54.
[51] Preston Thomas, op. cit., p. 227.
[52] Edwards, op. cit., pp. 68 ff.
[53] Smallburgh Guardians Minutes, 30 Nov. 1897.

humanity, custom, and self-interest combined in the countryside to buy off a degree of rural radicalism by allowing out-relief sufficient to keep the rural poor from the workhouse but which would still leave them dependent on and grateful for the charity of family, neighbours, and betters. As the Poor Law Commission of 1909 reported,

It is hard to say with any precision how the aged pauper spends the weekly florin or half crown. Rent, coal, bread, butter, tea, sugar, oil and the burial insurance are the recurring items. The garden supplies vegetables. Neighbours are kind. There are small charities at Christmas from the Church. Perhaps a married son pays the rent and a daughter in service sends a few shillings every month or half yearly. The clergyman leaves something when he calls . . . It is a lonely and precarious existence sustained by faith in Providence and the Poor Law. The latter can always be relied on for 2s. a week and the former for something in addition.[54]

Women guardians, whatever their private views, showed little disposition to challenge the system either by making levels of outdoor relief adequate, as labouring men demanded; or to limit its application to the principles of 1834, as the LGB and women on town boards continued to urge. Prudently, they kept their heads down and went along with prevailing attitudes.

So rural poor relief policies remained relatively unpoliticized by the battles over outdoor relief that divided city boards, and which were turning so many women guardians into the class enemies of working men. In any case, worries about its cost had been softened by the Agricultural Rating Relief Act of 1896. After a generation of rural emigration, farm labour was becoming scarce—Norfolk in 1891 had nearly 38,000 male workers, by 1911 only 33,000 and there were few able-bodied unemployed men needing outdoor relief. Ten years later, the Old-Age Pensions Act removed the need for new generations of the elderly to seek outdoor relief. With 5s. a week coming in, they were now welcome in their children's homes. Norfolk's relieving officers became redundant.[55]

---

[54] RC on Poor Laws, 1909, vol. xliii, p. 631.
[55] The 1896 Act de-rated agricultural land by half for poor rates, and by three-quarters for general rates, to cushion the depression; the deficit was made good by government grant. Bennett, *Problems of Village Life*, p. 145.

But on one issue rural guardians were particularly vindictive, and that was vagrancy. Women guardians and inspectors tried to stand between the half dozen tramps a night descending on the typical Norfolk workhouse, and the wrath of the farmers on their boards. With little success. Most tramps were offered a punitive diet, hard labour, forcible detention which meant they could not leave early enough to find work the next day, and a leaking lean-to shed. Tramps in Essex sensibly preferred the police cells to the casual wards; and the police were noticeably more humane than overseers in dealing with them. Not that it mattered much since guardians would turn able-bodied men out of the workhouse on the grounds that they should support themselves, and would then arrest them, in their capacity as JPs, for vagrancy and send them to prison.[56] Aylsham's women members tried to have their tramp shed made watertight and to provide vagrants with a fire and a bath. To their obvious distress, their proposals were flatly rejected. Mrs Edwards rightly counted it as one of her triumphs in Erpingham when the tramps were no longer required to pick oakum, and were given a better diet. At Docking in 1911 one guardian considered the tramps' wards 'as not fit to put a horse into. He would not put one of his into it in its present condition'. The board accepted that that was a powerful argument and had the shed roof repaired, but the master was forbidden to let them have a fire under any circumstances as that would only 'pamper them up'.[57]

Women joining RDCs to act as poor law guardians seemed to have less to show for their efforts than their sisters on urban boards. There was far less scope for policy initiatives in small and shrinking rural unions than in the large cities. There were other more subtle pressures at work, too, associated with the innate conservatism and caution of the countryside, and the lack of respect it had for any book learning expertise that members might acquire, and which stood women in such good stead elsewhere. Miss Blanche Pigott was a clergyman's daughter, a good friend of Mary Clifford with whom she worked on East End missions, and one of the few Norfolk rural women within the network of the national women's philan-

thropic circles. When she came to address the Norwich
women's Liberal association in 1894, she made it clear that

She was in no way identified with any political party . . . but was
only too glad to cooperate with all who were trying to right wrong
and make the world a little better than it is . . . Women must not
suppose for one moment that they could do the work of men on
Boards of Guardians, but there was plenty of work for them to do
that the men could not possibly undertake . . . She did not think it
was the function of women Guardians to speak much, but believed
it was better for them to place their ideas before their male
colleagues, and induce them to bring them before the Boards . . .
The less conspicuous they made themselves at the Board meetings,
the better it would be . . .[58]

If even Blanche Pigott, who was abler by far than most of her
colleagues on the RDC, was determined to practise such a
self-denying ordinance and to seek such a low profile for her
work, however manipulative and devious her methods, then it
is not surprising that women's voice and women's contribution on rural boards remains unrecorded and so hard to
recover.

For many years, poor law work dominated the RDCs.
While minutes of poor law business grace some eight or ten
pages of foolscap, the sanitary and highway work of RDCs
was often dispatched in a page or two. And while poor law
Minute-books are usually complete, a surprising number of
RDC Minute-books, in Norfolk at least, have been lost. Some
councils did not establish any committees for their sanitary
and highway work, a sure sign of lack of activity. Only slowly
did their officers, the MOH, the sanitary inspector, and the
highway surveyor become full time, as sanitary and highways
work grew in importance—though how fast and by how much
depended largely on the energy of the MOH himself. At first
the RDCs simply responded to prods and pokes about their
rubbish and their roads, from parish council, the county
council, or the LGB, but they slowly began to develop initiatives of their own.

The major new function was highways. The county council
retained responsibility for major roads, but the RDC took over
minor and some private roads. Highway-oriented RDCs, such

[58] *Norwich Mercury*, 1 Dec. 1894.

as Thetford or Henstead, undertook surveys and began programmes of repair and resurfacing, though Miss Buxton on Thetford RDC had some difficulty persuading her colleagues that they should not personally survey their own patch but should employ a professional surveyor for the lot. Individual councillors lobbied to have their roads done first, and those whose wards covered mainly water tried to get the RDC to pronounce on dykes and staithes instead. Parishes kept up the pressure. Paston, to take one example, complained that their roads were 'in a shocking state'; the surveyor retorted that they were '50 per cent better than last year', but when the parish threatened to go to the county, the RDC sent the surveyor back for a further report.[59] Bicycles and then motor cars were using mud roads for the first time, generating clouds of dust by day and driving dangerously without lights at night. Norfolk's RDCs called for tough speed limits, licenses (introduced in the 1909 Budget), and controls on horns and headlights. Motorists after all, *en route* to Poppyland, were seldom local ratepayers. Along the north coast, councillors faced the problems of sea defences, intermittent flooding, breakwaters, and groynes. Miss Pigott, though far more interested in poor law work, tried quite hard to get her RDC to build new timber groynes; and both Docking and Erpingham tried to stop builders stripping the beaches of flints.[60] After the initial flurry, highways work became routine road maintenance, and women councillors left the men to get on with it. In no Norfolk RDC did a woman councillor sit on a highways committee though they sometimes spoke in the full council debates.

Sanitary work was different. As Alice Busk had said, when trying to educate women councillors into wider responsibilities than merely poor law work: 'The RDC is the preventive agency, the Board of guardians is the curative. Common sense tells us which is the more needful—may I make bold to say, the more sacred work'.[61] In time women came to agree with her.

---

[59] Thetford RDC Minutes, 22 Nov. 1895; *Norwich Mercury*, 4 May 1904.
[60] Blofield RDC Minutes, 11 June 1907; Docking RDC Minutes, 7 Nov. 1906. The RDCs were encouraged by Selby's Royal Commission 1905–6.
[61] Busk, 'Women's Work'.

In their early years the new RDCs appear virtually indistinguishable from old sanitary districts. They regulated and abated those nuisances reported to them by ratepayers or by parish councils. On request, they would require property owners to provide clean water and adequate privies, or insist that smallholders no longer drain their manure heap into the road, or dump rubbish into the pond from which the village got its water-supply. Mrs Hoare of Aylsham RDC reported that cottages in Drabblegate were discharging sewage into the river. She was put on a committee of the RDC which recommended that the council spend £150 installing a new drainage system for Hungate Street. In Mitford and Launditch, Mrs Cobon complained that the worst offenders in her parish were the breweries, Morgans and Bullards, and the Earl of Leicester. To their credit, members slapped abatement orders on their property without fear or favour. Miss Henry, a councillor in Berkshire, commented 'the tenants cannot complain; if they do, they often get a weekly notice. Besides what good would a complaint be if made to a Railway Company or a Canal Company? But Companies have to listen to a district council.'[62] Sanitary committees visited sites, whitewashed and fumigated cottages, sealed polluted wells and bored new ones, cleaned ponds, closed churchyards, cleared rubbish, constructed new drainage and sewage systems, and tried to check gross overcrowding. (It was reported from Devizes that one woman summoned to abate the nuisance of overcrowding, insisted that her children 'were not nuisances, even if two of them did have to go out to another house to sleep'.)[63] Women members were the busiest of all, with the time to make endless site visits, the energy to take up complaints, and the independence to speak of what they had seen.

Within a year or two, members were encouraging their officers not simply to respond to complaints as they came in, but to begin a systematic trawl of their parishes. Individual officers were soon visiting a thousand cottages a year, and finding statutory nuisances of overcrowding, unfitness, and

[62] Aylsham RDC Minutes, 23 June 1896; Miss Henry, *Anglo-Japanese Congress*, 1910, p.15.
[63] *Daily Chronicle*, 24 May 1912, quoted Green, *The Tyranny of the Countryside*, p.141.

bad repair in a fifth of them. They began to test water-supplies on a regular basis, and to inspect dairies and cowsheds as well as slaughterhouses, since contaminated milk was a source of diphtheria and typhoid as well as TB. Members did not spare their fellow farmers. When Forehoe members learnt that the water-supply of one dairy farm was polluted with sewage, making it unfit for drinking, cooking, or washing out milk churns, the owner was given just fourteen days to remedy the situation. The inspector drily reported that the milkmen's caps 'were not a pleasant sight'.[64] MOHs also sought to identify and isolate infectious diseases. They were annoyed when women councillors refused to let them bring Reepham's smallpox or Coltishall's scarlet fever cases into the workhouse where the women feared they would infect inmates, insisting instead that the council pay for skilled nursing in the family's home. When a typhoid fever case was reported at Aylsham, members immediately agreed to provide a trained nurse, disinfect the cottage, analyse the water-supply, provide new drains and privies, and remove the square blood hole of the next-door slaughterhouse.[65] Women members on every board called for an isolation hospital to meet Norfolk's needs, but no one was willing to take the lead. Women councillors also tried to improve children's health by taking the sanitary arrangements of schools more seriously, cleaning the privies more often, and closing the schools during epidemics of measles.[66]

In other words, members and officers alike soon moved away from responding merely to reported nuisances, into a wider concern with public health in which the RDC itself initiated inquiries and reports. Revealingly, when one of the smaller RDCs, Forehoe, eventually decided to make its post of sanitary inspector full time, they hesitated to appoint their existing part time officer, just because he was 'casual' and only responded to complaints, instead of inspecting parishes on a regular basis as they wished him to do.[67] When at the turn of the century, and in the flurry over national degeneration, the MOH annual reports were given extensive publicity,

[64] *Norwich Mercury*, 15 Oct. 1904.
[65] Aylsham sanitary committee Minutes, 5 Aug. 1895.
[66] Henstead RDC Minutes, 10 Dec. 1895.
[67] *Norwich Mercury*, 19 Mar. 1904.

members began self-consciously to compare their findings with those of other RDCs. Blofield's MOH grumbled that his death rates were unfairly high simply because he had a large number of suicides, 'which were always plentiful in Thorpe, no doubt owing to the close proximity of the river for city people'. Given the filth of Norwich's rivers, members agreed that no one in their right mind would wish to drown themselves in Norwich.[68]

The most sensitive indicator of public health, and the one that most bothered women members, was infant mortality, the number of deaths among young children less than a year old. In 1899, infant mortality was nationally at its worst, 163 deaths per thousand live births, higher than at the beginning of the century, despite half a century of Chadwick-inspired sanitary engineering and public health measures. The reason why is not entirely clear, though it must in large part have been due to poor maternal health, engendering premature births and underweight and sickly children. Low-weight babies were vulnerable to convulsions if improperly fed, and had lowered resistance to whooping cough, bronchial diseases, and gastro-enteritis.[69] Infant mortality thereafter fell to 95 per thousand by 1912; but this concealed wide regional and class variations, with inner wards of London, Liverpool, and Glasgow having infant mortality rates ten times higher than the rural gentry. Most parts of Norfolk had infant death rates of 80 to 100 per thousand between 1903 and 1909. If it went much over this the MOHs felt obliged to defend themselves. When at Forehoe the figure for 1909 reached 137 per thousand (some 41 deaths in all), the doctor reported that this included nine premature deaths, eight due to 'injudicious feeding', eleven chest infections, nine cases of convulsions, and some cases of whooping cough and diarrhoea. 'If the nine deaths from premature birth were deducted, the death rate would be much more satisfactory', he reported irritably. And if mothers only

[68] *Norwich Mercury*, 26 Mar. 1904. Norwich had a long history of polluting Thorpe's water-supply, with sewage as well as corpses.

[69] Cf. today's W. European figure of some 10 per thousand. Norfolk's figures may have been affected by the opium tranquillizers which were mixed with milk; with Lincolnshire, it consumed half the opium imports into the country. A. Wohl, *Endangered Lives: Public Health in Victorian Britain*, 1983, p. 34.

learnt how to feed their babies properly, his statistics would be better still.[70]

What members and officers alike were coming to realize was that the public health question, the infant mortality question, and even the labour question of rural England was largely a rural housing question. Walter Crotch, journalist and Erpingham councillor, published his *Cottage Homes of England* in 1902, the same year as H. Rider Haggard's two-volume survey of *Rural England*.

Rider Haggard was chairman of Ditchingham parish council in Norfolk and a local landowner. Around him were cottages 'that look so pretty in summer with roses and ivy creeping about their crumbling stud work and their rotten thatch, but which often enough are scarcely fit to be inhabited by human beings'.[71] Damp earth floors, bulging walls, leaking thatch, broken windows, and no water-supply, meant that families suffered constant ill health. The cottages were too poor to be worth repairing, labourers too poor to afford anything better. The great estates led by Holkham preferred to clear bad cottages rather than repair or replace them, thereby reducing their rate bill and their repair bills, but increasing the pressure on the accommodation that remained. In any case, housing was a pawn in the power game that farmers played with their men. Insubordination could bring eviction. Walter Crotch watched such an event in Wroxham, then an unspoilt and beautiful Broadland village, where a landowner gave notice to six families. The men tramped miles looking for somewhere to live but could find nothing and faced the workhouse. One bright May morning Crotch watched the families with their few belongings turned out on to the roadside. Kindly but overcrowded neighbours did what they could. The children slept out in boathouses down by the river, and donations paid for an old railway carriage in which the families huddled for the next nine months. Eventually the local MP stepped in, bought land, and built them cottages. The hovels they had left remained empty. It had been a political eviction.[72]

[70] *Norwich Mercury*, 10 Apr. 1909.
[71] R. Haggard, *Rural England*, vol. ii, p. 520.   [72] Crotch, *Cottage Homes*, ch. 7.

As rural cottages crumbled, they produced Chesterton's 'horrible paradox' of 'over crowding even where there are not enough people'. The *Norfolk Mercury*'s survey in 1902 could not find an empty house, but found families of ten or more crammed into two- and three-roomed cottages. With wages at 13*s.* a week, half the industrial average, the labourer was not earning enough to feed his family by Rowntree's calculations, let alone pay the 5*s.* a week in rent which would encourage a builder to erect a cottage. Fewer than ten houses a year were being built in most of Norfolk's RDCs—Crotch and Haggard were joined by Progressives and Labour to argue that only government subsidy to rural local authorities would break the vicious circle of poor cottages, poor health, low wages, poor cottages, and rural depopulation.[73]

Mrs Ethelreda Birkbeck, a member of Blofield's RDC, brought the case of a young girl to her board. The child had advanced TB. Her parents had a two-bedroomed cottage. The sick girl slept in one bedroom; parents and five children slept in the other. Two daughters were boarded out with their grandmother, in whose single bedroom slept granny, girls, and an uncle of twenty-one. The guardians gave the family a further 5*s.* a week but they all recognized that what was presented to them as a poor law problem was actually a housing problem. Mrs Birkbeck pointed out in discussion that such overcrowding bred immorality, sickly children, worn-out mothers, and in the longer term the physical degeneration of the race.[74]

Medical officers and women members continued to press the housing question on their councils. In his 1907 annual report, Blofield's MOH 'respectfully submitted this difficult problem to the consideration of the council'. They did not consider it. Two years later he buttressed his report with detailed figures from Halvergate, a small marshland village with seventy-six cottages, of which fourteen with one or two bedrooms had between six and eleven people sleeping in

[73] Chesterton, quoted Bennett, *Problems of Village Life*, p. 74. By 1905, only 32 cottages had been built by local authorities under the 1890 Housing Act, nearly half of them by Jane Escombe in Penshurst. *Norwich Mercury*, 20 Dec. 1902. More generally see E. Gauldie, 'Country Homes' in Mingay, ed., *The Victorian Countryside*.

[74] Blofield Guardians Minutes, 20 Feb. 1906.

them. He 'respectfully' suggested that they adopt the powers of the 1909 Housing and Town Planning Act. They would not. A year later, the MOH was again telling members that he was required to report to the LGB what action, if any, the council was taking under the 1909 Act. The chairman commented, 'We cannot build except at a loss. Are we to put up houses from which we cannot get sufficient rents, and have the other money paid by the ratepayers?' No one replied to what was so obviously a rhetorical question. When, minutes later in the same meeting, members considered closing some unfit cottages, the chairman stated that 'there were two houses we should like to close, but we cannot because of the limited housing accommodation'. No one appeared willing to relate the one discussion to the other.[75]

Yet it was by now a cliché that housing in rural England was worse than in the East End, decorated though it was by roses. Miss Constance Cochrane of St Neots thought that 'it is doubtful whether many of the cottages of the poor have ever been entered by the various male members' of RDCs. She would cycle through country lanes distributing leaflets to cottagers on how to get their homes improved, but noted sadly that unless the parish or the district council took action, most tenants were too terrified to exercise their legal rights.[76]

The housing shortage in rural Essex appears to have been even worse than in Norfolk, since the RDC energetically closed unfit cottages while Londoners bought up fit ones as second homes. Homeless families were erecting shacks of orange boxes, tree branches, abandoned bits of tin and tarpaulin; or sleeping in tents. The MOH of Maldon was reported as saying, 'If met with in Central Africa it would be photographed as showing the poverty and squalor of the inhabitants'. These were all families in work.[77]

By 1911, not all Norfolk RDCs were being as obtuse as Blofield. Smallburgh was erecting six cottages at Horning,

[75] Blofield RDC Minutes, 14 Apr. 1908, 26 Apr. 1910; *Norwich Mercury*, 1 Apr. 1911. Burns's 1909 Housing and Town Planning Act made surveys of rural housing compulsory, simplified compulsory purchase and closing orders procedures, and offered cheap loans from the Public Works Commission.

[76] C. Cochrane, *How Women as Councillors may Improve Sanitation and Housing in Rural Districts*, Paper to Congress of the Sanitary Institute, Glasgow, July 1904.

[77] Cuttle, op. cit., p. 198.

Erpingham six at Briston and Edgefield (women members
checked the plans to ensure adequate larders and economical
grates), while Cromer UDC had started to build artisan
cottages back in 1897. The farmers fought all the way, fearing
not only that council houses would be a charge on the rates
but that they would reduce the villagers' dependence on tied
cottages. Only when district councils were required by law in
the 1970s to rehouse families from tied cottages when necessary, did generations of feuding over rural housing begin to
ease.[78]

Women district councillors were a sizeable group within
Norfolk. In poor law and sanitary work they carried a heavy
committee load, visiting sites, inspecting cottages, talking to
mothers, with far more assiduity than male colleagues. They
gave professional officers loyal support. Unlike women on
urban school and poor law boards or even on London vestries,
they do not appear to have been policy innovators: though
until the First World War it was hard to find any rural
members in Norfolk who were. Erpingham apart, there were
few Labour members on RDCs, and no one else with an
interest in change. Women were, however, accepted without
fuss and without apparently the overt sexual hostility many
women found in public life. To some degree this was due to
their social background, as they came from well-established
gentry or professional families, superior in standing to many
small farmers on the council. It was also due to their own
deviousness and diplomacy, since they carefully avoided
any threat or challenge to male hegemony. Beyond that,
they shared a common knowledge and perspective that
came from rural life. Progressive urban boards and councils
took pleasure in pioneering. In the countryside, change was
suspect and policy evolved only slowly. On urban boards,
women seem to come from one world, the domestic and the
philanthropic, men from another, business and commerce.
Their experiences could be complementary, could be conflicting, but that they were different, and that therefore women
had a distinctive contribution to make, was not usually

[78] Between 1909 and 1912, 15,000 improvement notices and 5,500 closing orders
were made nationwide; but local authorities added only 470 cottages to the stock;
Bennett, op. cit., p. 140.

denied. Such demarcation, such separate spheres, were far less obvious in rural life, where farming, parish work, and rural government interlocked more equally for women and for men. Many women, after all, owned and some farmed their own land.[79]

Women members, however, were still thought to be more familiar than men with the interiors of cottage homes; and far more than most men, they fought resolutely to repair the cottages and improve the lives of country women.

Women councillors were absorbed without stress and fuss into rural local government. They worked hard, they worked constructively, and they worked quietly. Much of what they did cannot be recovered as committees met in private and their minutes have been lost. But in Norfolk certainly, more women held elected office at district level, and took a larger and more effective share in local government in 1900, and the years before the First World War, than they were doing around 1980.[80]

[79] Jacob Bright reported in *WSJ* 1 July 1877; 10 per cent of farmers were women, five per cent of lay patrons of the Church of England, 14 per cent of owners of land of more than one acre.

[80] There were at least 40 women councillors and guardians in Norfolk in 1903, 36 in 1975 following the reorganization of local government, 49 in 1982–3, for the equivalent five district councils (Broadland, Breckland, West Norfolk, South Norfolk, North Norfolk). These are absolute numbers, not proportions. Rural school boards have been excluded.

8

Women and Town Councils 1907–1914

THE Women's Local Government Society was jubilant as it held its December committee meeting in 1905. Not only had a Liberal government been returned, but John Burns, long a sympathizer with the women's cause, had become President of the Local Government Board. In recent years, women had seen their opportunities for public service narrow; in 1899 they had lost their vestry work, in 1902 their school board work, when both were absorbed into the boroughs and counties for which women were not allowed to stand.[1] Now, off went their letter of congratulations to Burns, and with it their prayer that government would finally adopt their bill allowing women to stand for all local authorities. Government did indeed adopt it; and in August 1907 women ratepayers could at last seek election to the country's three hundred and fifty borough and sixty-two county councils.

School and poor law boards, vestries, parish and district councils, these were all authorities with limited, shared, or concurrent powers. The boroughs, particularly the great cities, were the glittering prizes of local government. They were forums of citizenship, comprehensive in function, strongly competed for, highly politicized, highly visible. Since 1835, town councils had acquired powers and sought responsibilities far beyond public order, public health, and public roads. A medium-sized county borough like Norwich not only exercised the functions of today's city and county councils for housing, education, public health, fire and police, amenities, highways, planning, economic development, libraries, and parks; but also those of the non-voluntary sector

---

[1] See above. Tories described themselves as Moderates on the LCC (a term acceptable to the Liberal Unionists and ratepayer organizations with which they were in coalition). From 1906, the term Municipal Reformer was often used as well. See K. Young, *Local Politics and the Rise of Party*, 1975, ch. 3.

of the later National Health Service, such as fever hospitals, mental hospitals, sanatoria, midwifery, and infant welfare; of today's Anglian Water Authority, for water and sewage; of the nationalized undertakings, gas, electricity, transport; as well as a considerable part of the social security system. Sidney Webb engagingly captured some of this sweep when he described the individualistic town councillor, staunchly *laissez faire*, who would none the less

walk along the municipal pavement, lit by the municipal gas and cleansed by municipal brooms with municipal water, and seeing by the municipal clock in the municipal market place that he is too early to meet his children coming from the municipal school hard by the county lunatic asylum, and municipal hospital, will use the national telegraph system to tell them not to walk through the municipal park, but to come by the municipal tram to meet him in the municipal art gallery, museum and library, where he intends to consult some of the national publications in order to prepare for his next speech in the municipal town hall in favour of the nationaliza-tion of canals and the increase of government control over the railways system ... 'Socialism, Sir', he will say, 'don't waste the time of a practical man with your fantastic absurdities. Self help, Sir, individual self help, that's what made our city what it is'.[2]

This growth in responsibility had been accompanied by, and made possible by, enhanced rateable value and profitable municipal undertakings. Norwich's rateable value doubled, and Newcastle's rateable value quadrupled, between 1870 and 1907. In London, increased rates creamed off most, if not all, of enhanced property values from the 1890s.[3] At a period when problems of urban deprivation stemmed, not as today from urban decline, but from rapid urban growth, ratepayers

[2] Quoted D. Fraser, *Power and Authority in the Victorian City*, 1979, p. 171.
[3] N. McCord, 'Ratepayers and Social Policy', in P. Thane, ed., *The Origins of British Social Policy*, 1978, p. 34. Newcastle's rateable value rose from £449,000 to £1,640,000; its population from 128,000 to 260,000. For Norwich, see *City of Norwich Abstract of Accounts, 1880–1915*, and Corporation Year-books. Its population and rates rose by a third, but its *per capita* resources doubled. In Leeds between 1895–9 and 1910–14, rates rose from £395,000 to £730,000, central government grants from £35,000 to £245,000, and profits from utilities from £28,000 to £103,000. See B. Barber, 'Municipal Government in Leeds 1835–1914', in D. Fraser, ed., *Municipal Reform and the Industrial City*, 1982, p. 107. For London, see A. Offner's splendid *Property and Politics 1879–1914*, 1981. Nationally, local authority expenditure rose from £30m. in 1871, to £55m. 1889, £101m. 1900, £161m. 1913.

could see that heavy investment of public capital in the urban infrastructure, in roads, sewers, and drains, in lighting and transport, paid dividends in more housing, more industry, and greater rateable value; that slum clearance, the cleaning of privies and rivers, the more effective removal of refuse, could help to halve infant mortality in fifteen years and help build a healthy and more productive labour force. Local government expenditure was visibly contributing to economic growth and urban prosperity.[4] In Birmingham, Chamberlain had long said that the city cleared its slums for public health but invested in gas undertakings for public profit.

Professional staff likewise multiplied. Newcastle in 1882 had seven inspectors and two clerks under its Medical Officer of Health. By 1907 the MOH supervised twenty-four inspectors, seven health visitors, and six clerks. Similarly, Norwich had an MOH, three sanitary inspectors, and a clerk in 1891, but by 1910 its MOH had under him a deputy, a public analyst, seven male sanitary inspectors, five lady inspectors, and three clerks.[5] Some of these officials became 'statesmen in disguise', inspecting, exposing, publicizing, cajoling, initiating.[6] Brought together in professional associations (the National Association for Local Government Officers was founded in 1905), backed by professional journals, the new sanitary inspectors, surveyors, education officers, and city architects were naturally to be found on the side of greater municipal activity.

Not, it should be noted, that this activity was always or even usually welcomed by its recipients. A reading of local government from below offers a very different perspective. School teachers imposed standards of discipline, punctuality, and cleanliness that poorer working-class families found hard to meet; school attendance officers denied families their children's earnings and brought recalcitrant parents to court; education committee members sent rebellious children to

---

[4] D. Dawson, 'Economic Change and the Changing Role of Local Government', in M. Loughlin *et al.*, *Half a Century of Municipal Decline 1935–1985*, 1985, p. 149. About half of local expenditure went on the material environment.

[5] McCord, op. cit., p. 34; *City of Norwich Abstract of Accounts 1890/1–1910/11*.

[6] O. McDonagh's description of civil servants in 'The Nineteenth Century Revolution in Government', *Historical Journal*, 1958; and G. Kitson Clark, 'Statesmen in Disguise', *Historical Journal*, 1959.

reformatories.. Sanitary inspectors were feared as an inquisitorial police, knocking in the night to see how many people were
sleeping in how many rooms; and as they abated the nuisance
of overcrowding, they brought homelessness, higher rents,
and further overcrowding with them. When trained midwives
replaced the untrained, they not only cost far more but did far
less; they were not so handy, obliging, and comfortable to
have around. Health visitors, eyeing every dark corner of the
living-room and every grubby child, seemed over-critical
of working-class mothers and their homes. Medical Officers of
Health sometimes confessed themselves baffled by the resistance of slum dwellers to slum clearance, even when shiny
new flats were on offer after the War, and could not
comprehend the attachment of the elderly to their homes and
their neighbourhood. Local government officers were indeed
often heroic, mostly well-intentioned, and remarkably honest;
but they were equally often impatient with opposition, dismissive of the difficulties facing working-class families and
mothers stretched all ways at once, and were regarded by
them as intrusive and oppressive. One of the very real
contributions of women and Labour councillors was to bring
somewhat greater sensitivity to the views as well as to the
needs of the urban poor, and to seek consent to local government interference.

Along with increased responsibilities and resources, local
government also acquired greater moral authority, among the
middle classes at least. T. H. Green, Oxford philosopher and
town councillor, was insisting that 'the whole body of citizens
ought to be called upon to do that as a body which under the
conditions of modern life cannot be done if everyone is left to
himself and cannot possibly be left undone without the whole
body suffering'. The Revd Robert Dale in Birmingham and
Canon Barnett in Whitechapel were investing local government with the dignity of religious commitment.[7] That ethic of
community service and social solidarity, also evident in
changing attitudes to poor relief, was toughened at the turn of
the century by the concerns of the national efficiency
movement, worried that Britain's military physique and
economic productivity were falling behind those of her

[7] M. Richter, *The Politics of Conscience: T. H. Green and his Age*, 1964, p. 363.

competitors; and was sharpened by the new perceptions of poverty drawn from the empirical investigations of Booth, Rowntree, and the Fabians, as well as from university social science departments and settlements. Social policy was becoming 'high politics'.[8]

The rise of the Labour movement, which demanded collectivist responses to destitution and unemployment, was giving social reform a new saliency and political urgency. On the Right, Chamberlain's social imperialism was linking social to tariff reform. 'New Liberalism', focused around the writings of Masterman and Hobhouse, was translating T. H. Green's idealist concepts of moralized capitalism and social harmony into practical and interconnected Progressive policies. Masterman's phrase, 'the *waste* of it all',[9] was being applied to slums, infant mortality, unemployment, ignorance, squalor, and disease. Such waste—women picked up the phrase—was cruel, costly, and inefficient. Ethics and economics, instinctive compassion and social audit, all pointed to an expanded role for local government: and to an emphasis on issues, such as infant mortality and housing, where women believed they had a particularly womanly contribution to make.

Candidates and Campaigns

The first town and county elections were to take place in November 1907, only three months after the passing of the bill. The WLGS co-ordinated the women's network, and brought forward seventeen candidates outside London. Five were successful. The frail Elizabeth Garrett Anderson was, appropriately, elected for Aldeburgh, a small fishing and retirement borough on the Suffolk coast. Miss Edith Sutton, a co-opted education member, and granddaughter of the Sutton Seeds civic élite, was returned unopposed in the biscuit town of Reading, as was Mrs Elizabeth Woodward for the township of Bewdly, near Kidderminster, popular as a hotel owner because she had erected public halls when the local council

---

[8] J. Harris, 'The Transition to High Politics in English Social Policy 1888–1914', in M. Bentley and J. Stevenson, eds., *High and Low Politics in Modern Britain*, 1983.

[9] C. F. G. Masterman, *The Heart of the Empire*, 1901, p. 48. On New Liberalism see M. Freeden, *The New Liberalism: an Ideology of Social Reform*, 1978; H. Emy, *Liberals, Radicals and Social Politics 1892–1914*, 1973.

would not. In Oxford Miss Sophia Merivale, daughter of the late Dean of Ely, was elected with the help of the Green–Toynbee circle; Miss Dove, who founded a girls' school at Wycombe, with the help of her staff. Mrs Lees, the lady squire of Oldham, came in a few weeks later on an aldermanic by-election.[10]

In 1908 they were followed by Miss Margaret Ashton in Manchester, well connected and well known for her work in Lancashire Liberalism and women's suffrage. The Liberal Mrs Hughes joined Miss Merivale, her brother, and her father on Oxford town council.[11] The 1909 elections raised the number of women councillors to eleven, with the election of Miss Eleanor Rathbone in Liverpool; of Miss Helen Hope, a guardian and daughter of a school board chairman, in Bath; and two conservative women, the suffragist Miss Coulcher and Mrs Chapman for Ipswich and Worthing. 1910 saw three more women elected, Miss Ada Newman, daughter of a wealthy iron master, and co-opted LEA member, in Walsall; Miss Elizabeth Bannister, a headmistress in South-end, and Miss Maud Burnett, daughter of a prosperous Tynemouth merchant, and co-opted LEA member. A year later in 1911 two women joined Birmingham council, Mrs Ellen Hume Pinsent, nationally known for her work with the mentally handicapped, and Miss Marjorie Pugh. Manchester elected Mrs Redford, a member of their Distress Committee. Miss Alison Ogilvy was returned in Godalming. 1912 brought two women doctors forward at Bromley and at Wimbledon, while in Staleybridge Mrs Summers, a wealthy ironmaster's widow and pillar of the charitable community, joined her council. Miss Mabel Clarkson, a solicitor's daughter, and guardian, was elected for Norwich in 1913 as was Miss Stancomb Wills in Ramsgate and Miss Clara Martineau in Birmingham. Meanwhile their numbers were growing on the London boroughs and in 1910 two women were elected to the LCC.

[10] *Kidderminster Times*, 21 Sept. 1901, *Kidderminster Shuttle*, 2 Nov. 1907 (Mrs Woodward). Women candidates were defeated at Bolton (Sarah Reddish), Newcastle (Ethel Bentham), Leicester, Liverpool, Hull, Manchester (Margaret Ashton).

[11] Miss Ashton, daughter of High Sheriff of Lancs., sister of an MP, sister-in-law of James Bryce. Women were defeated in Bristol (Helen Sturge), Yarmouth (Ethel Leach), Norwich (Annie Reeves, ILP candidate), as well as in Bolton, Hyde, Rochdale, Salford, Keighley.

Like school and poor law boards, women were not unsuccessful in London and its suburban towns;[12] and put up steady performances in the spa, sands, and spires towns of England.[13] They found it very much harder to win seats on the great city councils. In Manchester, Birmingham, and Liverpool only women with an outstanding reputation were successful. Women failed in Bristol, Bradford, and Sheffield, where hitherto they had a good track record of elected office, and seem not even to have sought a seat in Nottingham or Leeds. Plymouth, Portsmouth, and Southampton as ever were indifferent to the rights and wrongs of women. With the exception of Miss Mary Noble, a local landowner and local historian, who was returned unopposed for Westmorland county council, the shire counties remained beyond women's reach.

Together with a few women on Welsh councils, some fifty women were borough and county councillors at the outbreak of War in summer 1914. As elections were then suspended, these women found themselves the natural organizers of the women's voluntary and involuntary War effort in their towns, finding work for unemployed women, sanctuary for refugees, convalescent homes for the wounded, and pensions for their wives. They promoted War savings schemes, and pressed for women police.[14]

Their experiences inevitably were very various, as some sat on tiny and wealthy boroughs like Aldeburgh (2,200) and Bewdly (3,000) which met only quarterly, barely distinguishable from UDCs and spending most of their time on roads and drainage. Others joined substantial county towns of 50,000 to 150,000 people—Norwich, Ipswich, Reading, Oxford—which took their education, their housing, and increasingly their

---

[12] Seats won in Godalming, Bromley, Wembley, High Wycombe; lost in Ealing, Windsor, Croydon.

[13] Seats won in Bath, Oxford, Norwich, Worthing, Southend, Aldeburgh, Ramsgate, Tynemouth; lost in Yarmouth, Eastbourne, Scarborough, Cambridge, Folkestone, Bournemouth, Chester, St Ives, Cheltenham.

[14] By 1918, 15 women were elected or co-opted to 80 county boroughs; 11 on 245 town councils, 8 on county councils including the LCC, 23 on London boroughs, 19 on UDCs, and there were 1,585 guardians of whom 200 were also RD councillors. When elections were resumed in 1919, numbers soared—74 on county boroughs, 58 on town councils, 46 on county councils, 142 on London boroughs, 2,039 guardians. WLGS Annual Reports, 1914–1920 incl. See also Apps. B and C, pp. 484–5.

town planning functions seriously, though responding to them very differently. A few made their way on to great cities— Liverpool, Birmingham, and Manchester, with populations of over half a million and budgets of around £1½ million a year.

Even on the smallest borough they appear formidably well qualified. For the first time a new breed of university-educated, salaried, or self-employed professional women was offering educational and medical expertise to the community. Invariably they were already guardians, LEA co-opted members, members of distress committees where they sought public works for the unemployed, or of care committees ensuring the well-being of the schoolchild. Very few were now COS-trained, though many continued to offer philanthropic service in the settlement movement and in the less censorious Guilds of Help. Even the wealthiest, Mrs Lees and Mrs Summers, were already hard-working, self-disciplined, and experienced local authority committee members before their election. Almost always women councillors were suffragist, feminist, and whatever their politics, progressive in practice. Having already worked in local government, often with professional related occupations, and often active members of a political party, women councillors seemed to share a clear vision of what municipal enterprise could achieve. They were dismissive of individualism and doubtful of the efficacy of much so-called philanthropy. Many moved to the radical end of the political spectrum during their public life and, like Miss Clarkson of Norwich, Miss Sutton of Reading, and Miss Ashton of Manchester, found their homes in the Labour party after the War.[15]

As women were by now welcome and familiar figures on the local government landscape, why, said the *Municipal Journal*, were so few women coming forward as candidates and even fewer being elected? 'On the whole, women do not aspire to municipal honours. Perhaps they think they deserve better things than the dry routine of local government. But we warn them that St. Stephens is a Palace of Disillusionment. There is more work for women on municipal councils than in the

[15] For further evidence on this point, see O. Banks, *Becoming a Feminist: The Social Origins of 'First Wave' Feminism*, 1986, p. 23.

Imperial Parliament'. But with so few women standing, and only a tiny fraction voting, 'it hardly seems to matter whether they get the vote for other purposes or not'.[16]

This stung, as it was meant to. Outside London, the WLGS patiently explained, women had to be ratepaying electors before they could become councillors: just like guardians before 1894. All daughters and most married women were automatically barred from office. As Miss Ashton said, 'Picture to yourself a council from which all married men were excluded—who would be left?' As an unmarried woman herself, 'she wished to emphasize that it was the married women who were needed on town councils. They were the most valuable in public work'.[17] The Croydon women's movement petitioned John Burns to remove the ratepayer requirement: of their three LEA co-opted members and four guardians, from whom councillors might be drawn, none were householders; nor were any of Sheffield's eleven women guardians, three LEA members, and pension committee members.[18] Not until the summer of 1914 did government bring the borough and county councils into line with district councils, allowing residents as well as ratepayers to stand for election, and therefore wives and daughters as well as women householders. The War intervened, and elections were suspended.

Of those women who did stand, relatively few were successfully elected: perhaps one in three or four, in contrast to school and poor law boards where women were not only usually elected but often headed the poll.[19] School board seats, as has been suggested, were much prized, but women could break the male monopoly with the strategic use of the cumulative

[16] *Municipal Journal*, 4 Oct. 1907, 17 Sept. 1909, 28 Oct. 1911; *Englishwoman*, 21, 1914, pp. 14 ff. This was a point frequently made by the *Anti-Suffrage Review*; see Violet Markham's speech, reported May 1912, Lady St. Helier, Dec. 1912.

[17] *Municipal Journal*, 28 Oct. 1911; NUWW branch reports from Bradford, Occasional Papers, May 1914, p. 23.

[18] Petition of Croydon NUWW, Women's Suffrage Society, Women's Debating Society, WLA, Fabian Society, Croydon Education League, and the Croydon Ethical Fellowship, *Municipal Journal*, 29 Jan. 1909: petition of Mrs Wycliffe Wilson, 14 city councillors, 12 education committee members, and 22 guardians, *Municipal Journal*, 12 Mar. 1909.

[19] *EWR* 15 Apr. 1907, reported that of 86 women standing as guardians, 62 were elected and 29 came head of the poll.

vote. Poor law seats were less in demand, and over time, providing a woman did not insult the board by challenging a sitting member, a patient, tactful, and preferably silent woman could usually find a seat. City council elections, however, were of a different order. They were dry runs for general elections; and parties and government took delight or despondency from their outcome. Seats were in great demand. Just as men used poor law seats as a stepping-stone to the council, a council seat could be a stepping-stone to parliament. Vacancies were few, and ward committees naturally chose the business man with cash or the Labour man with a committed constituency before the woman with neither. To be told that women did not warrant the parliamentary vote because they were not winning local government seats was bitterly unfair, 'hypocrisy', wrote the suffragist Conservative Marion Chadwick. It was precisely because women did not have the parliamentary vote that they were not taken seriously at local level. Elections were run

almost exclusively on party lines ... the political associations and agencies are all-powerful and they make up their lists from amongst their party adherents, too often without much regard to the fitness of the candidates. Everything is done with an eye to the Parliamentary elections. A popular Mayor or County Councillor may be the next Member of Parliament. Thus it is of primary importance to strengthen the party and the efficient administration of local affairs is a secondary consideration. It is only natural, therefore, that a man who has the Parliamentary vote should usually be preferred to the woman who has no political value, however high her qualifications may be. If these political associations can be persuaded to adopt a woman as one of their candidates, well and good; but they will rarely do this, unless there is a shortage of candidates for some doubtful ward, or they have reason to dread a three cornered election.[20]

The most women could hope for at the start was the chance to fight an unwinnable seat and prove their pluck.

However, if women served their apprenticeship gallantly,

[20] M. Chadwick, *Women's Work in Local Government: A Reply to Mrs Humphrey Ward*, 1912, p. 2. Birmingham's 120 city council members in 1914 included 42 manufacturers, five merchants, 14 lawyers, 13 other professional men, 15 retailers, 10 working men, 5 builders, 2 farmers, 2 women, 12 miscellaneous. A. Briggs, *History of Birmingham*, 1951.

polled a good result, were determined to stay around, and made it clear they were loyal party workers, space might be found for them the second time round in a better seat. Miss Burnett changed wards and came in on the second attempt, threatening everyone with a third effort if she failed; as did Margaret Ashton of Manchester, Miss Hope of Bath, Miss Newman of Walsall, and Miss Clarkson of Norwich (who was allowed to transfer from the Tory ward of Coslany to the ever-marginal ward of Town Close). Women did a little better if, as often happened, they already represented part of their ward as a guardian, but even so, women like Mabel Clarkson of Norwich who were elected as guardians were often defeated when standing as councillors for the same patch.

Women's best chance came as members of a slate. Some smaller boroughs, under 6,000 in size, like Aldeburgh, were undivided; Elizabeth Garrett Anderson came home fourth of six. At Southend, the whole town was being re-warded and therefore all councillors had to seek re-election. The ratepayers ran a slate across the town, and included Miss Bannister when she answered their questions on economy satisfactorily if not audibly. She was given what was euphemistically described as a neglected ward but brought it safely home. She retired the next year.[21] In Birmingham a major boundary extension tripled the city's acreage; the city's wards rose from eighteen to thirty. This allowed Ellen Pinsent and Marjorie Pugh to stand in empty seats. Miss Merivale won her Oxford seat in an all-out election. Miss Ogilvy was one of eight candidates for Godalming's six seats. As befitted a Girton graduate she worked the borough hard. Alone of the candidates she issued an address, held public meetings, and personally canvassed a large number of ratepayers. She came in head of the poll. As the candidate with the highest votes, she refused to forfeit the traditional honour of thanking the returning officer for his expedition and his staff for their efficiency. Nor could she resist making a ladylike speech from the balcony. The next day was the sumptuous mayoral banquet, at which ladies appeared only after dinner for the speeches. Miss Ogilvy, an object of some interest to the guests, decided that she was a lady, not a councillor, and sat

[21] *Southend Standard*, 20 Oct. 1910.

alongside the Lady Mayoress. She was gently teased that she should be in the main body of the hall. By moving a motion for women's suffrage at the town council, she subsequently lost much of the early goodwill. The following year saw her absent from the banquet altogether.[22]

Women also seem to have done rather better at by-elections which did not usually disturb the political balance. Thus Eleanor Rathbone won her seat in Liverpool, Miss Newman in Walsall, and Miss Reina Lawrence in Hampstead, though two better known London Progressives, Mrs Bracey Wright, a temperance reformer and vestrywoman, and Mrs Evans, a popular vestrywoman in the Strand, could not win in a triangular by-election.

Mrs Sarah Lees likewise came in on an aldermanic by-election in Oldham in 1907. A wealthy widow, she and her daughters had decided not to retire to salubrious spas but to devote their money to the town. As it was thought to be the ugliest in Britain, this was a more altruistic gesture than it might seem. Mrs Lees had been an LEA member since 1902, was president of the local NUWW and of the Oldham Nursing Association, and helped to finance the suffragist *Common Cause*. She had spent money on schools, houses, nurses, and trees for the town. When the male Liberal candidate for an aldermanic vacancy fell through at the last moment, the Liberal leadership in some desperation approached Mrs Lees. To the chagrin of local Tories, who thought with her wealth she should be above politics, she agreed to stand. The Tories complained that her husband would not have approved, that she lacked business experience, and that the Liberals were manipulating a gentle charitable elderly lady to bribe the electorate with her private purse. Her meetings, reported the Tory newspapers, were attended by '287 ladies and somebody else'; her very willingness to 'fork out sovereigns with a shovel' showed she would not be economical with public money. They ran a local undertaker against her, who made much of the fact that he was a self-made man. Mrs Lees replied that as women were the main clients of local authority services, the council needed a

---

[22] *Surrey Advertizer*, 14 Oct., 4 Nov. 1911; Godalming Council Minutes, 9 Nov. 1912. (I am grateful to the Godalming Library staff for this reference.)

woman's eye in their provision. She won, narrowly, by 727 to 681 votes.[23]

The following year the Tories, still in control, ruefully accepted the inevitable, and to her amusement she found herself nominated by both political parties. As the Tories said,

There was no doubt that everyone present and Oldham generally appreciates Mrs Lees' very great services to the town . . . She did not however come out as Mrs Lees but as the nominee of the Liberal party . . . Mrs Lees was a lady who had done great things, she was a lady of whom Oldham people were very proud, she was a lady who had established the right to the name lady, and consequently conservatives were very diffident in saying anything against her from a political standpoint . . . What they did oppose was the nomination of a lady candidate. They did not believe in the principle of having lady councillors, but if they were to have lady councillors, they would sooner have Mrs Lees than anyone in the town.[24]

Her seat was never again challenged and passed without contest in time to her daughter, a poor law guardian. Two years later, however, the Tories nearly choked on her nomination by the Liberals as mayor, at a time when the office of mayor was closer to that of chief executive than today's ceremonial post. In the sour words of the *Oldham Standard*, 'She had come to the front of municipal life almost with a bang, as it were, and her liberal friends have receded behind her. This is due to some extent to her fine personality, but much more to the fact that her immense wealth enables her to do things which catch the public eye', like giving parks to the town. The Liberals 'paraded her position, munificence and influence as though it were the special preserve of the Liberal party'. So irresistible a candidate had she become—every time she proposed an improvement to the town, she offered to pay for it—that the Liberals fought and won the 1910 local elections largely on one campaign issue, that Mrs Lees be mayor. They won and she was. Lady Frances Balfour wrote in delight that 'the fierce masculine mind simply shuddered at

[23] *Oldham Standard*, 15 Nov. 1907, *Oldham Evening Chronicle*, 19 Nov. 1907. Her papers include a charming note from her ward secretary, a working man, tendering 'our heartfelt thanks for the work you have so diligently performed on the Oldham town council'.

[24] *Oldham Standard*, 30 Sept. 1908.

the idea'; the anti-suffragist Asquith found himself required to invent suitable honours to bestow on ladies who were serving as mayors while unable at parliamentary level to vote. Staunch Liberal ladies found themselves Dames after the War.[25]

To a lesser degree, Mrs Summers filled the same role in the much smaller borough of Staleybridge with its 26,000 population. She funded endless supplies of clogs for children, prizes for schools, trees for gardens, scholarships for students, toys for Christmas, clinics for the sick poor, playing-fields for the working poor, holidays for the convalescent, and rest and recreation for nurses and teachers alike. Like Mrs Lees, a lady bountiful, she was also a co-opted LEA member, a staunch suffragist, and a keen Liberal. When her husband died, the Tories were willing to allow her a casual non-party vacancy, but she refused, preferring to fight a sitting Tory for a full-term seat on the Liberal ticket. She won in an otherwise low-key election, and the following year the Liberals swept back to power after forty years in the wilderness.[26]

Local 'Queen Victorias' who should but would not remain above party politics, were one problem for the Tories. More commonly, the problem was the reverse. Women insisted on standing as independents, although they were known to be closet Liberals. The WLGS encouraged women to stand as independents in order to foreground their claims of gender rather than party or class, and to pick up Tory women and clerical votes. Miss Burnett of Tynemouth asked for votes as 'a representative woman' and wanted to avoid 'party prejudices closing doors to canvassers'. Many Tynemouth men also stood for election without party label, making her position much easier. Miss Bannister proudly reported that her nomination papers in Southend were signed entirely by women, and her campaign was organized and run by women. Miss Edith Sutton was asked by two wards to stand on a party ticket. 'I refused. It is possible for women to be elected and to serve on non-party lines, and I said I would rather not do the work at all than tie myself to either party . . . a vacancy

[25] *Oldham Standard*, 3 Nov. 1910; Lady F. Balfour to Mrs Lees 27 Nov. 1910, Fawcett letters; Asquith note, 18 May 1911, ibid.
[26] *Staleybridge Reporter*, 2 June 1938, 8 Nov. 1913.

occurred, and both the Liberals and Conservatives went about getting signatures for my nomination papers, and I was returned unopposed.'[27] Commentators, then as now, bemoaned the intrusion of party politics into local government, where issues of local concern could and should, they thought, be sorted out by mutual goodwill.

It was an honourable but increasingly inappropriate stance. The party system was already more open and transparent than the patronage system, less corrupt and more accountable than an oligarchic system, more accessible for poorer candidates than standing as a private individual. It encouraged political participation, higher polls, more scrutiny by an organized opposition, and fuller press reports. Though councillors cross-voted on issues of temperance, or feminism, they shared common party views on the role of municipal enterprise, how much, how fast, and by whom. Even the paradigmic siting of street lamps and planting of street trees represented a political choice—Marion Phillips was to complain bitterly in Kensington that Holland Park got the street trees at the expense of the poorer North Kensington. And though councillors might, irrespective of party, agree about providing public goods, such as sewers, drains, and roads, when it came to questions of rate-supported milk depots, swimming-baths for working-class children, the harrying of landlords to repair slum property, or the enforcing of contract compliance with trade-union wage rates, these were party political issues, because they redistributed public resources and the social wage from richer to poorer wards.

Indeed, the very reason that women were unhappy as co-opted LEA members was precisely because they could not take part in the more general council debates on the allocation of resources and the determining of financial priorities. As Miss Burnett had said, 'she wished to follow up that work in the council itself'. Women might win the argument in committee. To win votes in council they needed the support of the majority, and that came from group meetings and party

---

[27] WLGS Minutes, 17 June 1908 (Miss Burnett); Edith Sutton, speech at Anglo-Japanese congress, p. 9. More generally, see D. Fraser, *Power and Authority in the Victorian City*; G. Jones, *Borough Politics* (Wolverhampton) and H. Daunton, *Coal Metropolis: Cardiff 1870–1914*, 1977.

loyalty. School and poor law boards were essentially precept-ing and single-issue authorities. They could get what they asked for. Development in one dimension was seldom at odds with or at the cost of development in another. So women members may have been somewhat unsophisticated about the significance of the central establishment committees, finance, staff, general purposes, in determining who got what in local government, and the importance of party solidarity in deter-mining that anybody got anything at all.

Women, not surprisingly, tried to have it all ways round, to the immense and understandable irritation of party managers. Women would stand as independents, claiming that, unlike other people, they needed unfettered judgement and freedom of action, as though their consciences were more tender than the rest; while at the same time asking and expecting the established parties to make space for them in winnable wards, which would thus lose a reliable councillor of their own. Sarah Reddish, trade-unionist and Women's Co-operative Guild organizer, fought a Bolton ward in 1907 that was normally a Liberal stronghold. The other two parties each ran one candi-date for a two seat vacancy, the Tories shrewdly, the Liberals gallantly. The Liberal Councillor Scott said, 'there was a lady candidate in the field and he would frankly say that he would not put any obstacles in the way of the election of Miss Reddish . . . for ladies could do good work on public bodies'. Liberal aldermen graced her platform. In the event, the Tory vote held; but the Liberal vote split three ways, some voting Liberal and Miss Reddish, some plumping only for the Liberal, and some voting for the men, Liberal and Conserva-tive. Miss Reddish cost the Liberals a safe seat.[28] In 1909 Miss Burnett stood as an Independent. Though popular (she had got the LEA to provide spectacles for short-sighted children), she was defeated by the party candidates, a grocer and fishmerchant. The following year she changed wards, and as hon. sec. of the local women's Liberal association, had a better organization behind her. The official Liberal sensibly decided to retire, the Liberal party gracefully acquiesced. She for her

[28] Miss Burnett, *Shields Daily News*, 22 Oct. 1910; Miss Reddish, *Bolton Journal*, 25 Oct. 1907. (For references to Miss Burnett, I am most grateful to Mr Eric Hollerton, South Shields Library.)

part promised them covert support. 'They knew her politics. As her father was a Liberal, so was she; but she was perfectly determined that it was not right that politics should come into the council. Her opponent was putting too strong a meaning upon her being a liberal because she had not a vote. (laughter)'.[29] She beat her Tory opponent, an aerated-water manufacturer, in style. Likewise Miss Ogilvy made much of her independence in Godalming but her notice asking for canvassers was pinned up in the local Liberal club. Mrs Swann canvassed in Northampton on the dual ground that she was an Independent and her father a Liberal alderman. Margaret Ashton's independence was even more equivocal, reflecting her sense of betrayal. As the Liberal party refused to enact women's suffrage, she was standing for independent liberal principles and not for the Liberal party.[30] When she was given a straight fight with the Tories, she won.

In Liverpool, Eleanor Rathbone stood as an Independent in Granby ward where the sitting councillor had died. Her secretary, election agent, canvassers, party workers, and supporters were all women. As she could command the princely Rathbone influence, the energy of the women's Victoria Settlement, the standing of the University's School of Social Studies, and the philanthropic street knowledge of the Central Relief Society, all of which she or her father helped found, the parties bowed to the inevitable and would have allowed her the seat, until a working man, who had fought the seat for the socialists in 1907, refused to be intimidated and stood again. Radicalized by her study of dock labour, this was not the opponent Miss Rathbone would have wished to fight and defeat, but she had no choice. As she wrote to her mother:

They seem quite satisfied with the way things are progressing so far as can be told yet. We shall know better after tonight. People seem to think I ought to be safe to get in, but you can never tell: it is a queer sort of election. Both sides are very cross at not having a candidate of their own and are doing a great deal of grumbling at me and at each other.

The meeting last night was very successful, the room quite packed

[29] *Shields Daily News*, 9 Oct. 1909, 22 Oct. 1910.

[30] *Manchester Guardian*, 15 Oct. 1908; see also Miss Maud Lindsey, *Islington News and Hornsey Gazette*, 1 Nov. 1912; Miss Mary Beeton, *Paddington Mercury*, 25 Oct. 1912.

long before it began, and hundreds turned away. Everybody said it was very rare, almost unheard of, for a mere Municipal election candidate's meeting to attract so many, but of course that merely means curiosity at the novelty of a woman candidate, and doesn't really mean much as to prospects at the poll.[31]

She won by 1,066 votes to 516 and then made her peace with the Labour movement by helping the socialist candidate in a by-election, standing on a lorry, said the Tory council leader, Sir Charles Petrie, indignantly, 'spouting to working men at the dock'. At the next election the Tories gave her a rough ride. Liverpool election campaigns were notoriously virulent, and the Tories were known to orchestrate sectarian violence, dump cartloads of rubbish in enemy gardens, and hire heavies. She was now opposed by the local Tory chairman, proclaiming himself a true Protestant, a coded licence to beat up Catholics and Irish. Meanwhile, as the industrial unrest mounted, 75,000 men were on strike, and 2,300 soldiers and 4,000 special constables had to be drafted into the city to keep order. She won 'in the face of a most rancorous attack' and her Granby seat was thereafter her fief until in time it passed to Margaret Simey.[32]

With barely an exception, women councillors standing as Independents only won when the Progressives stood aside and effectively endorsed them. Even the Labour party, with a large natural constituency, found it difficult to win seats in a three-party fight; it was impossible for women. In Tory Oxford, Miss Merivale, almost certainly tory, stood as an Independent against three official Tories, and collared the women's votes, the latent Liberal vote, and the working-class vote. In her words, they thought 'a lady would have nothing to gain, whereas they suspected that each man councillor had his own axe to grind. They thought I would give more attention to their interests.'[33] On election day she was accosted by a 'well dressed man' (did that make it worse?) 'who began looking at my poster; when he had read it, he said with a sort of cheerful tone and knowing wink, "She won't get in, will

[31] Eleanor Rathbone Coll., xiv 1,2., 5 Oct. 1909.
[32] *Liverpool Courier*, 29 Oct., 1 Nov. 1910. *Liverpool Daily Post*, 27 Oct. 1910. More generally see P. J. Waller, *Democracy and Sectarianism: A Political and Social History of Liverpool 1868–1939*, 1981.
[33] WLGS annual meeting, 10 Mar. 1908.

she?" So I replied, "I am that she, and I hope she will."' And added, 'it requires strong nerve and a tough skin, besides plenty of self-sacrificing and zealous supporters to carry one through'. She came in head of the poll; but as she studiously distanced herself from the women's movement, women Liberals successfully ran Mrs Hughes under the Progressive ticket the following year.[34]

Over in Yarmouth, Mrs Leach, calling herself an Independent, fought a Tory in the Tory Gorleston ward. Her opponent had only one campaign issue, 'whether ladies should be allowed to participate in municipal work'. Election day was marked by rumours of a suffragette party descending from London to help her. Photographers gathered. She ended up on the front page of the *Daily Mirror* but not inside the town hall. That had to wait until after the War.[35] In like manner, Miss Pugh stood as an Independent Liberal in a Birmingham aldermanic by-election, challenged by a Tory opponent on the simple slogan, 'Do you want to be represented by a businessman or a suffragette?' Slightly short of content for his election manifesto but owning a new bicycle, he decided he was in favour of cheaper fares, more education, lower rates, and more cycling clubs. She won with the help of the WLGS and a strong women's movement amid the complexities of Birmingham's divided Liberalism. Mrs Ellen Hume Pinsent, a former member of the Royal Commission on the Feeble-minded and a co-opted LEA member, was elected as a Liberal Unionist the following year, on the safe anywhere programme of efficiency, economy, and progress.[36]

The WLGS still tried to encourage women to stand as independents, with or without the collusion of official parties; and some of their London members were in 1913 to join a fledgling Women's Municipal Party, led by the Duchess of Marlborough, which sought to place non-party women in council on an agreed programme, to safeguard the interests of women staff, and to mobilize the women's vote. It enticed

[34] *Oxford Chronicle*, 6 Nov. 1908.

[35] *Yarmouth Mercury*, 31 Oct. 1908, 7 Nov. 1909, *Daily Mirror*, 3 Nov. 1908.

[36] *Birmingham Daily Post*, 31 Oct. 1911; Birmingham Women's Suffrage Society, minutes, May 1912; Municipal election folios, Birmingham Public Library. Birmingham had two Liberal parties after Chamberlain's secession over home rule, and many distinguished families such as the Kenricks were split down the middle.

Eleanor Rathbone, the Fabian Mrs Pember Reeves, and the Whig Lady Frances Balfour on to its Executive; and Mrs Stanbury of the WLGS acted as its organizer, but it seems to have disappeared shortly after the War. As Marjorie Lees of Oldham said to would-be women candidates, 'In a town where the party system is as strongly entrenched as it is here, if you want to get women candidates *returned*, I should advise you to try and work it through your party associations. An Independent candidate might be of great educational value, but I fear it would not be successful in returning the candidate which is after all what we are out for.'[37]

The London Experience

The bigger the city, the tighter the party whip. Nowhere was this more the case than London. Few women even tried to stand here as independents. London women candidates were eligible to stand for office if they were residents, as, unlike provincial women, they did not have to be ratepayers;[38] but as they would insist on standing as Progressives when most of London was swinging Tory, their success rate was little more impressive. In 1909 only four of the twenty-eight boroughs were Progressive, in 1912 just three. Candidates who were women and Progressive, let alone ILP, were thus doubly disadvantaged. The Tories were better organized with fifty paid agents to the Liberals' five, and had therefore built up the electoral register and appear to have brought out the tory women's vote. The Liberals could neither register their lodgers nor chase their removals, and so lost perhaps a third of their support. The swing was also political, a local vote against a national Liberal government, a ratepayers' vote against heavy Liberal rate levies (spent, the Tories alleged, on

[37] Lees Coll., notes for a speech u.d., item 8. Cf. *Englishwoman*, 1 Jan. 1910, pp. 282–9; and leaflets, *The Wants of the Women's Municipal Party*, 1913, *The Reason Why it has been Formed; The Aims and Objects of the Women's Municipal Party* (Fawcett Library). Its programme included expanding education; a pure milk supply; housing reforms and women's municipal lodging-houses; more baths, washhouses, and public lavatories; more open space and playgrounds; better lighting and refuse collection; stricter enforcement of Acts against unfit housing, unsound food, and smoke pollution; and more women sanitary inspectors and health visitors. *Objects and Programme of the Women's Municipal Party*, 1915, p. 5.

[38] Because the London boroughs were under the LCC and were thus considered similar to RDCs and UDCs.

an inefficient Works Department and an extravagant river-boat service), while property values slumped. Liberal returns were not helped when Labour candidates also stood and split the Progressive vote.[39]

The first London women to win seats were Miss Reina Lawrence and a Quaker social worker, Miss Mary Balkwill, who captured casual vacancies in Hampstead. In the 1909 triennial elections, sixty-four women stood in nineteen boroughs. Two Conservative guardians won seats in Kensington; Mrs Salter took a seat for the ILP in Bermondsey, and four Progressives won seats in Islington, Paddington, and St Pancras. Three years later in 1912, twenty-two of the fifty-four women candidates were successful, eight Tory Moderates, eight Progressives, two Independents, and four Labour (two in Kensington, two in Woolwich).

It was all very difficult, but Labour women, like the Liberals, now had a support organization in the field, the Women's Labour League, founded by Margaret Ramsay MacDonald in 1906, reaching out to politicize the wives and daughters of Labour men. At its core were a cluster of remarkable women including Margaret Bondfield, Marion Phillips, and Ethel Bentham, all MPs after the War, as well as the shining and incandescent Mary MacArthur, who in 1911 found herself with twenty strikes in South London's female jam and confectionary trade, and with WLL help won advances in eighteen of them.[40] By 1910 the WLL had seventy branches, by 1913 it could claim six councillors, thirty-one guardians, many women on education and care committees, and the full support of the parliamentary Labour party.

The first ILP woman to win a seat was Mrs Ada Salter in

---

[39] *The Kensington Express*, 25 Oct. 1912, estimated that of 1,362 London borough councillors, 1,004 were Tory, 301 were Progressive, 57 were 'nondescripts'. In 1907, 100,000 women were thought to have voted, mainly Tory. Dr Blewett calculates that in 1911 30 per cent of adult males were not on the electoral register because they were removals or lodgers. N. Blewett, 'The Franchise in the U.K. 1885–1918', *Past and Present*, 32, 1965. This paragraph draws heavily on P. Thompson, *Socialists, Liberals and Labour: The Struggle for London 1885–1914*, 1967. For Miss Lawrence's views, see *Hampstead and Highgate Express*, 6 Nov. 1909.

[40] M. A. Hamilton, *Mary MacArthur*, 1925, p. 72. As she spoke on street corners, 'Keep on Miss', said a tired Lancashire woman with a rapt expression, 'It's better than t'seaside'; Annual Reports, WLL; J. R. MacDonald, *Margaret Ethel MacDonald*, 1920 ed.; L. Middleton, ed., *Women in the Labour Movement*, 1977.

Bermondsey in 1909. Her husband, Dr Salter, who held the LCC seat, had polled creditably for the ILP in the parliamentary by-election a couple of weeks before, which let in the Tory. She joined a hung council. When hustled by the Progressives, she insisted on her independence and withheld her casting vote, allowing the Tories to regain control by taking the aldermanic seats. She was never forgiven. Angry letters were exchanged in the press; and her husband lost his own LCC seat shortly afterwards. She was totally marginalized: by the Tories because they could not forgive her politics, by the Liberals because they would not forgive her 'treachery'. They would not place her on a single committee. At council debates she said little and could do nothing, and lost her seat in the next triennials when the Tories enhanced their majority. Throughout East London came reports of Labour and Liberals at 'daggers drawn'; in Poplar, Liberals and Tories arranged an electoral pact to keep out Labour candidates.[41] Mrs Salter's problems were those of third party rather than sexual politics, and were faced by many ILP candidates accused of splitting the Progressive vote.

By comparison, socialists in North Kensington had a more agreeable time. Here the ILP built up the register and worked just one ward, Goulbourne, which back in 1894 had returned women to the vestry. In 1910 the Women's Labour League baby clinic, started as a memorial to the early death of Margaret MacDonald, was opened under Ethel Bentham's supervision, and as Elizabeth Garrett found many years before, helped to build a loyal constituency. To this the WLL had added a co-operative trading store, a socialist Sunday school, and committed constituency work. In the 1912 triennial borough elections, Ethel Bentham, Marion Phillips, and four male comrades swept the ward, though the rest of Kensington remained solidly Tory. Marion Phillips indecorously put down motions, pressed amendments, and challenged rulings from the first meeting. She complained *inter alia* that the works committee were not promoting public

[41] *Southwark and Bermondsey Recorder*, 12, 19 Nov. 1909, 6 May, 8 Nov. 1910; *East London Observer*, 9 Nov. 1912; *Woolwich Gazette*, 5 Nov. 1912. See also Dr Salter's letter to WLGS, Minutes, 8 Dec. 1909.

works for the unemployed; tried to get more baby clinics, municipal milk, and school holiday meals; and policed the borough's Labour practices. She did succeed in getting more women's lavatories.[42] In all cases, she was to be found speaking for the dispossessed, though by cutting procedural corners, challenging the mayor's ruling, and failing to preserve proper decorum she, like Helen Taylor, caused offence. She too sought the publicity of confrontation rather than the possible negotiation of success. Not surprisingly, she and Ethel Bentham could point to much activity but little achievement for their efforts. As Ethel Bentham remarked to the WLL conference in 1913,

Whatever happens, we shall not want for work . . . to many, the round gets sometimes a little weary; the meetings and the committees and the discussion, the socials, the struggles on Boards against an overwhelming majority, the elections fought and lost and the ever-lasting difficulty of finding funds. But, after all we sometimes win, and even printers' bills are paid at last, and now and again we score our points . . .[43]

A handful of other Labour women won London seats, Miss Katherine Medley taking a seat in Poplar where the Liberals were fractured between Whigs and Progressives. In the old stronghold of Woolwich with its Royal Arsenal Co-operative, two women joined their comrades, but though they were armed with massive petitions, were unable to persuade the ruling Tory group to reopen the municipal milk depot.[44]

Spanning the London boroughs was the LCC, which in 1907 had swung to the Tories. At the 1910 triennial elections there were twelve women candidates. The Women's Labour League ran Margaret Bondfield in Woolwich, Ethel Bentham in Kensington, and Mrs Despard in Lambeth. Despite the chairmanship of Graham Wallace and the platform sparkle of George Bernard Shaw, the first two were defeated at the poll

[42] *Kensington Express*, 20 Dec. 1912, 17 Jan., 13 Mar., 25 Apr., 9 May, 18 July, 1 Aug., 18 Oct., 28 Nov. 1913; 6, 20 Feb., 1 May, 28 June 1914. Ethel Bentham had failed to win an LCC seat in 1910. Ensor had tried but failed to get Ethel Bentham and Margaret Bondfield co-opted on to the LEA in place of Susan Lawrence and Nettie Adler. Marion Phillips to R. C. K. Ensor, 9 Mar. 1910, Ensor papers.

[43] WLL 8th Annual Report, 1912–13, p. 36.

[44] *East London Observer*, 9 Nov. 1912, *Woolwich Gazette*, 5 Nov. 1912. Woolwich had been briefly held by Labour but had now 21 Tories, 15 Labour, and no Liberals.

and Mrs Despard was forced by the SDF to withdraw.[45] However, Henrietta Adler, Progressive daughter of the Chief Rabbi, was elected for Hackney with the help of the Revd Stewart Headlam, in his words as a 'banner-bearer'. A lady in the spirit of Mrs Surr, she had devoted her energies as a co-opted LEA member to checking child labour. Susan Lawrence, the clever, austere, and monocled Tory member of the old London school board and a co-opted LEA member, was elected for West Marylebone. Though both women said little at full council meetings, they worked hard on the asylums and education committees. During summer 1912 Nettie Adler was urging the LCC to feed hungry children as industrial disputes deepened.[46] Susan Lawrence, nominally a Tory, was becoming more and more uncomfortable. She had learnt that school charwomen were not only poorly paid but hired and fired at the whim of male caretakers. She could not persuade her party to amend their hours and conditions. In March 1911 she was writing to R. C. K. Ensor, the ILP Poplar member, that 'the charwomen's business is not going very well'. She turned instead to Mary MacArthur and Marion Phillips to help unionize women cleaners. On other issues, too, she was moving away from her party. By the end of 1911, she felt compelled to resign, writing to her Tory ward chairman that on matters of school feeding, poor law reform, and the standing of voluntary schools, 'I do not feel that I am in sufficient sympathy with the Municipal Reform [Tory] party to stand again'.[47] It was a courageous decision. She had little hope of returning to public office. She was soon addressing socialist conferences on child feeding and maternity benefits; speaking at suffrage meetings; and helping organize the charwomen's union. When Ensor decided not to seek re-election to the LCC in Spring 1913, she was adopted by the Poplar Executive in his place. She was launched as their

[45] WLL central branch Minutes, 12 Jan., 2 Nov., 7 Dec. 1909. The SDF made it clear that if Mrs Despard ran as an ILP candidate, they would oppose her.

[46] *Hackney & Kingsland Gazette*, 21 Feb., 4, 7 Mar. 1910; 19, 26 Feb. 1913; *East London Observer*, 27 July 1912.

[47] S. Lawrence to R. C. K. Ensor, 1 Mar., 26 Sept. 1911, Ensor papers; *The Times*, 24 Jan. 1912: LCC Minutes, 23 Jan. 1912, for tributes paid to her. Dr Sophia Jevrons tried but failed to take the vacated seat as an Independent with Miss Lawrence's support, *The Times*, 5 Feb. 1912. On Susan Lawrence, see C. Rackham's obit. in *Fabian Quarterly*, Spring 1949 (I owe this reference to Dr B. H. Harrison).

candidate at dock gate meetings, flanked by Will Crooks, their MP. 'Good meeting', Ensor noted laconically in his Diary.[48] In March 1913 she was elected for Poplar. Nettie Adler, in poor health and unable to canvass, held her Hackney seat with just eight votes.

Other women candidates were no more successful than in the past.[49] Three women held aldermanic seats, Lady St Helier, perhaps better known for her advocacy of rational dress than her devotion to education, the Progressive Katherine Wallas, and the Tory Mrs Wilton Phipps. Susan Lawrence on her return, and to her chagrin, was placed by her erstwhile Tory allies, not on the education committee, but on public health and on stores.[50]

Within London, women were handicapped by their political affiliation (they stood for the wrong parties), and outside London by their lack of property qualification and electoral leverage. It was also the case that where women did stand, they were often defeated by larger margins or won by less than had been predicted, the reverse of their experience on poor law and school boards. Once elected they met with little or no hostility from fellow councillors (again, unlike their poor law experience), but there was a deeper resistance to them from voters and therefore also from party agents. Women recognized this themselves, and searching around for explanations, blamed it on the militant tendency of the suffragettes. Margaret Ashton wrote angrily to Mrs Fawcett that women candidates were finding themselves damaged by 'the action of those few violent women who have so much injured the reputation of women politicians in Lancashire'.[51]

Of women councillors, Miss Newman of Walsall was anti-suffragist, Miss Merivale of Oxford was studiously neutral, and Elizabeth Garrett Anderson supported the Pankhurst

[48] Mrs Rackham to Miss Strachey, 12 Jan. 1912, Fawcett letters; Ensor diaries, entries for 1, 26, 29 Jan., 6 Mar. 1913. Miss Lawrence was imprisoned in the Poplar protest of 1921; in 1923 she became East Ham's MP.

[49] Mrs Miall Smith standing as an Independent Progressive, 'the men's association having again refused to adopt her', WLGS Minutes, 5 May 1910, cf. Minutes 13 Apr. 1910.

[50] *The Times*, 14 Mar. 1913. The Tory majority on the LCC rose from two to 15 after a highly professional campaign on the Tory side involving 12 million leaflets, 600 meetings, and 50 cinematographic films.

[51] 16 Jan. 1906, Ashton papers, Manchester Public Library.

Women's Social and Political Union. Indeed, on her mayor-making day, certain councillors asked whether she would agree not to invite Mrs Pankhurst to her home. She calmly ignored them and moved the next business. The remaining women councillors were suffragist, distancing themselves from the WSPU. In Reading Edith Sutton chaired suffragist meetings with a tactful hand, when Lady Frances Balfour, their speaker, was being attacked by both militants and anti-suffragists; and herself moved a suffrage resolution in council which attracted eight votes, mainly Labour. She carefully did not attend any WSPU meetings. At Tynemouth, the press reported that Miss Burnett was in favour of votes for women:

I don't think however, she is likely to agree with the antics employed by her militant sisters to further their cause. Miss Burnett is one of the sweetest women I know. She is sympathetic and kind and never tires of doing good work among the poor of the town. I heard her speak at a charity bazaar recently and I saw more than one of the audience slyly wiping a tear.

One of the women on her platform however, had still to ask the audience 'not to allow the tactics of the militant suffragettes to bias their minds against the work such women as Miss Burnett were doing'.[52] Helen Sturge, who twice fought a ward in Bristol, recorded that passers-by wanted to know whether 'the suffragette got in?' In Manchester, a voter was heard to pull down the blinds when Miss Ashton won in Withington, saying 'a suffragette is elected'. Mrs Hughes complained at Oxford that she was hampered by WSPU militancy. Miss Balkwill standing for the LCC blamed them for her defeat. Mrs Lees characteristically told Oldham voters that she sought a council seat—'I have no idea of revolutionizing municipal life but I hope to get some more playing fields for the children'. Eleanor Rathbone typically was robust. She deplored 'violent methods or the use of violence in political propaganda'; but the suffrage campaign was 'to her the greatest and dearest of all political causes' and would command 'all the time and energy she had to spare'.[53] Over in

[52] Quoted J. Manton, *E. G. Anderson*, 1965, p. 273; *Reading Mercury*, 13 Feb. 1909, 8 July, 1911. *Tyneside Weekly News*, 1 Jan. 1909, *Shields Daily News*, 18 Oct. 1909.
[53] E. Sturge, *Reminiscences of my Life*, 1928, pp. 64ff. O. Hertz, *Women Citizen*,

High Wycombe, Miss Dove as head of the poll was mayor-elect. A male councillor on mayor-making day asked whether this was the time 'to elect a woman to the mayoralty when the "shrieking sisterhood" were creating so much turmoil and trouble?' This was such a persuasive consideration that the formal motion that she be called to the mayoralty was defeated. Miss Dove handled the situation with dignity but was deeply upset by it. In 1913 she lost her seat. The *Anti-Suffrage Review* may not have been overstating its case when it reported that 'Everywhere one hears the same story. Canvassers for women candidates in London boroughs find the voters' doors shut in their faces. "We don't want no howling and screaming women on the Council". "No women for me— after the suffragettes".'[54]

Suffragette militancy hampered women in the eyes of men, but it may have helped to mobilize women for women. In Eleanor Rathbone's election, 74 per cent of the women but only 43 per cent of the men bothered to vote; in an election three weeks later with two men candidates, 56 per cent of the women and 41 per cent of the men voted.[55] However, even where the women ratepayer vote reached 20 per cent or more, this was not thought large enough to outweigh the disadvantages of a woman candidate. Mrs Lees apart, there is no evidence that parties ever sought a woman candidate for the town council, even though those same parties had sought women for school and poor law boards. The Progressives, who were sometimes helpful, had many candidates and few safe seats after 1906, and the Labour party who *were* supportive had even less.[56]

There was a deeper reason for hesitation and hostility, apart from suffragette hassle. Put simply, many people were unpersuaded that council work was suitable work for women.

Dec. 1937, p. 8; Mary Balkwill, WLGS Minutes, 14 Nov. 1908; *Oldham Evening Chronicle*, 5 Nov. 1910, *Liverpool Courier*, 5 Oct. 1909.

[54] Cuttings, Lees Coll., L. 151; *Manchester Guardian*, 23, 28 Oct. 1908; NUWW Occasional papers, May 1914, p. 35; Occasional papers, May 1913, p. 21; *Anti-Suffrage Review*, Sept. 1909, p. 1.

[55] *Truth*, 24 Nov. 1909. High polls in local government wards depend mainly on their marginality.

[56] G. D. H. Cole estimated that there were 420 Labour councillors under a variety of labels in provincial and London boroughs by 1914, *Municipal Decline*, p. 81.

School board and poor law work, yes. That stemmed from philanthropic roots; and as the care of children, the sick, and the elderly moved from the private into the public domain, so it was acceptable for their carers to move from domestic piety into domestic politics, and thus into local government. School and poor law work was conducted by specialist *ad hoc* boards, where knowledge of domestic detail, and care for female clients and staff, was helpful. Women were welcome to visit, inspect, and undertake after-care work, as long as men retained control of the budgets, the contracts, and the building work. In that sense, council *committees* were akin to school and poor law boards, specialist and service-centred. 'A woman was in her right place on the Education Committee', said one male councillor. Many councils were quite happy to co-opt women directly on to education and distress committees, where their presence was statutory, and occasionally on to library, pensions, midwives, and insurance committees, where it was not. Accordingly, Miss Burnett of Tynemouth council urged women to seek co-option to these committees 'because then men get accustomed to them, and it is easier to go from there on to the Councils'.[57] Miss Burnett notwithstanding, the city council itself was still held to be a very different animal.

The prime concern of the town council was the built environment of its community: drains, roads, refuse, lighting, and the like. The second prime task was the management of its utilities: water, gas, electricity, and transport undertakings. It was by no means obvious to electors that women without business experience or technical aptitude had anything much to contribute to either. Miss Merivale in Oxford found council work far more difficult, complex, and demanding than she had ever anticipated, and was subsequently 'quite ashamed' of her naive election speeches. Miss Ashton, a tough and forthright lady, stood in St George's ward in Manchester in 1907, claiming that she knew as much about building regulations 'as the ordinary man when he went upon the council'. She was stating a fact. When she found that a builder was erecting cottages for her father which departed from her designs, 'she

[57] *Englishwoman*, 16, 1914, p. 16; Councillor Tebb, *Shields Daily News*, 22 Oct. 1910; Councillor Miss Burnett, WLGS AGM, 12 Mar. 1913.

raised Hell and made the builder pull down what had been done and build them to her design'. She was defeated.[58] The following year she stood in Withington ward. She now suggested, ironically no doubt, that municipal council work 'was not entirely suitable for women but she was anxious to be allowed to try her small share of it'. No talk about her expertise in drains and ventilation this time, but about the need of women and children for homes, schools, clean water, baths and washhouses, adequate street lighting, and safe trams. 'Surely such things affected women even more than men . . . Who had to suffer if houses were ill-built, ill-drained, short of water, and dark, ill-paved courts and alleys outside them?' And a few days later, in case the electors had missed the point, she was insisting that 'if she were returned there would be no necessity for her to encroach upon the men's work'—not that it was entirely clear what she had left them to do—as there was so much only a woman could see to. 'Matters of great importance to women had been overlooked, not because men desired to neglect them, but because they never thought of them'. She was elected: and went on to obtain women's lavatories, municipal milk, and municipal lodging-houses for the women of Manchester.[59]

Even when women were elected, they were often quite careful to preserve the proprieties in office. Miss Merivale, who had perhaps the most exaggerated sense of women's sphere, refused to join civic deputations to Hastings to inspect their new electric trams on grounds of decorum, though whether it was Hastings or trams that were indecorous she did not say. It was thought rather odd that Miss Dove had sensible views about refuse destructors. Mrs Lees deliberately made her first council speech 'on the manufacture of water gas (laughter) which created a great nuisance in the ward I represent'. After which she was regularly invited to inspect the gasworks, and equally readily accepted. Edith Sutton was allowed her choice of committee at Reading, save only that the men thought the watch (police) committee improper for her. She persuaded them in time that it was even more improper for men alone to review the prosecution of prostitutes.[60]

---

[58] *Manchester Guardian*, 30 Oct. 1907.   [59] *Manchester Guardian*, 17, 21 Oct. 1908.
[60] *Oxford Chronicle*, 8 Nov. 1907; speech of Mrs Lees, WLGS Dinner, 21 May 1908.

Women councillors joined the service committees, education, health, and housing in particular, while men continued to run the finance, works, contracts, and trading committees. Reading the Minutes there is a strong sense of men raising the money and women spending it. But local authority concerns in the years before the First World War were widening, beyond the material infrastructure of the town and its utilities, to include child care, housing and planning, the promotion of employment: in all of which women had a singular contribution to make. It was, they said, a suitable job for a woman.

9

Council work 1907–1914
A Suitable Job for a Woman?

WOMEN found it hard to be selected for council seats and even harder to win them. Once elected, they also found it harder to make the impact that women of their distinction had done on school and poor law boards. Mrs Lees apart, they were not therefore seen as much of a political asset by their parties. Nor were they able to open up more seats for more women by virtue of their pioneering and performance. Why was it so difficult?

Unlike school or poor law boards, RDCs, or even vestries, city councils spoke for their communities. They did not confine their comment merely to matters within their technical jurisdiction, as school and poor law boards had to do, but could and did move motions on women's suffrage or tariff reform, as the collective voice of their locality. City councils were also generalist rather than specialist bodies, delivering a wide range of services. This meant that few councillors could work across the entire surface of city council work, as Miss Sturge of Bristol or Mrs Buckton of Leeds did in education, or Miss Clifford in poor law work. After all, the entire school board field was now just part of the work of one committee. Councillors had to specialize; but as they also protected their constituency interests by sitting on several committees, they had far less time or mental space to engage in the detail and the individual casework which were the hallmark of women on school and poor law boards. That was delegated to officers. One characteristic contribution of 'ladies elect', their unremitting attention to detail, the expertise that they thus acquired, their willingness to be 'officers-in-disguise' tackling executive work at a saving to the rates, was now somewhat less wanted or valued. Formal bureaucratic structures carried much of this load. The scope for any one individual to pioneer, as women had been able to do in the heroic days of the early school boards, had long passed, in education as in most other fields.

Because councils had wide responsibilities, this meant also that a commitment, say, to reduce infant mortality, could be pursued in different ways, by slum clearance, baby clinics, better refuse and sewage disposal, municipal milk depots, and by any of several committees. Which was the most cost-effective? The co-ordination of policy across many fronts, so that houses *and* schools *and* trams together developed suburban land, the pursuit of priorities, the allocation of resources, had necessarily to be done by party groups or by an inner caucus of committee chairmen, often aldermen, whose skills lay not so much in detailed knowledge as in administrative ability, polit-ical clout, and sensitivity to salient community pressure. Women arrived on town councils when resources were being trimmed by 'the urban scissors of rising needs and stagnant revenue'. The cost of local authority services rose (by 18 per cent between 1908 and 1913), government grant did not (by only 2 per cent), while in many areas property values actually fell.[1] Choices had to be made, and this gave somewhat greater significance to the central committees of general purposes and finance, staff, and property, where those choices were made, over the service committees on which women served. Women councillors pulled their weight, so to speak, held a watching brief on women's issues which on occasion they were able to foreground. But they were inevitably on the margins of city council work and at the periphery of power, just like other newly arrived, back-bench, or opposition members. Not until the 1920s were women promoted to full committee chairman-ships and the powerful aldermanic bench. Once on the council, they experienced little overt discrimination or sexual hostility, but not much deference or attention either. Some of them rather missed that air of assigned moral authority which they had found so useful on school and poor law boards.

Predictably, the first committee women sought to join was

[1] A. Offner, *Property and Politics 1870–1914*, 1981, p. 387, and chs. 18 and 23 for this point. The City treasurer of Birmingham calculated that property rated less than £18 p.a. cost more in services than its rates yielded; hence the battle for suburban property. See P. J. Waller's admirable book, *Town, City and Nation: England 1850–1914*, 1983, pp. 262–3.

education. Most were already co-opted LEA members, but as Miss Coulcher of Ipswich said, had found 'that when the women on that committee had given their views, they have no voice on the Council, and cannot take any part in the final stages of the Proposals they support'. They had had to watch helplessly while cherished schemes faltered. Nettie Adler's draft by-laws on child labour for the LCC were mangled in her absence, Susan Lawrence saw women teachers sacked and domestic science mismanaged but could not intervene. Now as fully elected members, they found seven hundred and fifty thousand London children clamouring for their attention. Margaret Ashton, an experienced education committee chairman of her small UDC, complained as a Manchester councillor that Manchester's teachers were poorly paid and trained. Her criticisms were bitterly resented, but the reports of HMIs substantiated her every point.[2]

In Reading Edith Sutton came on to the education committee where in her words she 'staved off' the resignation of the chairman, her cousin, by offering to do all the work for him by chairing the school management committee herself. With 13,000 children under her supervision, she ensured that Reading adopted the school meals act, and a generous school medical service. She started special schools for mentally and physically handicapped children, and, influenced by Margaret McMillan's work at Deptford, added an open-air school for tubercular children, as well as technical schools for less able children. When in 1911 Reading enlarged its boundaries, it fell to her to ensure that school standards and pupil-teacher ratios did not worsen. She sat on every committee and subcommittee, and her attendances were far higher than those of any man. Like Mrs Pinsent, she functioned as the classic, cool, competent school board lady. As she was virtually silent in full council, even when a women's municipal lodging-house was being discussed, her educational work was appreciated even more by the men.[3]

[2] Miss Coulcher, *Suffolk Chronicle*, 22 Oct. 1909; *Municipal Journal*, 10 July 1908; *Englishwoman*, 21, 1914, pp. 145–9. See also Mrs Homan, WLGS annual meeting, 15 Mar. 1906.

[3] Reading education committee Minutes, 1909–14, Annual Reports 1909–14, *Reading Mercury*, 3 Apr. 1909. Similarly, Miss Burnett tried to develop continuing education for 14 to 16-year-olds in Tynemouth.

Ellen Pinsent had trained with the NSPCC before being co-opted on to Birmingham LEA. Unusually for a co-opted member, she chaired a subcommittee, special schools, and discovered hundreds of mentally handicapped children in the city unknown to and unreached by the local authority, living out their lives, she said, in contaminated environments. She feared they were 'a source of racial degeneration'. She linked up with Mary Dendy of Manchester to hold conferences which called not only for special schooling but permanent protective after-care. She disaggregated epileptic children from their ranks, and built them a special boarding-school. She traced over a thousand Birmingham children, crippled with TB, polio, multiple sclerosis, and rheumatism, who were unknown to the education committee, and extracted the funds to build them special schools and sheltered workshops. In her spare time, she joined the Royal Commission on the Care and Control of the Feeble-minded and became, with Mary Dendy, one of the first commissioners of lunacy. By 1912 the full education committee seems mainly to have served as an echo-chamber for its special schools sub-committee; lethargy descended, and the Minutes shortened, when in summer 1913 she moved south. She had made an extraordinary impact by thinking large. She also had the wit to belong, as a Liberal Unionist working with the Tories, to the majority group. She surveyed, planned, phased, and implemented with irresistible efficiency. Other women LEA members distrusted her eugenic-inspired passion for segregated special boarding-schools, believing that all children needed families (in this sense, Mrs Pinsent was in the poor law rather than the school board mould). She, for her part, insisted that integration in practice meant neglect: if handicapped children were to get proper care and a decent quality of life, they needed her version of affirmative action.[4]

A second area of work to which council women were devoted was public health and sanitation. The experience of vestry and sanitary work in the 1890s had taught women that

[4] *Men of Mark, Manchester*, No. 64; Birmingham education committee Minutes, 24 June 1910, 19 July 1912, 27 Jan. 1911; *Birmingham Daily Press*, 24 Oct. 1911. In 1911, Birmingham had 41 Liberal Unionists, 45 Conservatives, 28 Liberals, and 6 Socialists on its council. See also Miss Clephan's work in Leicester, *Leicester: Its Civic, Industrial, Institutional and Social Life*, 1927.

healthy homes needed healthy streets, that private health still depended in large measure on public health. The Progressives, in one of their election leaflets to women voters, advised them sardonically, 'Always take a house in a street where a Borough Councillor lives if you can afford it, because that street will not be neglected by the officers of the council'.[5] Privies, drains, and sewers had to be kept wholesome and in good repair, dustbins regularly emptied, streets cleaned of refuse and horse dung. The women's press pointed out that much of the dust on furniture came from dried horse dung blown in from the street, and transferred 'impartially to our food, our clothing and our persons. The housekeeper and the sanitary authority must think of these; other people for the sake of their peace of mind prefer to ignore them'. It noted that when a woman householder or woman councillor complained, they were often met with the reproof, 'this is not a nice subject for a lady'—yet

the lady is likely to have this not-nice subject descend in clouds upon her superior person and her ultra-refined home. The male councillor would have her wash her hands and divert her mind from the subject; but someone's mind must remain undiverted if our children's food is to remain unpoisoned; and the mind of the dustmen, consecrated though it is to the matter, is an inadequate protection to city dwellers'.[6]

Newly elected women councillors went out and bought textbooks on drainage systems to study at home, others arranged to be given crash courses in the running of sewage farms.[7] They asked for common lodging-houses to be policed, smelly cemeteries to be closed, public WCs to be provided for women as well as for men so staircases and courts were not fouled, and sought by-laws to control the spitting which spread TB. Miss Burnett of Tyneside expected that her public health work would attract criticism for offending propriety, but 'she was sure when this prejudice died away, those men who agreed that it was the right thing, but could not quite make up their minds whether it was the nicest thing' would be glad of women's work. As Mabel Clarkson of Norwich said in

  [5] Progressive leaflet No. 5, *A Word to Women Electors about the Borough Councils*, 1906.
  [6] *Englishwoman*, 28, 1913, pp. 107–14.
  [7] *Anglo-Japanese Congress*, 1910, p. 13.

her 1913 election address, 'the best wealth of a city is in the health of its citizens'.[8]

One sub-branch of public health work was lunacy. Susan Lawrence and Nettie Adler conscientiously visited London asylums with an eye to the welfare of their eleven thousand women inmates.[9] Miss Burnett, Mrs Pinsent, Miss Pugh, and Miss Martineau of Birmingham, spent considerable hours visiting and inspecting lunatic asylums, because women patients need 'one of their own sex on the Management Committee'.[10] Eleanor Rathbone found herself on the Port, Sanitary, and Hospital committee in Liverpool in 1909, checking that ships brought neither infected goods nor infectious people into the port. Edith Sutton, overworked at Reading, still found time to give support to the matron of the town's isolation hospital and supervised the midwifery service. Miss Henry noted, somewhat sourly, that male councillors were usually married men whose wives did not like them visiting hospitals in case they brought infection into the family home. It was left to women councillors to uncover (so to speak) patients sharing beds, and other improprieties.[11] Invariably too it was women members who pressed for open-air sanatoria for TB patients and open-air schools for delicate children.

Education and public health were two predictable committee interests of women members. A third was what might be called urban amenities; the provision of baths and washhouses to make washday tolerable and indeed possible for families in upper tenements without running water; the acquisition of open space for parks and gardens, desperately difficult when land values were so high and streets so densely populated; and the adoption of permissive legislation to provide libraries, museums, and art galleries. Hannah Mitchell, a councillor immediately after the War, acquired a washhouse for her ward, where women could hire a stall, a deep trough with its own boiler, hot water, extractors, hot-air driers, and ironing

---

[8] *Shields Daily News*, 22 Oct. 1910; Norwich Election Address, Oct. 1913.

[9] LCC Asylums committee, Annual Reports; *Shields Daily News*, 8 Jan. 1918.

[10] Miss Pugh, 1911 Election leaflet; Annual Reports of the committee of visistors to lunatic asylums for the City of Birmingham, 1911, 1913.

[11] Miss Sutton, Annual Reports, MOH Reading 1909–14; Miss Henry, *Anglo-Japanese Congress*, 1910, p. 16; NUWW Conference, Glasgow, 1911, p. 92, 'The work of women as councillors'.

boards. It was a 'real public service appreciated by women' without coppers or drying space of their own, and faced with large families, long skirts, and no detergents or synthetics. She built it. Male colleagues insisted on the glory of opening it. She also tried to start a travelling library, but this was held to be hopelessly impractical, until a male councillor had the same idea eight years later whereupon it was greeted with enthusiasm.[12] In Walsall, Miss Ada Newman became vice-chairman of the free library and art gallery committee, on the apparent understanding that she would provide many of the books. Women's local groups rallied to support their women members. In Birmingham, the NUWW campaigned for free lavatories for women, and the Women's Labour League got the price of admission reduced for women at the baths.[13]

Chamberlain's concept of urban amenity had been building-based—the boulevard, the clock tower, and the town hall, the public and the prestigious. Women councillors brought with them a more domestic and family version of amenity, wanting open space rather than built space, seeking to reduce densities rather than decorate them with imposing frontages. Hannah Mitchell wryly noted that there were male and female versions of parks then as now. When men had an open space, they wanted to put down tarmac and put up bandstands, mark it out and parcel it up for energetic games. Women wanted gardens, she said, not activity centres, but trees, flowers, and rest.[14]

Education, public health, and amenities were identified by women in their election addresses as suitable work. They paid less attention to three other areas of council responsibility, that of municipal undertakings, the central resource committees, and the licensing or regulatory committees, all of them committees much sought after by men.

The first of these male committees was municipal trading. All local authorities had an interest in the quality, price, and convenience of water, gas, electricity, and transport, the power networks sustaining modern urban life. By 1914, most

---

[12] H. Mitchell, *The Hard Way Up*, 1977 edn., p. 208.

[13] NUWW Occasional Papers, May 1913; Women's Labour League MS branch reports, Birmingham, 20 Sept. 1906.

[14] Mitchell, op. cit., p. 207.

water was supplied by local authorities rather than private companies. Women soon understood that without major investment in waterworks, towns could not clean their lavatories, private homes enjoy bathrooms, children swim in public pools, or industry expand. Similarly, if women wanted better street lighting or convenient gas ovens, then the price and quality of gas had to be controlled. Electricity was a relatively new industry and local authorities bought into it from the beginning to ensure a reliable power-supply for trams. Women quickly learnt that if they were to reduce densities and clear the slums, a decent transport system was vital. Margaret Ashton indeed tried, but failed, to get cheap workmen's fares extended to working women. For the most part, women stood aside as men argued where to draw the line between municipal and private enterprise, the scale of the corporation debt and the size of the municipal workforce, and whether profits from utilities should go to reduce prices for consumers or cut the rates for citizens.[15]

A second set of male committees was the central resource committees, finance, staff, estates, assessment, and valuation. These were less powerful than they would be today. Finance committee, for example, served mainly as an audit committee, while financial policy was determined by a general purposes committee of all members. The works committee, however, was of key importance to Labour members. Labour women fought robustly to maintain contract compliance with trade-union rates and conditions of service for their male manual staff. Mrs Pankhurst was somewhat cool about women's local government work, regarding it as diversionary, but she, like Susan Lawrence, noticed immediately that councils employed women inspectors, clerks, and cleaners, many of whom were overworked, underpaid, and ignored. 'Much has been done to make the municipality a model employer of men, but little or nothing has been done for women', she pointed out. The plight of LCC cleaning women was to take Susan Lawrence on her journey from the Tory party to the ILP.[16]

Women did invade a final set of traditionally male commit-

---

[15] For local government services, see A. Briggs's classic account in *History of Birmingham*, 1951, vol. ii; Waller, *Town, City and Nation*, esp. pp. 302–7.

[16] *Municipal Journal*, 4 Oct. 1907.

tees, the licensing and regulatory duties imposed by central government. Some of these functions grew out of public health work—inspecting dairies, sampling food and drink, controlling nuisances. Registering midwives and inspecting baby farms had been womanly work since the days of Lady Sandhurst. Other duties developed from police work, regulating common lodging-houses, pubs, and places of public entertainment. This was work dear to the heart of many social and moral purity groups disturbed that children and decent women were exposed to indecent posters in shop windows, embarrassed by prostitutes servicing clients in the public parks, and accosted by men the worse for wear pouring out of pubs at night looking for trouble and women. Reclaiming the streets was always a feminist concern. Women coming from a temperance background and unconnected by business ties to the brewing trade, sought more stringent licensing laws. Others began to call for women police. Edith Sutton patrolled Reading's lodging-houses, suppressed its brothels, and started rescue homes for fallen women.

From the 1890s local authorities also began more vigorously to inspect workshops, laundries, offices, and shop premises, checking their hours, conditions, and sanitation. Women sanitary and factory inspectors begged women to come into public life, so that they would have women members to report to; women councillors for their part tried to get more women sanitary staff appointed. Together, women sanitary staff and women councillors such as Margaret Ashton, Eleanor Rathbone, and Marion Phillips, took the lead on their councils in exposing and denouncing sweated labour.

Women's concept of their work steadily widened. Margaret Ashton spoke for the others when she said that certain issues such as insanitary housing and infant mortality mattered particularly to women, but that as a woman councillor she also needed to hold 'what I may call a watching brief for the women on other matters'. The women's perspective had been ignored, not wilfully, but because there was no one on the council to say 'that women have the same requirements and pay the same rates as the better represented men'.[17] There was

[17] *Englishwoman*, Sept. 1910. See also *Bath Chronicle*, 24 June 1922, for Miss Hope's work devoted 'to the welfare and betterment of her sisters'. Eleanor Rathbone: 'as the

a step beyond that, not just to perform traditional womanly work for women and children, not just to hold a watching brief for women on other council matters, but also to work with men for all men and women. Eleanor Rathbone's 1910 election address put it neatly: 'she represents the men of the Ward as well as the women of the City'. As Miss Reddish said standing for Bolton town council in 1907, 'Work on town councils was human work, and why should work for humanity be confined to one sex?' Miss Balkwill describing her work on Hampstead borough council explained that the council had twelve committees, from assessment through to works, and added carefully, 'I do not wish to make out that any of these twelve committees are particularly suitable for women's work, because I hold that suitable women of experience and of capability can do good work on all of them'. In Mrs Rackham's words,

> Speakers and writers on women's work in local government usually dwell on those parts of the Council's work which specially appeal to women as women—to the mothering and housekeeping instincts of women . . . And indeed it is absurd that questions of this kind should be discussed and settled without the help of women. But women claim an interest as human beings in every question that comes before a municipal authority, whether it may be of finance or rating, or general policy. On some matters women have more to learn, but none belong exclusively to either sex . . .[18]

Beyond the areas of traditional local authority work of education, health, amenities, trading, and licensing, new areas of concern were coming to the fore, often at the interstices of traditional departments: and it is to three of the most important of these, infant mortality, housing and town planning, and unemployment, that we now turn.

Infant Mortality

Women in the education field had long insisted that much of only woman on the Council she of course considered it her duty to keep a sort of watching brief in the council for anything that especially concerned the interests of women or children', she said, in *Liverpool Daily Post*, 26 Oct. 1910. Her zeal was said to provoke mirth in town hall corridors.

[18] Election address, 1910 for Granby ward, Liverpool: *Bolton Journal*, 1 Nov. 1907; Miss Balkwill, *Anglo-Japanese Congress* 1910, p. 11; Mrs Rackham, *Englishwoman*, 21, 1914, pp. 145–9.

what passed as stupidity among working-class children stemmed from poor health. It was useless, they said, to try and teach a hungry, weary, or sickly child. Children must first be washed, fed, clothed, and doctored. Women school board members had mobilized their philanthropic networks to do precisely that.

Instinctive compassion had, by the turn of the century, become overlaid with new considerations of public health and national efficiency, and informed by new perceptions on poverty. Britain's economic performance was under pressure from American and German industrial competition. The productivity of British workers and the physique of British soldiers seemed clearly inferior to their rivals. At the same time, the surveys of Booth in London, Rowntree in York, and then A. L. Bowley in provincial towns, showed that some 30 per cent of people lived at or below the poverty line. The Fitzroy Committee on Physical Deterioration in 1904 though dismissing fears that urban health was deteriorating, highlighted the interrelation of urban problems, child life, and education. Slums were reproducing slums. Britain was physiologically as well as culturally two nations, its poorer third puny, stunted by malnutrition, misshapen from polio, rickets, and industrial injury, weak of sight, hard of hearing, inarticulate in speech. Mrs Townsend, a middle-class Fabian and school manager, was much distressed by the linguistic poverty of working-class children. 'One had to delve deep to reach a response. To receive an answer prompt, fearless and distinct is so rare as to be absolutely startling. The children have never been taught to speak and most of them make very clumsy attempts at it . . . [using instead] a code of half articulated sounds which serve to express their more urgent needs and emotions'.[19]

The most sensitive indicator of the nation's health, 'the thermometer of the City' in Margaret Ashton's phrase, was infant mortality. Despite advances in formal medicine, a rise in real wages and half a century of public health legislation, infant mortality was at its worst ever in 1899. As birth rates

---

[19] The Fitzroy committee called for school feeding, medical inspection, clean milk, more public health control of poor housing and pollution, and education for mothers. Mrs Townsend, *The Case for School Nurseries*, 1909.

were also falling, especially among the more prosperous (by a third between 1900 and 1911) this was regarded as an ominous sign of racial decline. Towns were baffled by their inability to reduce infant deaths. Ashton, for example, closed privies, cleaned up the milk supply, and sent out health visitors to educate mothers in baby care, but as their new WCs flushed into the river, infant deaths remained stubbornly high.[20] Infant mortality, more than any other issue, focused the work of women councillors, embracing not only maternal health and midwifery, pure milk and proper feeding for infants, but also nurseries and health visitors for small children, school meals and clinics for older ones; and beyond that a mounting concern with housing, the environment of poverty, and town planning. 'Clean air, clean streets, clean yards, clean houses, all work together to protect infant life', said the Fabians.[21]

The first step, Progressive women insisted, was to improve maternal health. Mrs Pember Reeves, of the Fabian Women's Group, worked with Ethel Bentham to study the daily lives and budgets of Lambeth women between 1909 and 1913, in her classic *Round About a Pound a Week*. Lambeth husbands, steady and respectable men, brought home up to 25s. a week. This bought rooms, food, and clothing, but not enough of all three to sustain family health. 'She can never feel certain that she has found the right solution. Shall they all live in one room? Or shall they take two basement rooms at an equally low rent, but spend more on gas and coal, and suffer more from damp and cold? Or shall they take two rooms above the stairs, and take the extra rent out of the food?' If they did, it was the mother's food (bread, scrape, and weak tea) and her health that suffered. Pregnancy was a curse, abortions were sought, confinement itself 'a time of most awful misery . . . a dreadful fortnight'.[22] Margaret Llewellyn Davies collected

---

[20] M. Ashton, u.d. newspaper cuttings (Nov. 1909), fol. 352, Fawcett coll. For a fuller discussion, see D. Jones, 'Local Government Policy Towards the Infant and Child with Particular Reference to the North West 1890–1914', Manchester MA, 1983, ch. 6; J. Lewis, *The Politics of Motherhood: Child and Maternal Welfare in England, 1900–1939*, 1980. Infant mortality figures were 151 deaths per 1,000 live births 1890, 161 in 1895, 154 in 1900, 128 in 1905, 105 in 1910.

[21] Fabian pamphlet, *What a Health Committee Can Do*, 1910.

[22] Mrs Pember Reeves, *Round About a Pound a Week*, 1913, pp. 155–6. See also her speech, WLL Conference, 'The Needs of Little Children', 1912, pp. 14–15.

letters from members of the Women's Co-operative Guild, published as *Maternity* in 1915. They make harrowing reading. For months before the birth, women worked harder than ever while they systematically starved themselves to put by for clothes and medical costs. They worked until the last hour, unable to buy proper medical care, and rose within a day or two to face with broken health an endless succession of pregnancies, miscarriages, still births, and live births, all barely a year apart. 'I am not grumbling', said one, 'but now I am nearly used up'.[23] These were respectable women whose husbands were in work. For other women, less competent perhaps, whose men were in and out of casual labour, given to drink or in poor health, their situation must have been very much worse.

Poor maternal health produced sickly, low weight, puny, and premature babies who failed to thrive, often dying within the first few weeks. The connection between the health of the mother and the new baby was not fully understood; and in any case, doctors could do little to remedy years of maternal malnutrition. Indeed, medical opinion, overlaid with eugenic considerations, doubted that much could, and should, be done to save the frail newborn, who would live only to become the puny adult. Not until the war years did official attention turn to maternal welfare, culminating in the Maternal and Child Welfare Act of 1918, enabling local authorities to provide clinics, nurseries, salaried midwives, food, and milk for mothers and babies at risk.

But even when babies were born healthy, 'the healthy infant at birth', pointed out Mrs Pember Reeves, 'is less healthy at three months, less healthy still in a year'. Progressive women knew perfectly well that this was a consequence of family poverty, and the poor food and poor living conditions that followed from it. Most male commentators would have none of that. Infant deaths in the first month might not be preventable, but infants dying at three or six months, especially of diarrhoea, died of dirt—dirty milk, dirty homes, dirty and ignorant mothers. It was 'not climate, not topography, nor municipal sanitation, but the lives and habits of the mothers

[23] M. L. Davies, ed., *Maternity*, p. 58; E. Ross, 'London's Working Class Mothers 1870–1918', in J. Lewis, ed., *Labour and Love*, 1986.

in the homes', said one, that determined whether infants lived or died, the race declined or flourished. As John Burns told the first conference on infant mortality in 1906, 'At the bottom of infant mortality, high or low, is good or bad motherhood. Give us good motherhood and good pre-natal conditions, and I have no despair for the future of this or any other country'.[24] Burns's comments were double-edged. They pointed towards maternity benefits, clinics, and midwifery care, but they also permitted other responses. Infant death rates were among their worst in Lancashire where an unusually large number of married women worked in the mills. As their hand-fed babies had three times the mortality rate of breast-fed babies, mothers who went out to work, who placed money above motherhood, were held to blame for their babies' death. The women's movement was bitterly divided, then as now, Margaret Bondfield for one having very little patience with working mothers, Mrs Fawcett recognizing that women's earnings could improve family diet and the family's health overall, while Mary MacArthur insisted that married women had a right to choose.[25] Staleybridge decided that their notorious infant mortality figures were the fault of delinquent mothers. The mayor, however, was able to announce on mayor-making day that 'a little society had been formed at Staleybridge with the object of teaching mothers their duty' under the guidance of Councillor Mrs Summers (as if that did not damn her efforts from the start). This was somewhat cheaper than attending to the equally notorious state of Staleybridge's privies and sewage system. So Mrs Summers's work notwithstanding, while national mortality figures fell to 100 per thousand, the infants of Staleybridge continued to die off at the rate of 189 per thousand live births.[26]

One committee on which women were prominent was the midwives registration committee. The 1902 and 1905 Midwives Acts required midwives to be registered and

[24] *Round About a Pound a Week*, pp. 155–6; Burns, quoted S. Alexander, Introduction, *Round About a Pound a Week*, p. x; Lewis, op. cit., p. 65.

[25] WLL Annual Report, 1907–8, p. 8; M. A. Hamilton, *Margaret Bondfield*, 1924, p. 113. More generally, C. Dyhouse, 'Working Class Mothers and Infant Mortality 1895–1914', *Journal of Social History*, 1978, pp. 247–66; E. Roberts, *A Woman's Place*, 1984, pp. 165 ff.

[26] *Staleybridge Reporter*, 15 Nov. 1913.

trained. Few women would defend the use of grubby, elderly, and untrained midwives. Miss Burnside, Inspector of Midwives, found herself cutting the nails and teaching the use of disinfectant to some of them; one illiterate midwife, she recalled, used her grandson to write up the case notes, but was very careful not to let him record anything indelicate. Untrained, they lost far more mothers and were more likely to produce a handicapped child. Some doctors were generous. Manchester city council sensibly appointed a woman doctor to do the training, but others, encouraged by their General Medical Council, were bloody-minded. They harried untrained midwives for their incompetence and competent midwives for filching their fees. The record and attitude of most doctors was not a pleasant one, unwilling to offer adequate and inexpensive medical care to poor women and unwilling to help or let anyone else do it instead. Gradually, most midwives (80 per cent by 1920) came to have nurse training. Standards rose but so did the fees, from ten shillings to a pound, placing medical care even further beyond the reach of those most needing it.[27]

A similar regulatory role applied to baby farms which child-minded the illegitimate, unwanted, or handicapped. Horror stories abounded of babies dying by inches in less scrupulous farms. From 1897 inspection became more effective. In 1908 local authorities were given the power to remove children at risk. Women made such work their own.

The problems of poverty were sharpened for many expectant mothers by their isolation, inexperience, and ignorance, not always knowing, as letters in *Maternity* reveal, how a baby was born. For over half a century lady sanitary visitors and mission women had gone from house to house with tracts and soap; local authorities, led by Manchester, now began to contribute towards their pay, and to detach them from the condescension of charity. Ladies committees brought forward suitable young women to train for the exams of the Royal Sanitary Institute. By 1909 it was estimated that there were

[27] Miss Burnside, *Anglo-Japanese Congress*, 1910. Untrained midwives lost 1 mother in 266, trained midwives 1 in 1,200, though untrained midwives were also working with the poorest mothers most at risk. More generally, see J. Donnison, *Midwives and Medical Men*, 1977.

two to three hundred salaried health visitors and ten times that number of voluntary workers in the field, sometimes welcomed, often resented, by working-class mothers. Many female sanitary inspectors, checking workshops and premises, added health visiting to their load. Women councillors like Miss Burnett of Tynemouth, fought hard to increase their numbers and their pay, especially as they were rather better qualified than most men who came from the building trades. Councillor Mrs Lees personally employed an itinerant professional health lecturer of her own.[28]

The 1906 conference on infant mortality had brought together over one hundred and seventy local authorities and public bodies, all anxious to explore new approaches. In June 1907 'mainly through the earnest voluntary work of ladies of that district' said the *Englishwomen*, the St Pancras MOH opened a school for mothers—a central clinic where doctors checked newly born babies, and health visitors followed them home. The St Pancras School caught the imagination of women in public life. Sarah Reddish, for example, a poor law guardian, guildswoman, recently defeated for Bolton town council, and worried by her town's infant mortality rates which were worse even than Staleybridge's, visited the School, and returned to open a similar clinic in Bolton. She had the help of local co-operative societies, women doctors, and the MOH; but staunch opposition from the chairman of the borough's sanitary committee, himself a doctor, who was, she said, 'not helpful in the matter, and his idea of the proposed scheme seemed to be that of a Church Mothers Meeting'. They opened in summer 1908 with tea and biscuits 'to show friendliness and good feelings'. Of their first thirty-five infants, twenty needed medical treatment and half needed reliable milk supplies. Miss Reddish pressed reluctant male Co-operative members to foot the bill. Within a year she had five clinics in operation, within four years had obtained financial support from the borough council, and was feeding needy mothers as well.[29]

[28] Mrs Lees, *Oldham Chronicle*, 6 Oct. 1911. More generally, see W. Downing, 'The Ladies Sanitary Association and the Origins of the Health Visiting Service', London MA, 1963, p. 259.

[29] 'Work in the Municipality', *The Englishwoman*, 16, 1914, p. 18; Minutes of Bolton WLGS, 1 May 1908, 31 Aug. 1908, 22 Mar. 1912, 21 Nov. 1913.

Margaret Ashton started the Manchester Guild for mothers at the same time, again with the help of the town's women doctors and its MOH, and soon had six clinics under her supervision. The Guild offered classes in elementary hygiene to which it tempted mothers by offering cut-price fresh and dried milk, and the loan of prams. Nurses checked the babies' health. Though a voluntary scheme, it was increasingly funded by the council, which by 1916 faced to its dismay an annual bill of £16,000 for cheap milk alone. Unable to get the council to release a wing of the Old Infirmary for a baby unit, Miss Ashton set up in 1914 a small twelve-bedded baby unit of her own, which became Manchester's Baby Hospital. In 1915 she persuaded the council to establish a special committee for maternity and child welfare, which naturally she chaired. Where infant mortality had been 188 per thousand in 1900 it had fallen to 77 per thousand by the time of her death in 1937. Dr Niven, the MOH, credited her with much of this achievement.[30]

Elsewhere, Councillor Miss Hope developed maternity and child welfare services in Bath as did Councillor Miss McGregor in Wimbledon. Over in Oxford Miss Merivale had joined Mrs H. A. L. Fisher in a local Sanitary Association, founded in 1902 to promote public health and to manage cottage property, Octavia Hill style. Its Baby Welcome club helped mothers to prepare for lying-in, and lent them equipment. Following the 1907 Notification of Births Act, the MOH sent them details of infants born to poorer homes; and their voluntary health visitors were soon making seven hundred calls a year. Miss Merivale's coup was to persuade the Oxford Dairy Company to provide modified milk for over a hundred of their most sickly infants at cost; only seven of the babies subsequently died. In a shrewd gesture, the Dairy Company offered a clinic room and hygienic feeding bottles. Oxford's infant mortality rate of 94 per thousand in 1908 was already good; by 1912 it was down to 69 per thousand, and Miss Merivale's Health and Housing Association was given

[30] Dr J. Niven, *Observations on Public Health Effort in Manchester*, 1923, p. 181; S. Simon, *Margaret Ashton and her Times*, 1949, p. 9; Annual Reports, Women's Citizen Association; A. Redford, *The History of Local Government in Manchester*, 1940, vol. iii, pp. 72 ff. Dried milk was held to have checked Manchester's outbreaks of summer diarrhoea which killed so many infants. Dyhouse, op. cit., p. 256.

credit for it.[31] In Staleybridge, Mrs Summers had quietly turned her School to teach mothers their duty, into a Maternity and Child Welfare centre. By 1910 there were some eighty infant clinics in existence, by 1918 nearly a thousand. Though many voluntary workers were moralistic and patronizing (for that reason, the Women's Co-operative Guild pressed for more municipal clinics with salaried staff), the clinics were, conceded Sylvia Pankhurst, a 'great boon'. So when the Women's Labour League wished to establish a memorial to Mary Middleton and Margaret MacDonald, both of whom died suddenly in 1910, they naturally established a baby clinic in North Kensington, supervised by Dr Ethel Bentham, where two women doctors and a daily nurse weighed babies, checked on their physical development, and performed small operations. Ethel Bentham sadly noted 'almost all the diseases are those of malnutrition'.[32]

Around 85 per cent of mothers breast-fed their babies; but women at work, in poor health, or with narrowly spaced babies, could not do so. Yet fresh milk was often contaminated and costly, tinned milk was both quickly contaminated and unsuitable, and dried milk was not widely available until the War. Feeding bottles which were hard to clean without sterilizing equipment added to the hazards. St Pancras's MOH tested fifty samples of milk in 1899 and found less than a third of them clean. The rest were from tubercular cows, flavoured with manure, diluted with dirty water, 'enriched' with chalk, or preserved with dangerous chemicals. Doctors and womens' groups called for milk depots in every town, selling clean fresh milk from inspected dairies, as well as modified or humanized milk (sterilized and sweetened) for babies. Mrs Salter, the Bermondsey ILF councillor, characteristically suggested that 'the proper way of humanizing cows milk is to give it to the mother'.[33] The MOH of St Helens

---

[31] Miss Hope, *Bath Chronicle*, 24 June 1922; Oxford Sanitary Aid Association reports 1908–9; Oxford Health and Housing Association reports, 1912, 1915–16; *Municipal Journal*, 3 Mar. 1908.

[32] *Staleybridge Reporter*, 21 Jan. 1944; WLL, 'The Needs of Little Children', pp. 19 ff. The N. Kensington clinic handled a thousand cases a year. Sylvia Pankhurst, quoted J. Lewis, op. cit., p. 100.

[33] F. Lawson Dodd, *Municipal Milk*, Fabian Tract 122, 1905; Mrs Salter, 'The Needs of Little Children', p. 7.

started the first municipal milk depot in 1899, modelled on continental examples, and a few other towns followed. One bitter moment came when Woolwich borough council swung Tory in 1909 and promptly closed the milk depot started by the previous Labour group. Women councillors held public meetings in protest, abetted by the MOH, and presented petitions from local mothers, but were outvoted.[34] As with midwifery care, the problem was cost. Mrs Pember Reeves recalled a lecture by a West End doctor to West End charitable ladies. If they wished to help the children of the poor, he suggested, the most useful thing they could do was walk through the East End with placards bearing the legend, 'Milk is the proper food for infants'. His audience, she said, were 'deeply interested and utterly believing'. Mothers were again to blame. But, added Mrs Reeve, the reason infants did not get fresh milk but were fed with pappy bread, poor quality tinned milk, and adult scraps, was its cost. At four pence a quart, the same in Lambeth as in Mayfair, none of her mothers could afford to spend 2s. to 3s. a week on one infant when they had just 8s. a week with which to feed the entire family.[35] Only where milk was subsidized, as in Manchester, could it help to improve infant mortality.

Women councillors were also to campaign for nurseries and crèches, though opinion, then as now, divided between those who thought this undermined maternal responsibility (Manchester's MOH) and those who were determined to give high-risk infants a better chance of life (Southwark's MOH).[36] Mrs Townsend of the Fabian Women's Group pointed out that France had abundant nursery provision where children were not only safely off the streets, but were toilet-trained, and learnt to eat, sleep, and wash properly, all aspects of child rearing that inevitably suffered in overcrowded homes but which made early school years so hard for both teacher and child. In Tynemouth, Councillor Miss Burnett established a Guild for mothers and a day nursery. Both were so successful that the first was taken over by the health committee and the second qualified for an Education Department grant, much to

[34] WLL 6th Annual Report, 1910–11, pp. 23–4.
[35] *Round About a Pound a Week*, pp. 98–9.
[36] Redford, op. cit., p. 136; MOH, Southwark, Annual Reports, 1895–6.

local gratification.[37] Most nursery places were provided by the voluntary sector, though women councillors were often able to attach nursery classes to existing infant schools.

Like school board and co-opted LEA women before them, women councillors were concerned for the welfare of the whole child, and had a long record of organizing school dinners, holidays, and boots for needy children. Margaret McMillan in the same spirit had pioneered medical care and swimming-baths for Bradford children. In 1902 she had moved back to London where with her sister Rachel she had started nursery schooling in Deptford which was to be highly influential as a model of child care. Reports from the Fitzroy committee on physical deterioration had uncovered a pool, up to 15 per cent it was thought, of underfed school children. Voluntary agencies were clearly inadequate.[38] Even before they had taken office, progressive Liberals within and without the Shadow Cabinet had been studying questions of school feeding and child welfare policies. Now, pressure from Sir John Gorst, the NUT and the Labour movement converged to enact the 1906 Education (Provision of Meals) bill which allowed LEAs to finance school meals from the rates. The women's movement, the NUT and the Labour movement pressed their LEAs to adopt the Act. In Jarrow, the local branch of the Women's Labour League crowded the visitors gallery to hear their petition discussed at the annual council meeting. Said one woman bitterly, 'The mayor was re-elected, gowned and chained—the councillors made nice speeches to each other, but when the question of the petition came up the Mayoral banquet was waiting in the next room and the starving children must wait'. When the council reconvened, well-fed themselves, they were confident that Jarrow had no hungry children but referred the matter to the education committee who referred the matter to the school attendances subcommittee who were confident that Jarrow had no hungry children.[39] Nettie Adler's mother had initiated free and penny dinners throughout London's Jewish schools, cooking

---

[37] *The Case for School Nurseries*; *Shields Daily News* 8 Jan. 1918.
[38] By 1905, 55 of 71 county boroughs had voluntary school feeding, but only a quarter of the other boroughs, and then only for 2 or 3 days a week in winter.
[39] Jarrow branch MS report to WLL, u.d., fo. 57.

fourteen hundred dinners a week. Miss Adler urged the LCC
to continue to feed needy children during school holidays as
distress and industrial disputes deepened; but to Progressive
fury her motion was defeated.[40] By 1911–12 a third of all LEAs
had adopted the Act, especially the larger cities, though other
towns remained fearful of the cost to the rates and to moral
fibre if they should feed schoolchildren. In 1914 the Act was
made compulsory.

Food was not enough. Women councillors also turned their
attention to school medical treatment. Schools were required
by the 1907 Education Act to inspect children, and had
uncovered a reservoir of ill health—of six million children, two
and a half million were verminous, two million had bad teeth,
six hundred thousand had weak sight, five hundred thousand
were adenoidal, and many again had skin diseases. It was
assumed that following inspection, parents would then seek
medical treatment; but hospitals could not cope with the work
and parents could not afford the fees. Ethel Bentham and
Councillor Edith Kerrison reminded women, 'inspection
without treatment was a farce and treatment could not be
provided by poor parents'. Philanthropic groups offered
spectacles, tooth brushes, and tooth powder, but were unable
to do much about the food, sleep, housing, and fresh air
children needed. As in the early days of the school boards,
women councillors insisted that the parents must be 'taken
into partnership . . . be given their full share in assisting in the
child's care . . . Parental responsibility is not merely a
question of payment. It is a question also of interest,
knowledge and power'.[41] Margaret McMillan established the
first school clinic in 1908. By 1910 there were fourteen; by
1914 most LEAs provided some medical treatment; by 1921
some nine hundred clinics covered the country.

The LCC, like the LSB before it, pioneered much of the new
work, and at the prompting of Miss Frere (a latter-day
Margaret McMillan), established a network of child care
committees for each elementary school, bringing together

    [40] A. Burdett-Coutts, ed., *Women's Mission*, 1893, p. 20; *East London Observer*, 27 July
1912. See also Ethel Bentham, WLL 4th Annual Report, 1908–9, p. 22.
    [41] *Municipal Journal*, 21 Jan. 1911, p. 57; WLL, 'The Needs of Little Children', pp.
4–5.

managers, voluntary societies such as the NSPCC and Country Holiday funds, health visitors, settlement workers, employment agencies, and GPs. They tried to offer a total concept of child care, embracing health and feeding, boots and savings clubs, recreation and unemployment. Care workers visited parents and children in their homes keeping case notes, co-ordinating aid. It was, wrote Miss Frere, pre-eminently women's work.[42]

Women councillors like Nettie Adler sought also to restrict child labour. Surveys undertaken by the Women's Labour League confirmed the official Return on Wage-earning children of 1899. In these years before the War, there were nearly 300,000 children of twelve and thirteen at work, members of the reserve labour force which sucked down adult wages, and another 35,000 northern children working half-time. Almost more worrying were the 300,000 plus children who were at school full-time, but worked before and after school hours, a third of them more than twenty hours a week. Nettie Adler quoted one lad working fifty hours a week for a greengrocer, three hours before and five hours after school, as well all day Saturday; and girls doing forty hours a week of household drudgery on top of school. As these children had to be fairly clean and neat to obtain jobs and therefore seldom came from destitute families, they were, she thought, the children of the greediest rather than the neediest.[43] Lydia Becker and Annie Besant had, years before, found themselves baffled by the plight of a family dependent on its child's earnings. Women councillors no longer had such inhibitions. Like Margaret McMillan they insisted that children must not be sacrificed to the family well being. They had rights of their own. They were entitled to their schooling, their rest, and their childhood. Women members wanted the school leaving age raised, and child labour restricted by by-law. And if that made the council unpopular with hard-pressed mothers who wanted their children's labour, then so be it.

Slowly, in town after town, infant mortality figures fell, and

[42] Miss Frere, NUWW Occasional Papers, Feb. 1911; H. Jennings, *The Private Citizen in Public Social Work*, 1935, chs. 5 and 6.

[43] H. Adler, 'Children and Wage Earning', *Fortnightly Review*, May 1903; Mrs Hogg, 'School Children as Wage Earners', *Nineteenth Century*, Aug. 1897. H. Adler and R. Tawney, *Boy and Girl Labour*, 1909.

child health improved. The key factor was probably rising wages and falling family size, together with the continued investment by towns in pure water, efficient sewers, slum clearance, and low-density suburban development. The key figure was the Medical Officer of Health, his energy, the publicity he sought, the support he obtained from doctors and salient public opinion, the degree to which he carried even Tory (indeed, mostly Tory) councils along with him. Women councillors were a tiny number. On most town councils, steps taken to reduce infant mortality and child morbidity were taken by men, though very often under strong pressure from a mobilized women's movement which was entirely familiar with the work of a Margaret Ashton or the policies of the Progressive platform. But it is also true that where a town had a woman councillor she kept the question on the political agenda (often helped here by her independent label), fought for more resources, was the lead speaker on public platforms, and established suitable committees and subcommittees to press the work forward. She also became a focus for the efforts of others: women doctors, health visitors, women sanitary inspectors, voluntary and philanthropic groups. On matters of municipal trading, the views of women members were unwanted and unsought. On issues of infant and child welfare most male councillors were willing to allow their women colleagues an influence, and a work load, disproportionate to their numbers.

Unlike most of the men, the women were far less judgemental, less willing to attribute infant mortality to bad mothering, more willing to attribute it to poverty and poor housing. For all sorts of reasons but also from her experience as a town councillor, Eleanor Rathbone was to campaign as an MP for family allowances. Women spoke in progressive accents and pursued progressive policies, in much the same way as Liberal men. But it was not quite the same. Men talked of the waste of infant life, women of the massacre of the infants. Progressivism bridged the languages of social audit and moral anger, but that note of moral anger, of outrage, and of compassion came particularly from its women.

Mrs Pember Reeves had asked whether slender wages were better spent on food, warmth, or space. On reflection she

thought that better housing, the extra room, dry and ventilated, had perhaps the edge. Yet with high land values, high
rents, and most work still tied to the inner cities, few could
afford that choice. Mrs Pember Reeves was probably right.
Birmingham's infant mortality figure was 125 per thousand;
that of Bournville, just a short distance away with seven
houses to the acre, was 37 per thousand.

Housing and Town Planning

When women joined town councils in 1907, there had been
some thirty years of housing legislation designed to clear the
slums,[44] but without cheap transport and with ever higher
rents, overcrowding remained severe. Only eight of London's
twenty-eight boroughs showed any improvement in the
Edwardian years;[45] over a third of central London was still
living at two or more a room in 1911. The East End particularly was receiving waves of Jewish immigrants from the
pogroms of Eastern Europe, but every city had its plague
spots. Mabel Clarkson, a quick, sharp, compassionate lady,
wrote in her Election Address of 1912 for Norwich,

I know many, in fact most of the Courts and Yards in our poorer
districts, and I do not hesitate to say that some of the dwellings in
which some families are living today, are a disgrace to the City.
Those of us who care for the purity of our homes, for the right of
little children to opportunities of health and development, for the
prevention of infantile mortality, and of all the unnecessary sickness
and suffering caused by overcrowding and bad housing, are bound
to make every effort to get rid of the slums.

Those inward-looking, secretive, gloomy enclosed courts and
alleys, rotting into disrepair, whose dead ends collected the
moral and physical rubbish of the town, should be replaced by
open-ended streets which allowed free movement of people,

---

[44] The Torrens Act of 1868 empowered local authorities to repair, close, or
demolish an unfit house; the 1875 Cross Act extended these powers to areas. The
Housing of the Working Classes Acts of 1890 and 1900 permitted local authorities to
buy land for long-term improvement schemes and to rehouse at least half of those
cleared. See J. Tarn, *Five Per Cent Philanthropy: An Account of Housing in Urban Areas
1840–1914*, 1973; J. Burnett, *A Social History of Housing*, 1978; E. Gauldie, *Cruel Habitations: A History of Working Class Housing 1780–1918*, 1974, M. Daunton, 'Public Space and
Private Space', in D. Fraser and A. Sutcliffe, eds., *The Pursuit of Urban History*, 1983.

[45] Waller, op. cit., p. 31.

air, and opinion. Women worked patiently at the routine of public health and housing, closing cellars, banning back-to-back development, compelling repairs, abating severe overcrowding. Said Edith Sutton longingly, 'If only there were more of us; the amount that could be done! If you only knew what power we had, and the way in which we could cleanse the streets and purify the surroundings of the people'.[46]

Town councils not only wanted to clear the slums but had to stop instant slums being built in their place. In most provincial cities by now, private builders had to conform to by-laws requiring them to construct cottages with at least four rooms, piped water, separate sculleries, privies, and often internal WCs; with higher ceilings and larger windows; soundly built with damp-proof courses below, a yard behind, and a decent respect for the building line in front. Arranged in neat if repetitive terraces, these are the cottages of today's improvement areas, now cherished and coloured courtesy of Sandtex. As Dr Slater put it to the WLGS annual meeting of 1908, health committees had done their work 'very enthusiastically according to a man's view as to what is necessary for health administration—which is essentially the view of the plumber—the making of sewers down the streets and connecting the houses by drains'. Local authorities, he said, also needed a woman's view for 'the building up of the health of each little boy and girl'. Women must persuade council committees not to confine their concern merely to construction but also to care about design, to consider not only sanitation but also architecture. For the home was women's 'shelter and workshop', and its design 'determines the fate of the wife and mother'.[47] In stories that were quoted by the women's press, only women councillors seemed to notice that larders should be on the coolest not the warmest wall; that rooms with too many doors were impossible to furnish; that to have the dustbins sited next to the larder window was unhealthy, that squared corners collected dust, that steep narrow stairs were dangerous, that living rooms should be positioned to get the

---

[46] M. Clarkson, 1912 election address for Coslany ward; E. Sutton, *Anglo-Japanese Congress*, 1910, p. 10.

[47] Dr. Slater, WLGS Annual Meeting, 10 Mar. 1908; *Municipal Journal*, 24 July 1908.

sun; that children living in tenements five floors up were cribbed and confined, denied the safe play essential to their growth. Marion Phillips and Ethel Bentham claimed that sitting on Kensington borough council they had become an architectural panel of two: they had, for example, redesigned the cottages being built by the Great Western Railway to include hot and cold water, a bath, and well-arranged windows and doors. Male colleagues, they reported, objected that the poor would only keep potatoes in the bath (coal was presumably too expensive); that they did not need a second bedroom as they would only sublet it to lodgers, nor want a larder since women enjoyed shopping trips, nor use a washhouse/scullery because women preferred to decorate their living room with laundry.[48]

Privately built by-law housing was increasingly within reach of the urban working class, but not the families studied by Mrs Pember Reeves. They needed, she said, municipally subsidized housing, adding, cheerily, that such houses would be self-financing from savings on poor law and medical bills. By 1907 a few local authorities, notably the LCC, Liverpool, Sheffield, and Manchester, had begun to build their own cottages and tenements, avoiding as far as possible, the bleak oppressive style of the Peabody blocks labelled by Masterman 'Later Desolate'.[49] The LCC's pioneering Boundary Street development, started in 1893 on the site of the Old Jago, had shown a new sensitivity to urban townscape and human scale, though the Tory victory of 1907 had checked LCC house-building and was to sterilize LCC landholdings until after the War. By 1914, LGB inspectors calculated that some 23,000 council houses had been built, mainly since 1911.[50] Manchester, for example, had established its city architect's department in 1902; from 1904 it was building two-bedroom cottage estates. Margaret Ashton became much involved with the work, at last able to use her expert knowledge of drains and ventilation systems. By 1910 the city was building semi-

[48] WLL, 9th Annual Report, 1913–14, p. 59; *Labour Woman*, Apr. 1914; WLL leaflet, Dec. 1912.
[49] C. F. G. Masterman, *The Heart of the Empire*, 1901, p. 16; Mrs Pember Reeves, Central WLL branch conference, May 1913.
[50] H. Adler, *London Women and the L.C.C.*, 1924, p. 7; 44th Annual Report, LGB, 1913–14.

detached cottages, many with three bedrooms and bathrooms. In Oxford, Mrs Hughes and the Progressives were calling for the council not only to clear slums, but as many of the slums were council owned, to provide other accommodation in its place. Liverpool's council building stretched back to 1869, and by the time Eleanor Rathbone joined the authority it owned some 2,250 flats with a rental income of £20,000. She served on the housing and dwellings subcommittee which she found 'extraordinarily interesting' as she watched self-respect, cleanliness, sobriety, and 'home pride' develop among former slum dwellers and immigrant Irish, offered decent housing for the first time. Housing became her love. More practically, her mother, worried that children would fall into fires, offered permanent fireguards to every new tenant, which the council accepted with thanks.[51]

In Norwich, one can trace the enlargement of Mabel Clarkson's views. At first, she thought the council's responsibility was limited to imposing compulsory repairs notices on property. Then she was to be found suggesting that the council should itself acquire and renovate properties. Finally she urged the Tory majority to build council housing. She graphically described families with half a dozen children living in two or three damp and insanitary little rooms crouched under the shadow of the Cathedral, families who were then criticized by the moral majority for their slovenly ways. The municipality, she said, should prevent 'such pitiable waste', should give 'every child in its care . . . room to grow', every adult the environment 'that would uplift and not degrade'.[52] In Birmingham, Miss Pugh was making a similar stand. 'The property owners had a right to a fair profit', she argued, 'but a profit which was made on the toll of lives was not a fair profit'. The council must intervene. Labour women were committed to council housing from the start. Mrs Annie Reeves, who had flown the ILP flag in Norwich in 1907 and 1908, as a poor law guardian called on the council to replace houses lost in the great floods of 1912. Council houses, she insisted, must always be built to the highest standards and not

[51] Redford, op. cit., pp. 125–6; Mrs Hughes, *Oxford Chronicle*, 30 Oct. 1908; *Liverpool Daily Press*, 26 Oct. 1910; Liverpool housing committee Minutes, 27 Jan. 1911.
[52] *Norwich Mercury*, 26 Oct. 1912; election addresses, 1912, 1913 (Colman & Rye).

down to the lowest incomes.[53] But without central government subsidy, and with much of local government Tory-controlled, such calls were ignored.

Housing, even well-designed and solidly built housing, was no longer enough. The MOH for St George's Southwark had confessed in the 1890s that desperate though the need for more housing was in his parish, it was even more important for public health that slum-cleared land should remain as open space. Town planning was of course nothing new, as Roman, Renaissance, and Regency building testified. London's new thoroughfares in the mid-nineteenth century— New Oxford Street, Victoria Street, Farringdon Street, New Cannon Street—were designed to run through, open up, and eradicate the slums.[54] Public health, public order, public amenity, and enhanced property values happily came together when the poor were 'shovelled out', in the felicitous phrase of the vicar of Cripplegate in 1861. The Cross Act of 1875 had helped other energetic authorities, headed by Chamberlain's Birmingham, to follow suit. Handsome new streets ('Corporation Street' in Birmingham; what else?) were driven through slum quarters, their fine frontages of Palladian banks, Gothic town halls, and baroque department stores impressing the visitor, concealing the poorer quarters, raising rateable values, and allowing the Corporation to make a profit.

Women noted that it was a very masculine and engineering version of town planning, focused on the public domain, commercial life, and the street, and in which the concept of citizen was monopolized by the male business middle class. In the quarter century before the War, however, many cities— Birmingham and Manchester were only two of the largest— extended their boundaries to take in adjacent land, some of it green fields, other bits already an unplanned suburban sprawl. As Masterman wrote,

People are complaining bitterly today of the blind alleys and crowded narrow streets that are in our city. 'Alas', they say, 'the

[53] *Birmingham Post*, 31 Oct. 1911; Annie Reeves, WLL 9th Annual Report, 1913, p. 58.
[54] G. Stedman Jones, *Outcast London: A Study of Relationships between the Classes in Victorian Society*, 1971, p. 167.

property is so valuable, what we can do to put right the mistakes of our ancestors'—and all the while there are added to London crooked and bad streets by the hundreds of miles every year, and these same people do not raise their voices to stop it . . . The Housing problem is not to be solved in the slums of Camberwell, or Whitechapel, but in the green fields of Harrow and Hendon, in Woodford, East Ham and Barking, and the suburbs of the South . . . the public conscience should realise that it must protect its suburbs . . . [Instead] here a few streets facing one way, there a few streets facing another; here a house or two that are really decent and comfortable, there a whole row of houses that are a disgrace to civilization.[55]

He welcomed municipal housing, not because it made any great quantitive contribution to solving the problem but because, then as always, it set standards for the private sector to follow. The development of the electric tram and somewhat shorter working hours permitted working-class families to live further out and to buy more space with their rent. John Burns's 1909 Housing and Town Planning Act was in part a response to a wider concept of housing than merely regulating it by by-laws.[56] It recognized the need to protect open space, separate residential from industrial sites, and ensure community facilities.

Boundary extension was one pressure for town planning. Birmingham, for example, was planning the development of some four thousand acres between 1906 and 1913; and a new vocabulary of land use policies, residential zoning, and green belt began to enter the language. Another pressure was the garden suburb movement. Ruskin's espousal of the picturesque; the models of corporate villages at Bournville and Port

---

[55] Masterman, op. cit., pp. 80, 99.

[56] The 1909 Act required local authorities to survey their housing stock, and repealed the requirement that any housing built by local authorities must be resold within ten years. The Act did not equip local authorities to lay out privately owned suburban land, even though an Association of Municipal Councils survey of 1907 suggested that three-quarters of town expenditure on street widening and open spaces could have been avoided had authorities had planning powers. Waller, op. cit., p. 279. Town planning in theory brought together both negative and positive strands of municipal intervention into the housing and land market. See A. Sutcliffe, 'The Growth of Public Intervention in the British Urban Environment during the 19th Century: A Structural Approach', in J. H. Johnson and C. Pooley, *The Structure of Nineteenth Century Cities*, 1982; A. Sutcliffe, ed., *The Rise of Modern Urban Planning 1800–1914*, 1980.

Sunlight in the 1890s; the opening in 1903 of Ebenezer Howard's Letchworth, and the promotional work of Geddes, Unwin, and Abercrombie, were to have a profound influence on the urban imagination. Town planning here was less about the theatrical presentation of city centre street frontages and commercial life; and rather more about bringing town and country, class and class, together into the tapestry of a living village community; where rustic cottages with mansard roofs and decorative brick, each with their own garden, clustered along gently curving, ever sunlit, tree-lined roads. Its concerns were family life rather than public life, its clients women and children rather than men. Hampstead Garden Suburb, conceived by Octavia Hill and Henrietta Barnett, was one example, the garden village of Councillor Mrs Lees at Oldham, another. Some sixty such garden suburbs were started before the War, the direct forerunners of the new towns. As John Burns said, his Housing and Town Planning Act sought 'to secure the home healthy, the house beautiful, the town pleasing, the city dignified, and the suburb salubrious'. And a year later, 'I conceive the city of the future as Ruskin, Morris, Wren and Professor Geddes wished a city to be—that is, an enlarged hamlet of attractive healthy homes . . . harmonizing so far as may be possible with the life and characteristics of the people'.[57]

A third strand was more American in origin, the aesthetic concept of the City Beautiful, and again it was often women-led. Liverpool's Kyrle Society was carried by women to many cities. The planting of trees, the provision of domestic flower boxes, generous grass verges, the control of bill boards, the gift of water fountains, the landscaping of parks, would all help to beautify the streets which served as the gardens and open space of the poor. Liverpool hosted a City Beautiful Conference in 1907. Women councillors everywhere quoted Ruskin in their 'campaign against ugliness'.[58]

Mrs Lees had perhaps the clearest vision of what a town council could do. An American woman journalist visited her

---

[57] Gauldie, op. cit., p. 305; W. H. G. Armytage, *Heavens Below: Utopian Experiments in England 1560–1960*, 1961, p. 382.

[58] W. H. Wilson, 'Ideology, Aesthetics and Politics of the City Beautiful Movement', in Sutcliffe, op. cit.

at Oldham and (obviously having read Dickens on the train)
wrote an article entitled 'Mothering a Municipality':

We ascended the long sloping hill and beheld through the smoke one
of the ugliest of England's industrial towns. The streets are long and
narrow, the houses are in straight lines, each house is the exact
reproduction of every other, and every street is the exact counterpart
of every other street—except that some were more tidy and showed a
more prosperous tenantry. No trees could be seen, and only in
higher grade streets a little box of flowers or a bed of blooming
plants. Surrounding the monotony of ugliness are valleys crowded
with textile mills, whose hundreds of tall smoke stacks look down
upon the thousand tiny chimney pots on the little ugly stone
cottages, as if to say, 'We support you, so do your duty and be
content'.[59]

This was the town of which Mrs Lees said, in authentic
Progressive accents, as she received the freedom of the
borough:

Is it not well to set an ideal before us, something to follow and work
for? Is it too much to think of a beautiful Oldham devoid of black
smoke and smuts? A town without slums, a town where all the
houses are in good sanitary condition and not overcrowded; where
there are fewer temptations to excessive drinking and the opportunities for healthy recreation and pleasure are greater; perhaps
even a town where the trams are convenient for every one? (laughter) I know these improvements will cost money in the first instance
but would it not pay in the end? We want the rate and tax payer to
realize that not spending money is not necessarily economy . . .[60]

In 1903 she had founded the Beautiful Oldham Society to
much local mirth. By 1909 it had four hundred members.
They held competitions for cottage gardens though the
number of entries were disappointing, offered trees for planting, exhibited spring flowers, and searched for playing-fields
to make Oldham 'less dreary, more habitable, less monotonous'. They campaigned against smoke pollution which veiled
the hills, and with trade-union help researched local housing
conditions. Oldham had few back-to-backs and no cellar
dwellings, unlike Leeds and Nottingham, but many houses

---

[59] M. McDowell, 'Mothering a Municipality', *The Designer*, Feb. 1912, p. 227.
[60] Mrs Lees, *Oldham Standard*, 3 Nov. 1911.

were in a bad state of repair, and they were notoriously small. At least ten thousand of its thirty-one thousand homes were badly overcrowded in 1908. Typically, when the council, at last under Progressive leadership, thought it would be daring and build its first twelve council houses, the MOH insisted that it would be absurdly extravagant to provide them with two bedrooms. One was clearly sufficient. The houses were not built.[61]

Baffled on that front, Mrs Lees tried to teach by example. Drawing on the model of Letchworth, she planned a garden suburb of some three hundred homes, with a variety of façades and frontages, steeply pitched roofs, and surface rendering. They were to have three bedrooms and a bathroom, be built at fourteen to the acre, and offered at rents of 4*s*. to 5*s* a week—within working-class reach. She formed a co-operative building society among better-paid mill workers, and six years later with one hundred and fifty of the homes complete, the American journalist saw 'a garden city grown out of a clay and cinder hole'. They are quite charming; but her hope that the town council would follow where she had led came to nothing.[62]

So she and Mary Higgs again turned their attention to Oldham's plague spots. They sent letter after letter to the press urging closure and clearance: they uncovered one row of houses without a single privy among them. What appalled them even more than the disrepair was the indifference, even complacency, of slum dwellers towards their homes. Once on the council, she immediately introduced motions welcoming the 1909 Housing and Town Planning Act, and urged members to adopt 'a full scheme drawn up by experts' rather than permit haphazard development as and when an infill site became vacant. Recognizing that Oldham was built up to its boundaries, she begged the council to discuss overspill development with neighbouring authorities. Elected mayor and with a Progressive majority behind her, one of her first moves in 1911 was to call a local town planning conference,

[61] Beautiful Oldham Society, Annual Reports, Lees cuttings; C. Bedale, 'Property Relations and Housing Policy: Oldham in the Late 19th and Early 20th Centuries', in J. Melling, ed., *Housing, Social Policy and the State*, 1980.

[62] *The Designer*. See illustrations.

with experts from other cities, and prizes for the best architectural drawings offered by herself. The conference recommended that the council should specify the roads, densities, designs, and open space of new developments. They formed an advisory committee, which she chaired, and after some ten years of coaxing, lobbying, prodding, promoting, and publicizing with the help of the faithful *Oldham Chronicle*, she persuaded the town council to set up a housing committee and adopt planning guide-lines. There was no point in clearing slums unless they were replaced with something better, she said; as that could not be done except at a loss, she was personally willing to subsidize new housing. She was slowly winning and buying support when the War intervened.[63]

It was an extraordinary record. She was channelling all the ethic of late Victorian philanthropy, all the deference owed to her social standing, much of her private purse, and all the energy of a close circle of able women, into municipal life. At least as remarkable was her insistence on working through local government forms and democratic modes. She refused to bypass town council bureaucracy, refused to ignore the lethargic and sluggish among her male colleagues, but tried to educate them while submitting herself to the discipline of the party system and the out-turn of a committee vote.

There was one housing question that women councillors made exclusively their own—municipal lodging-houses for women. Mary Higgs, friend of Mrs Lees and wife of Oldham's Congregational minister, and then Julia Varley, a Bradford guardian, had both gone on the tramp a few years before. They had found nowhere for poor but respectable women to stay while they searched for work. Common lodging-houses were of doubtful cleanliness and doubtful reputation. Salvation Army homes were always full. The workhouse casual wards where both women slept were 'demoralizing and degrading'; they damaged rather than sheltered the women who entered them. YWCAs and the like were far too expensive. Women ended up sleeping rough. Most larger towns had clean and reasonably priced municipal lodging-houses for men; yet women who had less money and were more at risk

[63] *Oldham Chronicle*, 6 Oct. 1911, 8 Nov. 1913.

had nowhere to go. One WLL branch reported that twenty of forty prostitutes in their borough refuge 'fell' because they could not find decent accommodation.[64]

So in 1907 Margaret Ashton led a deputation of the Women's Local Government Society and the North of England Guardians Society to Manchester's sanitary committee pleading for a municipal lodging-house for women. They agreed. It opened in 1910, gracefully named Ashton House, situated in the poorer quarter behind Victoria Station and supported by council subsidy. It provided cubicle dormitory accommodation for two hundred and ten women, who could also use sitting-rooms, kitchen, laundry, and bathrooms at a cost of 5d. a night, or 1s. a day with food. There were few rules except a ban on drink.[65] Her lead was followed across the country. Miss Reddish pressed for a woman's lodging-house at Bolton, Dr Ethel Mordant at Bromley; in Liverpool, Eleanor Rathbone extracted a lodging-house from her housing committee; and the NUWW reported that its branches in Bradford, Brighton, Bristol, Leeds, Leicester, and London were negotiating for women's lodging-houses with their town councils. When that doughty feminist, W. T. Stead, went down in the *Titanic*, municipal lodging-houses were adopted by the women's movement as an appropriate memorial.

Unemployment and the Local Economy

Infant and child welfare were recognizably suitable work for women; housing and town planning conceivably so. A third question that assumed increasing importance to local councils in the decade before the War was unemployment and the local economy. There too newly elected women councillors were to make a surprising and unpredictable contribution.

Elected women had always made the well-being of public sector female staff—teachers, nurses, cleaners—their especial care. Now they had also the task of supervising and supporting some of the one hundred and seventy women sanitary

---

[64] [M. Higgs], *Five Days and Five Nights as a Tramp among Tramps—by a Lady*; *Three Nights in Women's Lodging Houses*, u.d.; J. Varley, *Life in the Casual Ward*, u.d.; M. Ashton, NUWW conference, Liverpool, 1909; WLL leaflet, Sept. 1912.

[65] *The Woman Citizen*, Dec. 1937, p. 11; WLL 7th Annual Report, 1911–12; *Liverpool Daily Post*, 27 Oct. 1910; Liverpool housing committee Minutes, 14 Oct. 1909.

inspectors as they visited factories, offices, and laundries, and tracked down small grubby backstreet workshops that persisted in flouting the law.[66] Mrs Greenwood, a Finsbury sanitary inspector, described herself as a woman working for women in a department staffed by men, under a male chief officer and reporting to a male-only committee. Everything, she said, 'is regarded from the male point of view'. She begged educated, intelligent, 'broad-minded' women to stand as councillors and to serve on health committees where they might 'infuse more common sense and less red tape'. Two years later Sister Maud Lindsey of the Claremont Mission was elected on just such a programme in Finsbury, though how broad-minded she was is less certain.[67]

Margaret Ashton's route into public life is perhaps worth recalling. She started in 1895 by acting as unofficial treasurer to Manchester women's TUC as they unionized laundry, dressmaking, and clerical work. She arbitrated in wage disputes, and working with women sanitary inspectors helped to arrange the city's Anti-Sweating Exhibition in 1906. She later started trade schools for girls, determined like school board women that they should not finish up in dead-end jobs.[68]

New, however, was their concern with adult unemployed women. Chamberlain had deliberately bypassed poor law boards when in 1886 he had encouraged local authorities to undertake public works to relieve winter unemployment, rather than force respectable families into the workhouse. When distress worsened in the early 1890s, most of the London boroughs and some sixty city councils were operating such schemes,[69] but with the coming of good times they were put aside. The depression of 1903–5 produced Walter Long's Unemployed Workmen's Act of 1905 which required towns of

[66] See A. Anderson, *Women in the Factory*, 1922; *Economic Review*, 'The Inspection of Women's Workshops in London', Jan. 1901.

[67] Mrs Greenwood, *Anglo-Japanese Congress*, p. 24; Miss Lindsey, *Islington News*, 1 Nov. 1912.

[68] *The Woman Citizen*, Dec. 1937, pp. 5, 8; Councillor Miss Burnett, *Shields Daily News*, 22 Oct. 1910.

[69] See J. Harris's fine *Unemployment and Politics: A Study in English Social Policy, 1886–1914*, 1972, pp. 77, 84.

over 50,000 to establish distress committees and labour bureaux to aid the involuntarily idle.

The distress committees brought together city councillors, guardians, and co-opted voluntary workers, including at least one woman. They could levy a halfpenny rate in the pound or a penny rate with the consent of the LGB to be spent on public relief schemes. The committees had an impossible brief, with limited funds and machinery. Those boroughs, like Poplar, with the greatest need had, as ever, the least resources. In any case, rescheduling the city engineer's road schemes, advancing slum clearance, or landscaping open space, the usual public relief schemes as they neither cost too much nor invaded private enterprise, could make only a modest impact on the local economy, could do little to help skilled men or to decasualize unskilled men. Such ambitions had to await municipal house-building after the War which could indeed prime the pump of the local economy. But at least public relief schemes did decriminalize this version of outdoor relief by detaching it from the stigma of the poor law.[70]

Manchester's distress committee included Miss Dendy, and they authorized storm relief and street works. Staleybridge was too small for an official distress committee, so Councillor Mrs Summers funded an unofficial employment centre of her own, mainly by keeping open an unprofitable section of her ironworks. Mabel Clarkson joined Norwich's distress committee in 1911. Half-time shifts running for sixteen weeks were offered to the unemployed, who included some fifteen hundred boot and shoe workers, to lay out roads, dig the water ponds (today replete with ducks and lilies) in Wensum Park, lay down tennis courts, trim paths and churchyards—just as in the 1930s the unemployed were to lay out Eaton Park. Three hundred Norwich men emigrated to Canada. The experience affected her deeply; she said, as a Labour lord mayor in 1931, that 'no action on our part can make up to the unemployed for the enforced waste of their powers, their loss of independence, and the awful feeling of insecurity continually haunting their

---

[70] SC on Distress from Want of Employment, pp. 1895, vols. viii, ix; Harris, op. cit., pp. 108–9; H. Emy, *Liberals, Radicals and Social Politics, 1892–1914*, 1973, p. 154.

lives. Whatever economies may be laid upon us, I hope the very last to be considered will be any that will accentuate their position'.[71]

Women councillors soon noted that the thrust of relief works was directed towards men. Unemployed women were invisible. Three thousand working women had at the end of 1905 marched from the East End to Westminster in silent protest at the plight of their unemployed sisters. The Women's Labour League found just three workshops for women run by the London boroughs, which could place two hundred and sixty-nine women, yet twelve hundred and forty were on the official register, and many thousands more were in need. They begged Burns to finance a farm colony where women might regain their health, but he refused.[72] Outside London, women were somewhat more successful. In Birmingham, Councillor Clara Martineau chaired the subcommittee and kept women's issues on the agenda. Mabel Clarkson persuaded Norwich to open clothing workshops for women, and started cookery and laundry classes for some two hundred unemployed factory girls, turning most of them, apparently with their consent, into handy domestic servants. Eleanor Rathbone, with a study of casual dock labour behind her, and a study of working-class budgets in progress, joined her distress committee in 1905. They were soon dealing with a hundred applications a week, rescheduling road works and finding allotments. Like Margaret Ashton, she worked hard to prevent boys and girls drifting into dead-end casual labour, and instead sought to persuade them and their parents to take up apprenticeships. As the annual reports noted, 'lady members of the committee have made the female cases a special care . . . owing to their efforts it has been found possible to permanently ameliorate the lot of many women in the poorest circumstances'.[73] She helped nearly two hundred women, mostly widows with children, by opening sewing rooms in a disused Wesleyan chapel, quite undaunted by the

[71] *Birmingham Mail*, 30 Jan. 1932; *Staleybridge Reporter*, 2 June 1939; Miss Clarkson, *EDP* 10 Nov. 1931.
[72] WLL 4th Annual Report, 1909–10, p. 23; WLL central branch Minutes, 16 Feb. 1909.
[73] Liverpool distress committee Minutes, 13 Nov. 1905; report of the distress committee, 1906, p. 17; Annual Report, 1910.

fact that, like Ethel Leach of Yarmouth, she could not sew a stitch. They made creditable quilts, bags, cotton clothes for the workhouse, curtains, and the like. Most women filtered through into waged work. None went to the workhouse.

Manchester was the commercial, Oldham the manufacturing centre of the cotton trade. Unemployment, which was less than 4 per cent in 1907, had in Oldham doubled the following year. Councillor Mrs Lees, with her daughter Marjorie, a poor law guardian, and Mrs Higgs representing the COS, supported a deputation of working men demanding public works and a Mayor's Fund. If council manual workers had their hours cut from 53 to 48, if they repainted their schools, extended the tram system, advanced the road works and slum clearance programme, and turned the churchyard into an open space, they could, she said, create eight hundred jobs. The *Oldham Chronicle* headed its editorial '*Ladies in Council*'. Experience and knowledge, 'blended by womanly sympathy', enabled the ladies 'to put before the council facts, arguments and reasons worthy for the closest attention'. The council as requested also established a mayor's committee on which, Mrs Lees insisted, working men should serve.[74]

The distress ebbed; but when it deepened three years later, it was made more savage by industrial disputes, lock-outs, and strikes. Mrs Lees was now mayor, and in August 1911 she was faced simultaneously with a railway, a carters', and a tramcar strike. Most of the mills closed. She collected relief funds; and in a long day of high drama brought masters and men together, acting for thirteen hours as their conciliator. She ended the strikes. As the *Chronicle* reported, 'She closed that day of successful endeavour by driving the first tramcar out after the strike down to Hollingwood [her ward] late at night', to public cheering.[75] What gave her as much satisfaction was that troops were stationed in many Lancashire towns. She kept them out of Oldham. Her obituary notice said:

She accepted without any reluctance at all the obligations as she saw them, of her Christianity, her sex, her wealth, and her convictions on political and social questions . . . Her great wealth was a trust, and

[74] *Oldham Chronicle*, 3 Oct. 1908.
[75] *Oldham Chronicle*, 26 Aug. 1911.

so . . . were her leading positions in the town, her strong constitution, her clear mind, and abundant energy. As a Christian she gave liberally out of her abundance. As a woman she felt called upon to take a part in unpaid public service . . . [as a Liberal] that the best service which can be rendered to men and women and their communities is the removal of any obstacle that lies in the way of a fuller, happier and more useful life for them.[76]

[76] *Oldham Evening Chronicle*, 15 Apr. 1935.

10

Local Government
and the Women's Movement

Women's Work for Women

PHILANTHROPIC women and suffrage women, earlier chapters have argued, converged in late Victorian England on local government.

Philanthropy was instinctively conservative. It sought to reconcile givers and receivers not only to each other but also to the status quo. It redistributed a modicum of social wealth without disturbing the social structure, and linked rich and poor together in 'an invisible chain . . . of sympathy'.[1] But clouded though charity was by class and cultural assumptions, some of its practitioners developed a more radical perspective, determined in the words of Emily Janes of the National Union of Women Workers, to 'fence the precipice at the top before she provides an ambulance at the bottom'.[2] Fencing that precipice took many women into temperance and social purity work, others into local government. Such work should not be dismissed as narrow-minded and corseted puritanism, intrusive and resented though it often was. The father brutalized by drink, terrorized wife and children alike, as older women recalling their childhood will confirm. The young servant girl, seduced, pregnant, and abandoned, unable to return to her family, did not always want to take to the streets. Women reformers invaded every avenue open to them. Miss Foster Newton banned drink in her workhouse, sold coffee in the streets, and campaigned for a better water-supply. Over in Leicester, Miss Fullagar roamed the streets to round up stray dogs for the RSPCA and stray children for the

[1] J. R. Kay, *Four Periods of Public Education*, 1862, pp. 26–7.
[2] E. Janes, 'On the Associated Work of Women in Religion and Philanthropy', in A. Burdett Coutts, ed., *Women's Mission*, 1893.

school board at the same time. Mrs Wycliffe Wilson of Sheffield sent out election addresses in one parcel, bibles for atheists and Africans in the next. Women who in their philanthropic life remonstrated with the inebriate and rescued the fallen would as elected authority members restrict public house licenses, and set the workhouse prostitute on the straight and very narrow. It was all of a piece. Women's philanthropic network was a valuable resource, providing information, cash, and contacts with which to amplify their work as elected members. It formed their political capital. In their turn, women members could ensure more stringent licensing hours, could obtain official recognition for voluntary work, and integrate it into the map of provision. School dinners, care committees, health visiting, and baby clinics are all instances in which voluntary work was first aided by and then absorbed into local authority committees. Women were particularly adept at knitting together the public and voluntary fields of social service.

Suffrage women were equally insistent that politics needed a womanly perspective. If the statutes of the State were to reflect the moral order of the nation, then women must concern themselves with war and peace, the slave trade and the drink trade, child labour, animal welfare, slum life, the wrongs of Ireland, and the rights of prostitutes. To turn moral concern to effect, they needed the vote. Suffrage women shared many of the same assumptions as their philanthropic sisters—they came from the same milieu, after all—but they had perhaps a greater respect for the formal power of the State to elevate or discipline its members. Far more than philanthropic women, they were feminists, insisting that women as women should possess equal citizen rights alongside men. They turned to local government not only as a place of political power in its own right but as a stepping-stone to national power. They hoped that if women were seen as responsible local electors and valuable local authority members, then men would reward them with the parliamentary vote. They were wrong. Suffragist MPs certainly quoted women's local government work in suffrage debates in the House of Commons, but to little effect. Women's suffrage was not a moral or rational issue for men and could not therefore be won by women in moral or rational ways. (It could not be won by irrational or violent ways either.)[3] Deep Tory chauvinism allied itself to cynical Liberal party calculation to bar women from the vote. Indeed, anti-suffragists like Mrs Humphrey Ward were quite willing to argue that as women had the local vote they no longer needed the national vote. Women's public virtues, their moralism, their freedom from corruption, their independence of caucus and clique, were, one suspects, private handicaps in the eyes of men. Many MPs who supported women's suffrage were utterly hostile to the notion of women MPs. Whenever local government ladies highlighted their work as members, they stirred dark fears, that when they were given the vote they would sit in the House. Women were for a long time and quite illogically denied seats on town and county councils precisely because MPs were drawing a cordon sanitaire between local and parliamentary politics. When women's suffrage came in 1918, it owed little or nothing to women's local government work.

'Ladies elect' stood as women for women, stressing that they had abilities that were in no way inferior to men's, and aptitudes that men had not. They pointed to the clientele of school and poor law boards, mainly women and children, who required feeding and clothing, teaching and training, nursing and rescuing. This was work grounded in the detail of their daily lives, a far cry from the commercial office or customs house in which men passed their working hours. They could offer, women said, full time service, conscientious attention to detail, tender loving care to each individual. Few men had the time or taste for such work. It was all very seemly, unthreatening, and appropriate. In time most men conceded that school and poor law work was suitable for women. Town councils were rather different, as they addressed themselves as much to the business community as to families. Women reflected the confusion, not quite sure after 1907 whether they were standing as women councillors or as councillors who happened to be women, whether they should emphasize their equal competence or their differing experience. Sensibly, they usually had

---

[3] See B. H. Harrison, *Separate Spheres: The Opposition to Women's Suffrage in Britain*, 1978; R. Evans, *The Feminists: Women's Emancipation Movements in Europe, America and Australia, 1840–1920*, 1977.

it both ways, allowing transport, utilities, and sewage farms to be managed by men, but recasting most of the rest, such as housing, health, planning, and amenities into a domestic vocabulary.

In the opening elections to school boards and in the early decades of women's poor law work, women stood as independent candidates. This was not because they saw themselves as free-standing and unaligned individuals, as the men seemed to think; but because they wanted support as representative women, tacit members of a hidden women's party with its own agenda of family values, individualized care, moral reform, and fair and equal treatment of female clients and staff.

Increasingly, however, women were members of a political party. In the late 1870s as church–chapel conflicts gave way to party conflicts, school board women came to share the platforms and write the manifestos of their male colleagues. From the mid-1890s, poor law boards were politicized, as working men and Labour candidates arrived ready to battle for generous outdoor relief. Town council elections were usually, though not invariably, fought on party lines; if wards were not contested (a practice that continued into the 1970s) this only meant they were not considered marginal. An irritating habit of the Labour Party was the fielding of candidates in unwinnable seats, to the annoyance and expense of everyone else.

With more or less reluctance, most women came to embed their elections into the party system. For Labour women this presented no problems. They knew that local government was about political priorities. They were socialists first, and their feminism was limited to the well-being of working-class women, and had little to do with middle-class rights. Tory women had come in from parish work to serve the poor. They operated within a party that did not really believe in local party politics, and they often continued to stand as independents. When required to come in behind the party line on municipal enterprise, for example, they did not let this stand in the way of their committee voting—which was usually progressive in everything except Anglicanism, imperialism, and unionism, none of which had much to do with local government. Liberal women, however, found themselves in real difficulties. If local male Liberals were suffragist and supportive, women like Miss Emily Sturge of Bristol, or Mrs Cowen of Nottingham were proud to be part of the slate, believing that this was the greater triumph. More hesitant male caucuses were often persuaded to incorporate women when they experienced the effect of the school board cumulative vote; or when they needed to call on the efforts of women's Liberal associations to build up the register, canvass the electorate, chase the removals, bring out the women's vote and generally come to the aid of the party. But where the caucus remained anti-suffrage and antagonistic to women's entryism, women Liberals were in disarray, torn between conflicting loyalties. Some tried to stand as independents, holding fast to liberalism while abandoning the Liberal party, others withdrew from the fight. Whether women candidates were welcomed, neglected, or deplored, depended, as previous chapters have suggested, on the local political scene and the chauvinism of the local caucus. Yet in so far as women were adopted by their slates, they were still primarily women candidates, who expected to pass their seats down the female line. Other women would follow where they had ploughed.

Their numbers grew and quickly. By 1900 most town and city school boards of eleven to fifteen members had a couple of women, as healthy a proportion as would sit today on an education committee. Although women were proportionately fewer on poor law boards, their presence of around a thousand in 1900 and over fifteen hundred in 1914 formed a 'critical mass' of some visibility. However, on town councils women were almost as scarce before the War as the first generation of women MPs were to be after the War. Why? As with parliament, what mattered was not selection for a ward but which ward you were selected for. Women were handicapped by a property qualification outside London, and standing for the wrong party in London. They were seeking winnable seats that men were hungry to have, expected to hold, and in a simple majority system were more likely to win. Only at a by-election, boundary revision, or all-out contest, where something akin to a slate might operate, did women have much chance. As it was not clear to party agents or voters that women had much to contribute to town hall business, and as they had neither the local cumulative vote nor the parliamentary vote to give them electoral clout, then the scarcity of their numbers is not entirely surprising. When after the War these disabilities largely disappeared, their numbers rose sevenfold.[4]

Within education, women kept the quality questions to the fore. In town after town, men built the buildings, sought grants, awarded contracts, maintained the fabric, and levied budgets. Quite properly they brought their business skills to bear on public service. They seldom talked about children, except in the aggregate. It was women who 'read' the school population as a community of groups with special needs, who persuaded male colleagues to value kindergarten teaching for small children, who protected older girls from over-pressure, who befriended delinquent children and supported handicapped ones. They sought to cultivate the entire child and did much to turn Huxley's dream of a liberal education into the minutiae of daily lessons, broad curriculum, and practical care. They insisted that instruction was not education; and resisted payment by results, rote learning, and 'dinning it in' by corporal punishment, with equal fervour. They insisted, too, on caring for the whole child, recognizing that you cannot teach a hungry child, cannot reach a tired child, and cannot discipline a sickly one. They mobilized medical opinion, their philanthropic network, and the women's movement to provide clothing and swimming, school meals and medical inspection. They also understood that poorer children came from families and streets deeply hostile to regimented and coercive formal schooling, and they showed far more sensitivity than most men to the pressures that compulsory attendance hand in hand with compulsory fees placed on fragile working-class family economies. Women were among the most outspoken in attacking school fees and while never wavering in their commitment to compulsory attendance were among the more hesitant in imposing it in an arbitrary way. They preferred to win consent, spending endless hours counselling parents and visiting tired mothers in their homes. They tried to employ sympathetic staff to bring reluctant

[4] See Appendix B.

children to school. They wanted to root the school into the local community and not merely impose it on the neighbourhood. They asked for buildings and books that parents could use as well, a syllabus more relevant to the children's life, mature trained teachers that parents would respect, and the addition of working men as school managers. Women were often among the first to declare for trade-union rates and a direct labour organization. It was not by accident that in town after town women won the vote and loyalty not only of other women but also of working men. When in 1902 elementary education passed to town councils, they were required to co-opt women on to their LEAs, a back-handed recognition of the value of their work.

Like school board women, poor law women also read their work through the eyes of its recipients, in this case the pauper inmate. Female guardians faced a general mixed workhouse designed to institutionalize and intimidate its occupants. They worked to subvert its purpose. They substituted domestic values for disciplinary ones, as they sought to deconstruct the moral ideology and the very bricks and mortar of the Victorian workhouse. They brought children out into cottage homes and foster care; the sick into purpose-built infirmaries with trained nurses; the imbecile and fallen into small training homes (though there were never enough places for them); and they helped to provide the elderly with more comfortable surroundings where they might gossip, brew tea, and avoid baths, just like home. Along with prisoners, tramps, and lunatics, paupers were among the most marginal members of society, dismissed as worthless, denied most civic rights. Yet over some twenty-five years, the stigma softened and their quality of life immeasurably improved. Women did it.

There was another side to their work. Instinctive economic liberals, trained and toughened by the COS and temperance work, independent of party manifesto and relatively distanced from working-class life, women guardians were ruthless in their opposition to outdoor relief, believing it to be moral leprosy. Labour men and working-class communities trusted school board women; they very often detested poor law women. After 1894 many were voted off the board, and their place taken by women who were married, often active party workers, and looking to the NUWW, Guild of Help, or the settlement movement for philanthropic support rather than the censorious COS. More and more, they lived north of the Watford line. They were to support pensions for the elderly, public works for the unemployed, outdoor relief for the widowed; and were to work with Labour men under a Progressive ticket.

Philanthropic, education, and suffrage work brought women into local government. So did their sanitary work with its roots in the health lecturing of the 1860s, and their concern to build healthy homes for healthy families within healthy streets. The 1894 Local Government Act brought a few women on to London's vestries, and some one hundred and forty women on to the rural district councils, where they not only served as guardians, but also repaired cottages, surfaced roads, found allotments, and did something about drains and dustbins.

The final citadel fell when in 1907 women came on to borough councils. They had little to say about major labour contracts, a new waterworks scheme, the acquisition of a transport undertaking, or the virtues of boundary extension. However, as full education committee members they chaired sub-committees and built a more generous school system for fit and handicapped children alike. On health committees they sought to repair property and replan slums, building garden suburb cottages designed with women and children in mind. They promoted baths and washhouses, parks and gardens, street trees and safer street lighting, a womanly version of the built environment. Above all, women's council concerns converged on the matter of infant mortality. What, in other decades, might have been a marginal woman-only question, had been given salience by current eugenic debates and the driving energy of Medical Officers of Health. Maternity and infant mortality had become too important to be left to mothers. For a while it occupied the ground of municipal high politics. As all women naturally knew about children, on such matters they could command authority and attention normally denied to opposition back-benchers. They 'fronted' for their male Progressive colleagues.

Fifty years after women obtained the municipal vote, they

acquired the parliamentary vote. Led by Lady Astor, they also became MPs. How far was the experience of women breaking into local government shared half a century on by women breaking into parliamentary politics?[5] From what background did women MPs emerge? How easily did they win seats? To what extent were they women MPs or MPs who happened to be women? To what degree were they incorporated like local government women, or marginalized by male colleagues?

Seventeen women stood for parliament in 1918, well-educated, childless, often trained in local government. By 1929 the parties were fielding nearly seventy candidates, a number that remained steady until in the 1970s it doubled. The number of women actually elected did not. From the late 1920s, between twenty and thirty women were elected to each parliament. More women stood; no more were elected. Their problem was not finding a seat to contest, but a seat they could win. Tory women did relatively better at national than at local level, since from Nancy Astor on, nearly a third of Tory women MPs inherited the seats of elevated or absent husbands, and could therefore enjoy a secure parliamentary career. Labour women MPs by contrast sought seats on their own account, and like most women councillors found themselves with marginals held by the opposition or carved out by redistribution. They came in at one election, and, like Susan Lawrence or Dorothy Jewson of Norwich, went out at the next. They could keep their seats, they could not pass them on. 'Each time, it seems, the ground has to be ploughed anew'.[6] Left to themselves, constituencies almost always choose the safe candidate, the middle-aged man from the 'articulate' professions.

Parliamentary women also faced, even more sharply than local government women, the question of where their loyalties

[5] This account draws on P. Brookes, *Women at Westminster: An Account of Women in the British Parliament 1918–1966*, 1967; E. Vallance, *Women in the House: A Study of Women M.P.s*, 1979; B. H. Harrison, 'Women in a Men's House. The Women M.P.s, 1919–1945', *The Historical Journal*, 29. 3 (1986), pp. 623–54, and more generally, J. Lovenduski and J. Hills, eds., *The Politics of the Second Electorate: Women and Public Participation*, 1981; O. Banks, *The Faces of Feminism*, 1981; V. Randall, *Women and Politics*, 1982; J. Siltanen and M. Stanworth, eds., *Women and the Public Sphere*, 1984.

[6] Vallance, op. cit., p. 58.

lay. Local authority women before 1914 addressed women's questions, and stretched the municipal language to take in sewers and streets as well. Sometimes these women questions were already an integral part of Progressive manifestos—slum clearance, school meals, and medical inspection—but men accepted that women possessed privileged experience and at least in public welcomed their expertise. Other questions were pressed by women on men with considerable success—bringing children out of the workhouse, for example, or developing special schools for the handicapped—and men allowed women to reshape these local government policies and to add on further services, as long as it rebounded to the credit of the party. Women had only partial success with a third cluster of issues, usually requiring building-based capital expenditure, such as washhouses, lavatories, and lodging-houses, items to which most men were indifferent most of the time, though if women councillors were backed by a strong local women's movement, they could sometimes extract the occasional success. A final set of women's questions related to women staff—equal opportunities for women teachers, inspectors, doctors—and here women were vigorously resisted by men. Women's gains would be at the expense of men. They were trespassing on male space. In whatever board or body of town, most elected local women would have recognized this agenda, and would have accomplished a somewhat similar portfolio of achievement.

Women MPs were far more ambivalent. They were (mostly) party MPs first (Eleanor Rathbone a striking exception),[7] constituency MPs concerned with coal, shipping, or the shoe trade second, and women MPs only third. Parliamentary time was dominated less by social questions than by fiscal, foreign-policy, and industrial issues unfamiliar to local government. This did not stop men from pressing women MPs on to women's issue committees when they arose, nor women MPs from devoting their private bills to women's matters.[8] Eleanor Rathbone made family allowances her life's

---

[7] Miss Rathbone sat as an Independent MP for the Combined Universities Seat. See M. Stocks, *Eleanor Rathbone*, 1949.

[8] Of 25 private bills introduced by women, 3 were on drink, 3 on animals, 9 on women and children, 4 on consumer interests. Brookes, op. cit., Appendix C, p. 280.

work; and women MPs could occasionally be found operating as a woman's party on topics such as womanpower during the Second World War, Castle's Equal Pay legislation, or abortion. However, many women MPs resisted attempts to confine them to women's issues, fearing with some justification that in the male clubland of Westminster this would send them into a career cul-de-sac. Instead, they accepted male versions of what mattered, and as junior ministers sought 'hard' portfolios in transport, pensions, foreign affairs, rather than the 'soft' portfolios of education, health, culture, and welfare. Only recently, have some younger women MPs come to challenge the vocabulary of hard and soft, male issues which are mankind's and women's issues which are only women's.

It was all much simpler for local government women. Local government had more women's content in it than national government, and women councillors had less to fear from cul-de-sacing than women MPs. They were not, after all, bidding to become Cabinet Ministers. They occupied, and clearly felt comfortable in, a semi-detached sphere of their own.

The language of separate spheres has had a bad press. It seems at first sight irredeemably conservative, validating a male stereotype of women's nature (feeling rather than thinking, being rather than doing) which was used to confine women to the private and domestic realm. They were denied the vote in their own true interest. As the anti-suffragist Mrs Humphrey Ward deployed a vocabulary of separate spheres when founding her Local Government Advancement Committee in September 1911 to deflect suffrage demand, this seems evidence enough.[9]

Yet in a world where few middle-class women had salaried work, and where few men in public life had experience of caring or parenting work, women knew that they did indeed possess different attributes and skills from men. To have sought access to local government mainly as surrogate men would have seemed to them to be not only absurd, but morally and socially wasteful. Local government needed—needs— both management skills and client-centred skills. Before the

---

[9] Harrison, op. cit., pp. 134–6; *Anti-Suffrage Review*, Aug., Dec. 1911.

First World War, these skills were gender specific. (They do not have to be, but they were.) Hence women could argue with both accuracy and conviction that local government needed a communion of labour. Separate spheres, the work that only women could do for women, was a language in which level-headed local government women talked about their practical and relevant experience and not in some more mystical way about their essential nature. They thought they were diluting male hegemony and male notions of what was important, not reinforcing it. Mrs Ward's Committee was recognized as the theatrical device it was; though it ran one or two women in hopeless wards, there is no evidence that it elected a single woman to office.[10] Local government women ignored it.

The language of separate or semi-detached spheres was deployed by women because they found it helped them, not because it helped men. It was a supportive language. It valued women's domestic background and showed how it could strengthen civic life. 'Enlarge a household and it becomes a workhouse . . . '[11] It gave women the confidence to come forward, it allowed them to avoid conflicts between their private and their public values and tensions between their private and their public work. It allowed them to avoid public criticism that they were competing with men in shrill and unseemly ways. (Elizabeth Sturge after all had been turned out of the house of one old clergyman 'for my unwomanly conduct' in supporting Mary Clifford's election as a guardian in the 1880s.)[12] It also carried a radical cutting edge. Standing for election, selling one's virtues, seeking votes, was itself a profound challenge to ladylike notions of propriety and if that challenge was softened by ladylike clothes and ladylike language, that was only sensible. At their bravest, women were well aware that they were seeking to reshape the priorities of local government, that they were refusing to accept male definitions of what was central and what was

---

[10] See Dorothy Ward's description of running Miss Alice Willoughby in Hoxton, *Anti-Suffrage Review*, Apr. 1912, who polled 158 votes in a Progressive seat. Canvassers were appalled by 'the degraded and degrading poverty' they encountered, and found themselves inspired by a 'burning spirit of the crying need for women's work'.

[11] See above, pp. 230–1.

[12] E. Sturge, *Reminiscences of My Life*, 1928, p. 56.

marginal, that they were asserting that women's needs counted for as much as men's. The language of separate spheres helped to give them the public space to say so.

Josephine Butler understood this very well.

I wish it were felt that women who are labouring especially for women are not one-sided or selfish. We are human first; women secondarily. We care for the evils affecting women most of all because they react upon the whole of society, and abstract from the common good . . . When men nobly born and possessing advantages of wealth and education had fought the battles of poor men, and claimed and wrung from parliament an extension of privileges enjoyed by a few to classes of their brother-men who are toiling and suffering, I do not remember ever to have heard them charged with self-seeking; on the contrary, the regard that such men have had for the rights of men has been praised, and deservedly so, as noble and unselfish. And why should the matter be judged otherwise when the eyes of educated and thoughtful women of the better classes are opened to the terrible truths regarding the millions of their less favoured countrywomen, and they ask on *their* behalf for the redress of wrongs, and for liberty to work and to live in honesty and self reliance?[13]

There was a price to be paid. By making an issue of their sex, women exposed themselves to sexual hostility. Men inevitably played the gender card, seeking extra votes as men of business at election time by dismissing the value of women's work. But most women, once elected, thereafter met with nothing but courtesy. Poor law women, more than any other group, faced dirty jokes and dubious pleasantries at their boards. A few were kept off the committees of their choice, at least initially, but over time they won a grudging acceptance and even admiration for their work. The antagonism that Miss Ricketts incurred in Brighton, Miss Richardson met in Bristol, Mrs Charles found on Paddington vestry, or Mrs Salter generated in Bermondsey, was clearly political rather than sexual in origin. That met by Miss Lupton and Mrs Byles in Bradford, Mrs Leach in Yarmouth, Miss Ryley at Southport, or Miss Taylor in London, was equally clearly sexual. To a degree they had provoked it, even relished it, as they sometimes admitted and as other women kept telling them. The worst

[13] J. Butler, *Women's Work and Women's Culture*, 1869, Introduction.

offenders seem to have been ill-equipped male staff, assistant teachers, workhouse masters and instructors, who did not take kindly to supervision by moralistic and better-educated women. Professional men, doctors, inspectors, senior clerks, Local Government Board staff, made women welcome and regarded them as allies.

It could not have been easy for the men either. Customarily, men offered and women expected a degree of social deference that was at odds with the tussle of electoral contest. Men disliked fighting women for seats—they had to be so careful in what they said and did—and much preferred women to be co-opted committee members without an independent base of their own. Women were well aware of the tensions their arrival generated. Prudently, they worked hard, kept quiet, and bore with clumsy male condescension. For the most part, women's good sense, men's good manners, and enough goodwill prevailed.

Local Government and the New Feminism

Since Victorian days women had been told that biology is their destiny. Just like cows, said Emily Davies crossly. Child rearing made women dependent on their husbands and confined them to a family life that was essentially private and non-public. The languages of biology, culture, and economics interlocked to justify and perpetuate a subordinate status for women. As Mill said over a century ago, the customary is regarded as the natural, the natural as the innate, and the innate as inevitable.

It was Mill's stepdaughter, Helen Taylor of the London school board, who claimed for the first feminists 'the right to belong to herself'. The circle around her, Elizabeth Garrett, Emily Davies, and Barbara Leigh Smith Bodichon, fought to provide legal rights for women within marriage and to permit them economic self-sufficiency outside it. Over all arched their claim for the vote. With the vote won, feminism fragmented between the Wars, some women working in the birth-control movement, others in the peace movement, and still others for their political parties. The two decades after the Second World War saw women adopt the full skirts, overstuffed houses, and privatized domestic values of a previous age. Men, if not women, had never had it so good.

Not until the later 1960s did feminism re-emerge to plough yet again the same ground marked out by Victorian feminists, to challenge yet again the biological, cultural, and economic subordination of women. The pill, and then the 1967 Abortion Act which followed the thalidomide tragedies, taught many women that their bodies belonged to them. Defence of the Abortion Act against persistent attacks in parliament and the press rallied women who would not otherwise have considered themselves feminist. Battered wives' refuges, women-centred maternity care, rape crisis centres, and well-woman clinics have since then been provided by women for women. Women councillors in the face of male hostility often helped to find them premises and obtain grant aid.

The American civil rights movement taught the same generation of women to 'read' cultural oppression. If some of the feelings of frustration, powerlessness, and inadequacy experienced by women were not peculiar to them as individuals but widely shared by other women, then they were structural and social in origin, not personal and pathological. In an oft-repeated phrase, the personal was political. Domestic violence, the concept of the family wage, the apportionment of housework, sexual harassment, the burden of caring for the elderly dependants, the sense of vulnerability on the streets, these were private experiences shaped by the public sphere; just as women's experience of 'public' paid work, as part-timers for example, was patterned by their 'private' responsibility for child care. The public sphere in its turn—the school, the workshop, the media, the union, the party, the social security system, the professions, the Church, the law—was not culturally neutral as between the sexes but presumed that male ways of seeing and doing things were the norm. They constantly signalled women's second-class status. In language that Josephine Butler would have recognized and sometimes used, women's groups in the late 1960s and 1970s began to offer alternative readings of the world around them, which emphasized women's consciousness rather than men's. Such groups wanted to teach their women members to value the authority of their own experience, to have confidence in their own skills, to favour political forms that were open, informal, and participatory. They viewed with grave distrust 'male' politics with its hierarchy, bureaucracy, centralized decision-making, its minutes, standing orders, and rule books.

The third dimension in forming women's lives has been their economic inferiority. Most working women are in women-only jobs, in retailing and catering, clerical and personal services, which are ill paid, often part-time, under-trained, and insecure. Women in the trade-union movement began in the 1960s to display a new assertiveness that was channelled into the equal pay and opportunities policies of the 1970s. From the later 1970s, working wives, unemployed husbands, and broken marriages, have all made the cultural and economic norm of working husband and dependent wife and children rather unusual.

As in the late nineteenth century, this second women's movement has multiple voices. Apart from ideological divides between Liberal, Radical, and Marxist feminists, women (like men) continue to be divided by social standing and social class; by race; by education, age, and affluence; by whether they are in work, at home, in marriage, with children; by political loyalties; and by place, north or south, city or countryside. The skills, views, and life-style of the comfort-ably-off home counties woman voluntary worker will be more than miles apart from the newly politicized miner's wife or the young black woman, estranged and unemployed in a midlands town. Yet many women have come, via voluntary work, support groups, tenants' committees, and peace camps, into formal politics. Local government, as it did a century before, has again experienced the backwash of the woman's movement.[14]

Local government, let us remind ourselves, touches the lives of women in at least three ways. It employs women; it

---

[14] See S. Bristow, 'Women Councillors: An Explanation of their Under-representation in Local Government', *Local Government Studies*, May/June 1980, pp. 73–90; id., 'More Women in County Government', *County Councils Gazette*, Feb. 1982; J. Hills, 'Women Local Councillors: A Reply to Bristow', *Local Government Studies*, Jan/Feb. 1982, pp. 61–71; J. Gyford, 'Our Changing Local Councillors', *New Society*, 3 May 1984; S. Goss, 'Women's Initiatives in Local Government', in M. Boddy and C. Fudge, eds., *Local Socialism*, 1984; S. Button, *Women's Committees: A Study of Gender and Local Government Formulation*, SAUS Working Paper, 45, Bristol, 1985; J. Loven-duski, *Women and European Politics*, 1986.

provides services for women; and it is a place of political power and public advancement for women. Take jobs. About half of all local government staff are women (1¾ million). Apart from education, they are overwhelmingly in the clerical and manual grades, archetypically, the home help. In 1986, 521 local authorities could muster three women chief executives and around fifty chief officers, about 1 per cent of the field between them and far less than in law, medicine, higher education, commerce, and industry.[15] Local councils also provide services that largely determine the social wage of ordinary families. Decent housing, schools and nurseries, day centres and meals on wheels, swimming-baths and sports centres, parks and theatres. Only the affluent can buy these for themselves, yet they are essential to the quality of women's life. More recently, other local authority services have become politically salient. The safety of street lighting, the visibility of local policing, the reliability of public transport, now determine whether women can move around the streets without fear or favour. When local government services contract, women are hurt twice over. If a nursery closes, or school dinners are stopped, women lose services and jobs. To take a less obvious example, the current concepts of community care bring the handicapped, the damaged, and the frail out of building-based institutional care and back into the family within the community. Highly desirable. However, women lose their paid work as nurses and professional carers within hospitals and homes, while at the same time they are expected to provide the same nursing care, unpaid and often unaided, within their own families. 'Women fill a space in public policy provision and their labour is assumed to be free . . . When the family has failed to fill the policy gap, women as mothers, daughters and wives have been blamed.'[16]

Local government matters to women for a third reason. It provides a platform for the latter-day version of the philanthropic and suffrage woman seeking to offer service, seeking access to decision-making. Their numbers have grown. The 1,800 women holding elected office at the outbreak of the

---

[15] Button, op. cit., p. 24. See also Local Government Operational Research Unit, *Women in Local Government, the Neglected Resource*, 1982.

[16] Button, op. cit., p. 30.

First World War had risen to 2,700 in the 1919 elections. By
the 1930s women nationally formed some 12 per cent to 15 per
cent of elected local members, a figure that remained steady
until the 1970s, though it varied sharply city by city. Liverpool, for example, had more women councillors in 1930 than
in 1970, Newcastle more in 1950 than in 1930 or 1970,
Plymouth more in 1970 than ever before. In 1974 local
government was reorganized, aldermen were abolished, older
councillors retired. Within a year women formed 17 per cent
of council members, within a decade nearly 20 per cent, far
more than in parliament.[17]

Why have women done so much better in local than in
parliamentary politics? Some of the reasons are fairly obvious.
Women have found it hard to obtain safe parliamentary seats.
Vacancies come seldom and are often a long way off. The
lucky few elected to Westminster may find themselves with
parliament in London, their family in one county, and their
constituency in another. Local government on the other hand
is local, and for most people part-time (between ten and
twenty hours a week). Women can integrate, rather than
splinter, their family, working, and political lives around a
sense of place. They do not have to make the divisive choices,
take the risks, or offer the sacrifices demanded by parliamentary politics. Indeed, women who help to develop local
services may well consider that they are enhancing rather
than diminishing the quality of their family's lives.

Local government is more accessible. There are many more
seats and far less competition for them. Multi-member wards
permit a slate to operate, though this probably matters less as
women have grown in number. It is true that women still sit
for the more marginal wards, but this may reflect urban
geography rather than discrimination. Ward committees like
candidates to live in the ward. Labour women are often in
teaching or public administration, where the hours are flexible
and the employer helpful; and they tend to live not in safe
Labour council estates but in residentially more mixed and
politically more marginal wards. Impressionistic evidence
from all-out elections suggests that women candidates are

---

[17] Municipal Year-books, quoted in Appendix C. *The Widdicombe Report*, research
vol. ii, 1986, p. 19.

welcomed by the electorate, as they have a reputation for conscientious constituency work.

Local government is less intimidating than national government. The Conservative Party handbook of 1975 encouraged women to stand for election in words redolent of 1875, 'Women are extremely well-equipped for local government. They have a vested interest in, and immediate knowledge of, the schools, services, housing, care of children, and the environment, which are the responsibilities of the local authorities.'[18] Conservative women continue to come into local government from voluntary work, having governed schools, delivered meals on wheels, or advised at the CAB, as generations of women have done before them. Labour women are more likely to have gained their confidence and experience from local party work, canvassing, and holding minor office. If they are in paid work, that is likely to be in the public sector, where they acquire a feel for how such institutions function. Men as ever find their way into local government from the world of full-time work, as trade-unionists, estate agents, and small business men, or on retirement.[19]

Those women with experience of both worlds often find local government the more satisfying, and not just because it is 'domestic'. Whether women find fulfilment in helping people with problems at ward level, developing a housing or education service at committee, or determining overall financial policy, local government offers more of it than in any other field. Parliamentary politics consists mainly of the patch-work of a constituency back-bench MP, three years as a parliamentary private secretary, and a once-in-a-lifetime chance of a private member's bill. Only those making it to ministerial rank experience a qualitative change in political life-style. Because local government is organized not on the cabinet system but on the committee system, power is more

---

[18] Quoted in Bristow, 'Women Councillors: An Explanation', p. 76.
[19] The British Federation of University Women, *Participation in Public Life at Local Level*, 1985, p. 31; men's routes, work 55 per cent, voluntary work 33 per cent, political work 12 per cent, 'life' 0 per cent; women's routes, work 12 per cent, voluntary work 41 per cent, political work 17 per cent, 'life' 26 per cent. John Gyford suggests that 'Conservatives . . . cast their net wider than their active membership in seeking candidates, drawing in those of sympathetic persuasion who have made themselves known to the social and charitable life of the community'. *Local Politics in England*, 1976, p. 68.

widely diffused. All councillors are members of committees, helping to formulate policy and moving the business along. There is less call for confrontational politics and adversarial style in the committee room, to the relief of those women members who have little liking for the combative cut-and-thrust of Westminster.[20] Councillors are also members of a party group that meets frequently and shapes policy collectively along manifesto lines. If she happens to chair a senior committee, a woman can probably do more, influence more, and deliver more than any MP outside government. She has the power to make a difference. Pedestrianizing a street, restoring an old building, developing an alarm call system for the elderly, let alone establishing an enterprise agency, or negotiating a major shopping complex, these are tangible gains. The sense of working for a knowable community, whether it be a rural district pioneering peripatetic services perhaps, an historic city with its heritage in trust, or a run-down borough with its economy to rebuild, this can be a heady aphrodisiac, even if it also demands unremitting attention to drains and dustbins. Governments come and go. Communities endure.[21]

Over the last decade, local government has become more open to the women's movement. More women have sought election; town halls are actively promoting equal opportunity policies for women staff; and women's issues are firmly on the agenda. Norwich, to take an example to hand, had one woman councillor, Mabel Clarkson, in 1913; three in 1920; eight in 1950 and 1970; but eighteen in 1986, or over 35 per cent of its members. Women also chaired 45 per cent of its committees and working parties. Women are fewer in the north and in Wales. In cities, the south, and in Conservative counties they have done rather well, 30 per cent in Winchester and 26 per cent in Buckinghamshire, to set against the 5 per cent in Welsh counties and 13 per cent in Cheshire.[22] As regional

---

[20] E. Vallance and E. Davies make the same point about women in the European Parliament, *Women of Europe, Women M.E.P.s and Equality Policy*, 1986, p. 8.

[21] J. Stewart, G. Jones, R. Greenwood, & J. Raine, eds., *In Defence of Local Government*, Inlogov Studies, 1981.

[22] *Municipal Year-book*, 1985, British Federation of University Women, op. cit., p. 11, whose figures for 1984 suggest that 19 per cent of county councillors, 22 per cent of city councillors, and 25 per cent of district councillors are women. Overall they comprise 5,300 of 24,000 elected members.

variations narrow, we may expect women to form some 25 per cent to 30 per cent of local authority members across the country within the next decade. Since women also vote in the same proportions as men, are members of political parties and take part in politically related organizations to the same degree as men, this might be thought to falsify the theories of earlier political scientists that women are indifferent to politics, that they are non-citizens.[23]

Victorian board women and Edwardian councillors would have shared with today's elected women the same concern with women staff, women's services, and women's political strength. They would have eyed appreciatively a recent initiative that is networking women's local government efforts in progressive cities, the women's committee. With the phrase 'Don't agonize—organize', younger women councillors in London pressed women's committees, women's officers, women's units, and open women's meetings on their Labour groups. The GLC women's committee led the way in 1982, to be followed by many London boroughs and a score of provincial cities. Male Labour politicians greeted them with more or less grace, Tory politicians with more or less mockery, older women councillors with considerable doubt. At the same time, local government conferences started to hold separate workshops to discuss women's issues, and women's sections within the Labour Party have revived. The newly formed Local Government Information Unit appointed women officers, who have co-ordinated the women's committee network following the demise of the GLC.[24]

[23] V. Randall gives the figures for women's participation rates, op. cit., p. 37; in 1964 90 per cent of women and 92.5 per cent of men voted, in 1970, 83.3 per cent to 87 per cent, in 1974 76 per cent to 79 per cent. The difference is explained by women's longevity; some women are too frail to vote. 40 per cent of Labour party members and 52 per cent of Conservative party members are women. This is not to deny that even at local government level, women coping with dependent children and waged work find it hard to make space for political activity.

[24] The London boroughs included Lewisham, Haringey, Hackney, Camden, Greenwich, Islington, Southwark; the provincial cities included Birmingham, Sheffield, Leeds, Norwich, Newcastle, Basildon, Wolverhampton, Nottingham, Manchester. I am not aware that any Conservative-controlled council has a woman's committee. For the former GLC, see their *Bulletins of the G.L.C. Women's Committees*; for the LGIU, their *Women's News Bulletins*; see also newsletters and minutes of women's committees, Camden, Hackney, and Haringey. (I owe these references to their women's officers, and to Ms V. Wise of the former GLC women's committee.)

Despite grant-aiding of a few voluntary groups that the media have professed to find provocative, most women's committees have worked to a solid agenda that any earlier elected woman member would have recognized. They have grant-aided child care centres, regulated sex shops, promoted training and employment schemes for women, organized women homeworkers, campaigned for safer streets and better public transport, built additional women's lavatories, scrutinized housing policies, funded caring schemes for the elderly and handicapped, and held open forums and workshops on policing policies, domestic violence, and sexism in education. (Though one Merseyside woman councillor sadly reported to a conference of women's committees that her proposals were received 'with incomprehension and resistance so far'.)[25]

Women's commitees have met hesitation from some other women councillors, who understandably fear that male comrades will push women's issues off the main agenda into a ghetto of their own. If a women's committee funds women's theatre, will the council's arts subcommittee with its much larger budget then only fund formal 'male' theatre? With that in mind, many women's committees have declined a more generous budget, confining themselves to publicity, small grants, and awareness-training within the town hall.

Separate spheres have been one problem, issues of accountability and structure have been another. Women's committees have a dual and sometimes divided loyalty, both to their party group and district party, and to the wider women's movement. A deeper tension has been that of structure. The women's movement has been experiential, open, and informal. Attendance and decisions shift from week to week in ways that local government people find impossible and, more importantly, exasperating. Women's groups which take part in public policy making or receive public money have had to adopt the male town hall procedures they so much distrust, such as the structured agenda and the respect for formal qualification. By so doing, they become incorporated, 'add-on structures' in their own words, which they regard as a

---

[25] *Minutes of the Standing Conference of Women's Committees*, 26 July 1984.

betrayal.[26] Yet in so far as this second wave of feminism is generating new modes of policy-making, out-reach consultation, and greater openness and publicity, it may help to change local government methodology itself.

Women's committees have coincided with other initiatives that Victorian women would have recognized and welcomed. The commitment in many cities to 'go local', to bring housing management, for example, out to area offices run by tenants committees, to decentralize decision-making, to share power, and to make local government more porous to community pressure, is one such. The respect on the Left for the new political palette of herbivores, greens, and peace protesters, or black pride and gay rights, is also within the tradition of women's support for hidden or marginalized minorities.

Politics, the process by which people determine the allocation of resources, matters as much to women as to men, and interests women as much as men. Where political systems are open to women, there they take part. However, the formal political system with most power, Westminster, has least women. Although parliamentary short lists usually now include at least one woman, and although between 15 per cent and 25 per cent of parliamentary candidates are women, very few are elected. As each constituency makes its own choice, without reference to the common good, it seems unlikely that the percentage of women MPs will reach double figures in the foreseeable future.[27]

There are two other possible ways of enlarging women's role in formal politics. Each involves constitutional change. Experience both of local government and of European elections shows that women (and blacks and working men) gain from large multi-member divisions, whose representatives are elected by proportional representation from a party list. Parties cannot afford to go to the electorate without a balanced slate. Yet the price may be thought high, aside from

---

[26] S. Roelofs, 'G.L.C. Women's Committees—femocrats or feminism', *London Labour Briefing*, Apr. 1983, p. 18; *Spare Rib*, 'Takeover Town Hall', Apr. 1983, pp. 6–8.

[27] The 300 Group, and the more recent Women into Public Life Campaign, sponsored by the 300 Group and the Fawcett Society, may help to bring more women forward. See Polly Toynbee, 'Going Public', *The Guardian*, 27 Oct. 1986. The 300 Group works for the election of 300 women MPs. See their newsletter, *Three Hundred Group News*.

party political considerations. Proportional representation would detach the member from the constituency; in an all too literal sense, MPs become party placemen. There is an alternative.

The boundaries between central and local government are man-made, as any comparison with European constitutional structures will confirm, or as any comparison between the county boroughs of the 1930s and of the 1980s will show.[28] Local government could do far more and central government far less if we so chose. The 1979 general election brought to power a government determined to curb public expenditure, limit local autonomy, and shift the boundary between municipal and private enterprise. To some extent, national party battles were relocated as a struggle between the central government of one party and the urban local government of another. The reorganization of 1974 was re-reorganized in 1986. The old London school board was reinvented, the old LCC/GLC abolished. Further reform of local government must follow. Its functions, finance, and structures face perhaps a decade of change. Rates? Regions? Full-time salaried councillors? Given the vast and perhaps irremediable disparity between the number of women in parliament and the number of women in local government, and given that local government reform is on the agenda, then the most effective way of strengthening women's participation in politics may be to devolve power to where women are, that is to local government, rather than to seek to being women to where power currently resides, at Westminster. Constitutional reform that devolved power from the centre to the locality, from London to the provinces, from south to north, from larger to smaller, from the upper-middle class to lower-middle class, from older to younger, from whites to some blacks, and from men to rather more women, surely cannot be bad. Since opinion polls show[29] that local government is more highly regarded than central government, such a reform would even be popular.

[28] See Vallance and Davies, *Women of Europe*; Lovenduski, *Women and European Politics*; Council of Europe, 'Women in Local and Regional Life', Conference papers, Sept. 1986.
[29] See e.g. 'Attitudes Towards Local Government', Oct. 1985, MORI research study for the Association of Metropolitan Authorities; MORI opinion poll conducted for the Audit Commission, July 1986, quoted in *Local Government Chronicle*, 1 Aug. 1986.

APPENDIX A

Local Government Functions and Structure 1894–1914[1]

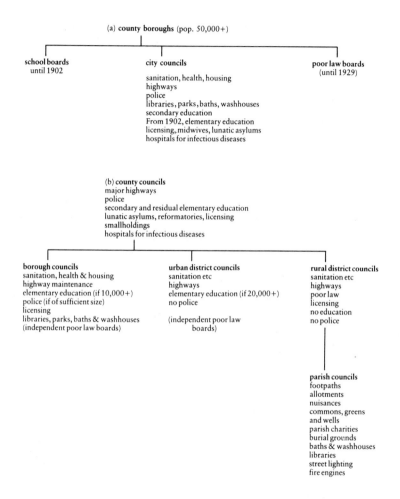

(a) county boroughs (pop. 50,000+)

school boards
until 1902

city councils
sanitation, health, housing
highways
police
libraries, parks, baths, washhouses
secondary education
From 1902, elementary education
licensing, midwives, lunatic asylums
hospitals for infectious diseases

poor law boards
(until 1929)

(b) county councils
major highways
police
secondary and residual elementary education
lunatic asylums, reformatories, licensing
smallholdings
hospitals for infectious diseases

borough councils
sanitation, health & housing
highway maintenance
elementary education (if 10,000+)
police (if of sufficient size)
licensing
libraries, parks, baths & washhouses
(independent poor law boards)

urban district councils
sanitation etc
highways
elementary education (if 20,000+)
no police

(independent poor law
boards)

rural district councils
sanitation etc
highways
poor law
licensing
no education
no police

parish councils
footpaths
allotments
nuisances
commons, greens
and wells
parish charities
burial grounds
baths & washhouses
libraries
street lighting
fire engines

[1] Local government structures outside London remained unchanged between 1914 and 1974; but services such as planning, council house building, personal health, social services, and consumer protection developed largely after 1918. See *The Royal Commission on Local Government in England*, 1966–9 (The Maud Report), Vol. i, Annex 3, pp. 323–30.

Women Elected Members
(England and Wales) 1870–1920

	School Boards	Poor Law Boards	RDCs	UDCs	Town & County Councils (incl. London)
1870	3				
1875	17	1			
1880	71	8			
1885	78	37			
1890	c.100	80			
1895	128	893[1]	140	2	
1900	c.270	1147	170	2	
1905	594[2]	1157	112	4	
1910	641	1310	147	4	22
1914–15	679	1546	200	15	48
1919–20	c.680	2039	263	67	320[3]

[1] From 1895, the poor law figures include RDC members.

[2] From 1903, women were co-opted to education committees, and not elected to school boards.

[3] The 320 women included 46 county councillors, 74 county borough councillors, 142 London borough councillors, and 58 borough councillors.

Sources: *Women's Suffrage Journal, Englishwoman's Review,* Annual Reports of the Women Guardians Society, Women's Local Government Society.

Municipal Councils:
The Representation of Women 1920–1985[1]

	1920	1930	1950	1970	1985
GLC[2]	9	24	39 (26%)	16 (14%)	14 (15%)
Bradford[3]	2	4	8	11 (14%)	19 (21%)
Bristol	0	4	17	18 (16%)	16 (24%)
Brighton	2	3	8	12 (16%)	13 (27%)
Birmingham	6	9	17	28 (18%)	26 (22%)
Liverpool	4	12	11	10 (6%)	14 (14%)
Manchester	4	8	13	22 (14%)	18 (18%)
Norwich	3	7	8	9 (14%)	15 (32%)
Leeds	0	14	17	22 (18%)	19 (19%)
Plymouth	3	5	9	15 (17%)	12 (20%)
Sheffield	2	12	8	20 (19%)	21 (24%)
Newcastle	1	5	18	16 (21%)	19 (24%)
Yarmouth	1	4	6	5 (10%)	4 (8%)

[1] The figures to 1974 include aldermen, subsequently abolished.

[2] London boroughs and smaller district councils have not been included as they were reorganized in the 1960s and 1970s respectively. On average, smaller district councils have more women members.

[3] Most cities changed the size of the council, upwards before 1950, downwards after 1970, over and beyond the loss of aldermen.

Sources: *Municipal Year-books*.

Summary of Relevant Local Government Legislation[1]

Poor Law Amendment Act, 1834 (4 and 5 Wm IV c. 76) as amended by the Poor Law Amendment Act of 1844 (7 and 8 Vic. c. 101)

Provides for election of Guardians in Poor Law Unions of Parishes.

Franchise for elections of Guardians:
All owners of land in the parish, and ratepayers rated for one year.

Scale of voting — for owners and ratepayers:
For property assessed up to   £50 — 1 vote
£100 — 2 votes
(and by £50 stages to £250 or more — 6 votes).

Qualification for election:
To be fixed by Commissioners, but not exceeding assessment for rates of £40.

Ex-officio members:
All JPs resident in the Union.

Method of voting:
In writing.

Municipal Corporations Act, 1835

Franchise for election of councillors:
All burgesses, i.e. male occupiers of any house, warehouse, counting house, or shop within the borough during the past three years, having been rated and paid all rates. Paupers excluded.

Scale of voting:
One vote for each vacancy to be filled.

Qualification for election:
Burgesses,
(a) in towns divided into four or more wards, owning real or personal property worth £1,000, or occupying land assessed for rating at £30 or more, and
(b) in other towns, real or personal property worth £500 or land assessed at £15 or more.

[1] Reproduced with kind permission from B. Keith-Lucas, *The English Local Government Franchise*, 1952. Appendix A, pp. 227–35.

Method of voting:
   By signed voting papers.

Co-opted members:
   Aldermen, chosen by the council.

The Municipal Franchise Act, 1869 (32 and 33 Vic. c. 55), amending Municipal Corporations Act, 1835

(*a*) One year's occupation of premises substituted for 2½ years, and list of qualifying premises extended to include 'other buildings'.

(*b*) If qualified as burgesses in all ways except residence, those resident within 15 miles of the borough are eligible for election as councillors or aldermen.

(*c*) Words importing the masculine gender held to include females.

Elementary Education Act, 1870 (33 and 34 Vic. c. 75)

Provides for the election of school boards.

Franchise:
   (*a*) In boroughs, by burgesses.
   (*b*) In London, as for vestrymen under Metropolis Management Act, 1855.
   (*c*) Elsewhere, all ratepayers.

Scale of voting:
   Each voter, one vote for each vacancy to be filled; voter can distribute votes as he will; may give all to one candidate.

Qualification for election:
   None.

Method of voting:
   In accordance with regulations to be made by the Education Department; in London, in accordance with the Metropolis Management Act, 1855.

Ballot Act, 1872 (35 and 36 Vic. c. 33)

Makes the secret ballot compulsory in all parliamentary and borough council elections, but not applicable to other local government elections.

County Electors Act, 1888 (51 and 52 Vic. c. 10)

1. Defines the franchise for county councils as:
   (*a*) The same as the burgess qualification in boroughs (12 months' occupation of a house, warehouse, counting house,

shop or other building in the borough, 12 months' residence in the borough or within 7 miles, 12 months' rating, and having paid the rates up to the previous 20th January), and

(*b*) Ten pound occupiers entitled to the parliamentary vote under the Registration Act, 1885 (12 months' occupation of any land or tenement in the area of £10 yearly value, six months' residence within 7 miles, 12 months' rating and having paid the rates up to the previous 5th January).

2. Extends the franchise for borough council elections by the inclusion of ten pound occupiers.

Local Government Act, 1888 (51 and 52 Vic. c. 41)

Provides for establishment of county councils.

Franchise for election of county councillors:
As under the County Electors Act, 1888.

Qualification for election:
As for borough councils, but including also
(*a*) Ministers of religion.
(*b*) Peers owning property in the county.
(*c*) Parliamentary voters in respect of the ownership of property.

Method of election:
As for borough councils (i.e. under the Ballot Act, 1872).

Co-opted members:
Aldermen, as in borough councils.

Local Government Act, 1894 (56 and 57 Vic. c. 73)

1. Parish meeting to consist of all parochial electors, i.e.:
(*a*) Local Government electors under the Local Government Act, 1888, and
(*b*) Parliamentary electors.

2. Parish councils.

Franchise for election of councillors:
Elected at the parish meeting by the parochial electors.

Qualification for election:
Parochial electors and persons resident in the parish or within three miles for previous 12 months (including women).

Method of voting:
By show of hands or by poll under the Ballot Act, 1872.

3. Rural district councils and urban district councils.

Franchise:
Parochial electors.

Qualification for election:
Parochial electors, and persons resident in the area for the previous 12 months (including women).

Method of voting:
As under Ballot Act, 1872.

4. Guardians.
(*a*) Ex-officio guardians and plural voting abolished.
(*b*) In rural districts, the rural district councillors to be the guardians.
(*c*) Elsewhere, to be elected in the same way as rural and urban district councillors and with the same qualifications.

5. London vestries.
The franchise, qualifications, and procedure for the election of vestrymen are made the same as those for urban district councils.

London Government Act, 1899 (66 and 63 Vic. c. 14)

Establishes Metropolitan Borough Councils, elected as the London vestries under the Local Government Act, 1894, but women ineligible for membership. Aldermen elected as in boroughs.

Qualification of Women (County and Borough Councils) Act, 1907 (7 Edw. VII c. 33)

Provides that a woman shall not be disqualified by sex or marriage for election as a councillor or alderman of a county council or borough council (including metropolitan boroughs).

County and Borough Councils (Qualification) Act, 1914 (4 and 5 Geo. V. c. 21)

Any person of either sex who has resided for twelve months in a borough or county is eligible for membership of the council.

Representation of the People Act, 1918 (7 and 8 Geo. V c. 64)

1. Establishes one standard franchise for the election of borough councils, county councils, urban and rural district councils, boards of guardians, parish councils and metropolitan borough councils.
For men:
Six months' occupation of land or premises in the area.

For women:

    Six months' occupation of land or premises in the area, or the wife of a man so qualified, on account of premises in which they both reside, if she is 30 years old.

2. Abolishes disfranchisement of paupers for all local government purposes, thus making them eligible for membership of county and borough councils, but not of other local authorities.

3. Owners of land within the area qualified to be elected but not to vote.

Representation of the People (Equal Franchise) Act, 1928 (18 and 19 Geo. V c. 12)

Amends the Act of 1918 in order to assimilate the franchise for men and women, the following to be entitled to be registered and to vote:

    (*a*) Any person of full age who has occupied land or premises in the area for six months, and

    (*b*) the husband or wife of a person so entitled to vote on account of premises in which they both reside.

Select Bibliography

PRIMARY SOURCES

i. Manuscript Collections

Ashton family papers, Manchester.
Blackburn collection, Girton College, Cambridge.
Bodichon papers, Girton College, Cambridge.
Cobden papers, West Sussex Record Office.
Courtney collection, vols. 22–30, LSE, London.
Elmy MSS, British Museum, London.
Emily Davies papers, Girton College, Cambridge.
Ensor papers, Bodley Library, Oxford.
Fawcett collections on suffrage, Manchester Public Library (Becker papers).
Garrett letters, Fawcett Library, London.
Howell collections, Bishopsgate Institute, London.
Lees collection, Oldham Public Library.
Lees letters, Fawcett Library, London.
M'Ilquham letters, Fawcett Library, London.
National Union of Women Workers papers, London.
Rathbone papers, University of Liverpool.

Eleanor Smith papers (Pattison papers, Pearson papers), Oxford, Bodley Library.
Strachey papers, Fawcett Library, London.
Helen Taylor papers, LSE, London.
Women's Emancipation Union papers, Fawcett Library, London.
Women's Labour League papers, Labour party Library, London.
Women's Local Government Society papers, GLC Record Office, London.

ii. Voluntary Societies, Papers and Printed Reports

Birmingham
    City of Birmingham Aid Society, Objects and Organisation, 1911.
    City Aid Review, 1906–14.
    Annual Reports, 1907–14.
    A Retrospect, 1907.
Birmingham Women's Suffrage Society, Annual Reports, 1869–1914.
Birmingham Women Workers Quarterly Magazine, 1893–1914.
Bolton Guild of Help, Annual Reports, 1908–14.
Bolton Guild of Help Magazine, 1906–14.
Bolton Women's Citizenship Association, Reports.
Bolton Women's Co-operative Guild, School for Mothers, Minute-books.
Bolton Women's Local Government Society, Minute-books, 1897–1915.
Bolton Women's Suffrage Society Minute-books, Annual Reports, 1909–19.
Bradford Charity Organization Society, Annual Reports.
Bradford Girls' Grammar school, *Jubilee Chronicle*, 1875–1925.
Birkdale (Southport) Association to Promote the Return of Women to Local Government Bodies, Reports.
Birkdale (Southport) Women's Liberal Association papers, Minute-books, cuttings, circulars, scrap-books.
Cambridge Charity Organisation Society, Reports.
Fabian Society, Annual Reports, 1888–1914.
Fabian Women's Group, *Three Years Work*, 1911.
    Annual Reports.
Fabian Society, *The Workers School Board Programme*, 1894.
Liverpool Central Relief Society, Annual Reports.
Liverpool Women's Citizen Association, Reports.
London Reform Union, Annual Reports, 1892–1914; leaflets.
Manchester & Salford Central Association of Societies for Girls and Women, Annual Reports, 1893–9.

Manchester & Salford Women's Citizen Association, Annual Reports.

'Manchester women's organizations' (u.d. typescript, Manchester Public Library).

National Union of Women Workers, conference reports, 1888–1914.
    Handbooks, 1892–1914
    Minute-books, 1891–1901; 1904–14.
    Occasional papers, 1890–1914.
    Threefold Chord, 1891–4.
    Tracts, 1896–1900.
    (J. M. King), *Women and public work*, NUWW pamphlet, 1902.
    Women candidates in local elections. Useful hints. How Cheltenham branch of the NUWW procured the return of a lady guardian, 1904.

Northern Workhouse Nursing Association, Annual Reports.

Norwich District Visiting Society, Annual Reports.

Nottingham & Notts. Association for Promoting the Election of Women Guardians, Reports.

Nottingham Town and Country Social Guild, Annual Reports.

Oldham Charity Organization Society, Minute-books, 1905–13.

Oldham Guild of Brave Poor Things, Minute-books, case books, 1903–14.

Oldham Liberal Bazaar souvenir, *Land Marks of Local Liberalism*, 1913.

Oldham National Union of Women Workers, Minute-books, 1910–14.

Oldham Women's Suffrage Society, Annual Reports, 1910–14.

Oxford Anti-mendicity & Charity Organization Society, Annual Reports, 1894–14.

Oxford Benevolent Society, Reports, 1909–13.

Oxford Girls Central School, Minutes.

Oxford Health and Housing Association, Reports, 1912, 1915–6.

Oxford Provident Dispensary, Rules and Reports, 1875–1896.

Oxford Sanitary Aid Association, Reports, 1908, 1909.

Oxford, Sarah Acland Memorial Home for Nurses, Annual Reports, 1879–97.

Oxford Women's Suffrage Society, Membership Lists, 1906–7.

Reading Guild of Help, Annual Reports, 1909–14.

Reading Health Society and School for Mothers, Annual Reports, 1912, 1913.

Society for the Promotion of Women as Poor Law Guardians (Women Guardians Society), Annual Reports, 1882–1904.

James Stansfeld Memorial Trust, Papers and Reports.

Vigilance Association for the Defence of Personal Rights, Annual Reports.

Women's Co-operative Guild, Annual Reports, 1892–1911.
    Annual congress proceedings, 1905–13.
    Coming of Age Congress, 1904.
    (M. L. Davies), *District Work*, 1908.
    Lillian Harris papers, LSE.
    Present Day Poor Law Administration, papers.
    Scheme of Winter Work, 1905.
Women's Emancipation Union, Annual Reports, 1891–9.
    Epitome of Eight Years Effort for Justice towards Women, 1899.
    Its Origin and its Work, 1892.
    Report of Bedford Conference, 1894.
    Report of Inaugural Meeting, 1892.
Women's Franchise League, *Report of the Executive Committee*, 1889–90.
    (Mrs Fenwick Miller) *On the Programme of the Women's Franchise League*, 1890.
    Woman and the Local Government Act, u.d.
Women's Labour League, Annual Reports, 1906–15.
    Minute-books, correspondence, and accounts, 1908–11.
    League leaflets, 1911–13.
    Labour Woman, May 1913–July 1914.
    Pamphlets, *The Feeding of School Children*.
        The Green Sprig Party.
        Help the Babies.
        Labour Women and Town Councils.
        Labour Women as Guardians of the Poor.
        Why Women want a Labour League.
    (M. MacDonald), *Women and the Labour Party*, 1909.
    —— *Women workers, souvenir of women's labour day*, July 1909.
    —— *Children at work, c.* 1914.
    (M. Phillips), ed., *Women and the Labour Party*, 1918.
    —— *The School Doctor and the Home.*
    (L. E. Sim), *The Working Woman in Politics*, 1914.
Women's Labour League, central branch, Annual Reports, 1910–14.
    Minute-books, 1906–14.
    Conference report, *The Needs of Little Children*, 1912.
Women's Labour League, MSS Annual Reports from Birmingham, Jarrow, Leicester, Liskeard, Kirkdale, Barrow, Portsmouth branches.
Women's Liberal Federation, Annual Reports, 1896–1906.

Women's Local Government Society, Annual Reports, 1888–1925.
  Minute-books, correspondence, and accounts, 1888–1914.
  Reports of the annual general meetings, 1905–14.

Women's Local Government Society leaflets:
  Why Women are wanted on Town Councils.
  Why Women are wanted as County Councillors.
  Why Women are wanted as Parish Councillors.
  Why Women are wanted as Poor Law Guardians.
  Why Women are wanted as Rural and Urban District Councillors.
  Why Women are wanted on London Borough Councils.
  From Westminster Women to the Women of Westminster, 1894.
  Women and Technical Education, 1894.
  Dr. Clifford on Women and the Education Bill of 1896.
  The London Government Act. The Latest Disqualification of Women, 1899.
  Local Education Authorities and the Need for Women as Members, 1901.
  Opinions . . . of the London School Board on the Work of Women Members, 1902.
  Women and the London Education Act, 1903.
  (Lady Trevelyan), *Women on County Councils,* 1904.
  —— *Women and Education.*
  The Position of Women in Respect to the Administration of the Midwives Act, 1902, 1904.
  Address to Miss Leigh Browne, 1907.
  Report on Reception, Lord Brassey's, 14 Feb. 1907.
  Report of the Trocadero dinner speeches, 1908.
  Seventeen Reasons why Women are wanted on the London County Council, 1908.
  Seventeen Reasons why Women are wanted on Town Councils, 1908.
  Some Reasons why Women are wanted on Town Councils, 1908.
  Speeches at the Anglo-Japanese Congress, June 1910.
  The Local Government Bill: An Explanation and an Appeal, 1912.
  Women on Local Authorities, 1912.
  The Franchise and Registration Bill, 1912.
  The Local Government Qualification Bill: An Explanation and an Appeal, 1912.
  Nineteen Reasons why Women are wanted on the London County Council, 1913.
  An Appeal to Electors, 1913.
  (Mary Beeton), *The Help and Influence of Women in Local Government,* 1916.
  (Clara Martineau), *The Work awaiting Women on County Borough Councils* 1920.

Women's Local Government Associations and How to Form Them.

(Miss Harston), *The Work of Women's Local Government Associations*, u.d.

Women's Local Government Society Branches, Annual Reports, Bolton (1914), Tynemouth (1913), Hampstead (1909, 1910, 1913).

Women's Municipal Party, pamphlets. *The Wants of the Women's Municipal Party*, 1913.

How All Women can Help to Govern London, 1913.

The Reason Why it Has Been Formed, 1913.

The Aims and Objects of the Party.

Objects and Programme of the Women's Municipal Party and Annual Report, 1915.

Handbook on Local Government.

Women's National Anti-suffrage pamphlets:

No. 3, *Speech by Mrs Humphrey Ward.*

No. 8, *Women's Suffrage and the National Welfare.*

Women's National Liberal Association, quarterly leaflets, 1896–1907.

Women's Suffrage, Annual Reports of the Executive Committee of the NUWSS 1869–1914.

Annual Reports of the Central National Society.

Workhouse Infirmary Nursing Association, Annual Reports, 1879–89.

iii.   Public Bodies and Local Authorities, Papers and Printed Reports

Birmingham, Minutes of Education Committee 1903–14.

Annual Reports of the committee of visitors of the lunatic asylums 1908–14.

Municipal year-books, 1908–14.

Bolton School Board, Minutes, 1900–3.

Year-books.

Bradford School Board Reports, 1880–1903.

Year-books.

Brighton School Board Reports, 1870–1903.

Year-books.

The Council of Europe, *Report to the Parliamentary Assembly on Women in Politics*, 22 Feb. 1983, Doc. 5370.

Women in Local and Regional Life, papers of the Standing Conference of local and regional authorities of Europe, Sept. 1986.

Manchester School Board, Reports, 1870–1903.

Norfolk rural district councils:
    Aylsham Rural District Council, Minutes, 1890–1914.
        Finance committee, Minutes, 1894–1914.
        Highways committee, Minutes, 1894–1914.
        Sanitary committee, Minutes, 1894–1914.
        Board of Guardians, Minutes, 1894–1914.
        Board of Guardians, House Committee, Minutes, 1894–1914.
    Blofield Rural District Council, Minutes, 1907–14 (incomplete).
        Board of Guardians, Minutes, 1894–1914.
    Docking Rural District Council, Minutes, 1904–8 (incomplete).
        Board of Guardians, Minutes, 1894–1914.
    Erpingham Rural District Council, Board of Guardians, Minutes, 1894–1914.
    Freebridge Lynn Rural District Council, Board of Guardians, Minutes, 1894–1914.
    Henstead Rural District Council, Minutes, 1894–1914.
        Board of Guardians, Minutes, 1894–1914.
    Mitford and Launditch Rural District Council, Minutes, 1894.
        Board of Guardians, Minutes, 1894–1914.
    Smallburgh Rural District Council, Board of Guardians, Minutes, 1894–1914.
    Thetford Rural District Council, Minutes, 1894–1914.
    Wayland Rural District Council, Minutes, 1907–14 (incomplete).
Norwich *Corporation Year-books*, 1880–1915.
    Abstract of Accounts, 1880–1915.
Norwich Incorporation of the Poor, Minutes, 1885–1914.
    Workhouse visiting committee, Minutes, 1890–1900.
    Relief committee minutes, 1894–1900.
    Instructions for the Foster Mothers of the Children's Homes, 1912.
Norwich School Board, Minutes, 1880–1903.
    Annual Reports, 1871–1903.
    Year-books.
Nottingham, School Board Reports, 1871–1903.
Oxford City Council, Year-books, 1907–14.
Oxford School Board Reports, 1870–1903.
Poor Law conferences: Reports of central conferences, 1880–1912.
    Reports of district conferences: West Midlands, North Midlands, South Wales, North, North-west, South Midlands, Yorkshire South-west, Eastern, South-east, and Metropolitan, 1880–1912.
Reading Corporation: Council minutes, 1911–13.
    Education committee, Minutes, 1907–14.

Sanitary committee, Minutes, 1907–14.
Watch committee, Minutes, 1907–14.
Annual Reports of the Chief Constable, 1908–14.
Annual Reports, Reading Education Committee, 1908–14.
Annual Reports, Medical Officer of Health, 1904–14.
Sheffield, School Board, Minutes, 1880–1903.
School Board Reports, 1871–1903.
Year-books.
Southport Corporation Year-books, 1905–14.
Tynemouth Borough Council Minutes, 1910–14.
Yarmouth School Board Minutes, 1880–1903.
School Board Reports, 1880–1903.

iv. Parliamentary Papers

Hansard reports
Select Committee on Poor Relief, PP 1861 ix; 1862 x; 1863 vii; 1864 ix.
School Board Attendance Committees, 1870–1889, PP 1889 lix.
School Board Attendance Committees, 1889–1895, PP 1895 lxxvi.
Select Committee on Poor Law Guardians, PP 1878 xvii.
Minutes of Evidence on School Board Elections, PP 1884–5 xi.
Cross Commission on Education, PP 1886 xxv; 1887 xxix; 1888 xxxvii; 1888 xxxv.
Select Committee on Poor Law Relief, PP 1888 xv.
Local Government Board Annual Reports, 1894–1914.
Select Committee on Distress from Want of Employment, PP 1895 viii, ix.
Royal Commission on Aged Poor, PP 1895 xiv, xv.
Royal Commission on Poor Law and the Relief of Distress, PP 1909 xxxvii, xl, xli, xlii; 1910 xlvii, xlviii, xlix.
Royal Commission on Local Government in England, 1966–9 (The Maud Report), 1969.
The Committee of Enquiry into the Conduct of Local Authority Business, (the Widdicombe Report), 1986, Report; Research, vols. i–iv.

v. Journals

Anti-Suffrage Review
Berkshire Chronicle
Birmingham Daily Gazette
Birmingham Daily Post
Birmingham Mail
Bolton Chronicle
Bolton Evening News
Bolton Journal
Bradford Daily Telegraph
Bradford Observer
Brighton Daily News
Brighton Examiner
Brighton Gazette
Brighton Guardian

Brighton Herald
British Women's Temperance Journal
Cambridge Chronicle
Cambridge Daily News
Charity Organisation Reporter
Charity Organisation Review
Contemporary Review
Daylight (Norwich)
Eastbourne Chronicle
Eastbourne Herald
Eastern Argus
Eastern Daily Press
Eastern Weekly Leader
East London Observer
Edgbastonia (Birmingham)
Englishwoman (1909–14)
English Woman's Journal
Englishwoman's Review
Englishwoman's Year Book (1894–1914)
Fortnightly Review
GLC Women's committee bulletins 1983–6
Hackney and Kingsland Gazette
Hampstead and Highgate Gazette
The Harvest (Manchester)
Islington News and Hornsey Gazette
Jackson's Oxford Journal
Kensington Express
The Lady's Realm
Leeds Mercury
Liverpool Courier
Liverpool Daily Post
Local Government Chronicle
London
Macmillan's Magazine
Manchester Faces and Places, 1889–90
Manchester Guardian
National Reformer
National Review
The Nineteenth Century
Norwich Mercury
Nottingham Evening Post
Nottingham & Midland Counties Daily Express
Oldham Evening Chronicle
Oldham Standard

Oxford Chronicle
Oxford Magazine
Paddington Mercury
The Parish Councillor
The Parish Councils Journal
Parish, District and Town Council Gazette
Poor Law Officers Journal
Primrose League Gazette
Primrose Record
Reading Mercury
Richmond & Twickenham Times
School Board Chronicle
School Board Echo (Bradford)
Shafts
Sheffield Daily Telegraph
Sheffield Independent
Shields Daily News
Shields Hustler
Southend Standard
Southend & Westcliff Gazette
Southport Guardian
Southport Visitor
Southwark & Bermondsey Reporter
Staleybridge Reporter
The Standard
The Times
Transactions of the National Association for the Promotion of Social Science
Westminster Review
Western Daily Press
Wings
Woman's Gazette
Woman's Herald
Woman's Penny Paper
Woman's Signal
Woman's Suffrage Journal
Woman's Trade Union Gazette
Woolwich Gazette
Woolwich Herald
Yarmouth Mercury
Yarmouth Times
Yorkshire Evening News
Yorkshire Observer
Yorkshire Post

vi. Pamphlets, articles, and memoirs

ABBOTT, Mrs G., *How as Guildswomen, Shall We Most Fully Use Our Powers under the New Local Government Act?*, 1894.

ADLER, H., 'Children and Wage Earning', *Fortnightly Review*, May 1903.
—— *London Women and the LCC*, 1924.
—— & TAWNEY R., *Boy and Girl Labour*, 1909.
ANDERSON, A., *Women in the Factory*, 1922.
ANDERSON, L. G., *Elizabeth Garrett Anderson*, 1939.
ARCH, J., *The Autobiography of Joseph Arch*, 1966 edn.
BALFOUR, Lady FRANCES, *Ne Obliviscaris*, 2 vols., 1930.
BALKWILL, M. E., *The Work of Women as School Managers*, 1914.
BANKS, E., 'Electioneering Women', *Nineteenth Century*, Nov. 1900, pp. 791–800.
BARKER, Mrs, *Parish Council Work for Women*, 1896.
BARNETT, H., 'Young Women in our Workhouses', *Macmillan's Magazine*, June 1879, pp. 133–9.
—— *Canon Barnett, His Life, Works and Friends, by his Wife*, 1921.
BAYLEY, E., *The Work of the School Board for London 1888–1891*, 1891.
BECKER, L., *The Rights and Duties of Women in Local Government*, 1879.
BEDDOE, Mrs A. M., *The Early Years of the Suffrage Movement*, 1911.
BENNETT, E. N., *The Problems of Village Life*, 1914.
BESANT, A., *An Autobiography*, 1908 edn.
BLAKE, M., 'Women as Poor Law Guardians', *Westminster Review*, Jan. 1893.
BOSANQUET, H., *Social Work in London 1869–1912*, 1914.
BRABAZON, R., 'A Woman's Work', *National Review*, Sept. 1884, pp. 80–90.
—— *Memories of the Nineteenth Century*, 1923 (Lord Meath).
BROWN, W. H., *Cooperation in a University Town*, u.d.
BROWNLOW, J., *Women's Work in Local Government*, 1911.
BUCKLEY, J., *A Village Politician*, 1897.
BUCKLEY, M. E., *The Feeding of School Children*, 1914.
BUCKTON, Mrs C., *Address on Leeds School Board: The Elementary Education Question*, 1882.
BURDETT-COUTTS, A., ed., *Women's Mission: A Series of Congress Papers on the Philanthropic Work of Women*, 1893.
BUSK, A., 'Women's Work on Vestries and Councils', in Revd J. Hand, ed., *Good Citizenship*, 1899.
BUTLER, J. *Women's Work and Women's Culture*, 1869.
CADBURY, M. C., *The Making of a Woman Guardian*, 1937.
CARPENTER, J. E., *The Life and Work of Mary Carpenter*, 1879.
CARPENTER, M., *Reformatory Schools for the Children of the Perishing and Dangerous Classes*, 1851.
—— *Juvenile Delinquents*, 1853.
—— *The Case Against Diggleism at the London School Board*, 1894.

CHADWICK, M., *Women's Work in Local Government: A reply to Mrs Humphrey Ward*, 1910/12.

CLIFFORD, M., *The Relation of Women to the State*, 1894.

—— *Out Relief*, 1896.

COBBE, F. P., *The Workhouse as a Hospital*, 1861.

—— *The Philosophy of the Poor Laws*, 1865.

—— *The Life of Frances Power Cobbe as told by Herself*, 1904.

COCHRANE, C., *How Women as Councillors may Improve Sanitation and Housing in Rural Districts*, 1904.

COLLET, C., *The Social Status of Occupiers*, 1908.

COOTE, W. A., *The Romance of Philanthropy*, 1916.

CRAVEN, F., 'Servants of the Sick and Poor', *Nineteenth Century*, Apr. 1883.

CROTCH, W., *The Cottage Homes of England*, 1901.

DALE, Revd R., *The Laws of Christ for Common Life*, 1884.

DAVIES, M. L., *The Women's Cooperative Guild, 1883–1904*, 1904.

—— ed., *Maternity: Letters from Working Women*, 1915.

—— ed., *Life as We have Known it*, 1930.

DE PLEDGE, J., 'Nursing in Poor Law Institutions', *Westminster Review*, Aug. 1894.

DESPARD, C., *Women in the Nation*, 1907.

DIXIE, Lady FLORENCE, *Women's Position, Social, Physical and Political*, Women's Emancipation Union Paper, 1892.

EDWARDS, G., *From Crowscaring to Westminster*, 1922.

ELMY, Mrs WOLSTENHOLME, 'Women in Local Administration', *Westminster Review*, July 1898.

FABIAN TRACTS:
  The Case for School Clinics, 1911.
  The Case for School Nurseries, 1909.
  Municipal Milk, 1905.
  Parish Councils and Village Life, 1908 (repr. 1920).
  Parish and District Councils: What they are and What they can do, 1926.
  What a Health Committee Can Do, 1910, repr. 1919, 1922.

FARNINGHAM, M. (Hearn), *A Working Woman's Life: An Autobiography*, 1907.

FAWCETT, M. G., 'The Women's Suffrage Question', *Contemporary Review*, June 1892, pp. 761–78.

—— *Home and Politics*, 1894.

—— *What I Remember*, 1924.

FORD, I. O., *Letter to a Friend Urging her to stand as a Parish Council Candidate*, u.d.

FORDHAM, E. O., *Women and Parish Councils*, 1906.

—— *Women as Poor Law Guardians*, 1907.

GALLOWAY, M., 'Women and Politics', *Nineteenth Century*, June 1886, pp. 896–901.

GANTREY, T., *Lux Mihi Laus: School Board Memories*, 1937.

GREEN, F. E., *The Tyranny of the Countryside*, 1913.

HAGGARD, H. RIDER, *Rural England*, 2 vols., 1902.

HASLAM, M., *Women's Suffrage in Bolton, 1908–1920*, 1920.

HEATH, F. G. *British Rural Life and Labour*, 1911.

HEATH, R., 'The Rural Revolution', *Contemporary Review*, 1895, pp. 182–200.

HERTZ, O., *Why We Need Women Guardians*, 1907.

HIGGS, M., *Five Days and Five Nights as a Tramp among Tramps — by a Lady.*

—— *Three Nights in Women's Lodging Houses*, u.d.

HIGGS, M. K., *Mary Higgs of Oldham*, 1954.

HILL, F. D., *The Children of the State*, 1868.

—— 'The family system for workhouse children', *Contemporary Review*, 1870.

'The boarding out of Pauper Children', *Economic Journal*, 1893, pp. 62–73.

HILL, G., *Women in English Life from Medieval to Modern Times*, 1896.

HILL, J., 'Workhouse Girls', *Macmillan's Magazine*, June 1873.

HILL, R. D., 'Technical Education in Boarding Schools', *Contemporary Review*, May 1888.

—— *Women on School Boards*, 1896.

HOGG, E. F., 'School Children as Wage Earners', *Nineteenth Century*, Aug. 1897, pp. 235–44.

HOMAN, R., *Women as Candidates: An Open Letter*, 1908.

HOPKINS, E., 'The Training of Pauper Girls', *Contemporary Review*, July 1882.

How the Destitute Live; and the *Reply by Norwich Guardians*, 1912.

HUTCHINS, B., *Working Women and the Poor Law*, 1909.

'The Inspection of Women's Workshops in London', *Economic Review*, Jan. 1901.

JAGGER, Mrs, *The History of Honley*, u.d.

JAMESON, A., *The Communion of Labour*, 1856.

JEFFERIES, R., *Hodge and His Masters*, 1880.

Memorials of Agnes Elizabeth Jones by her Sister, 1871.

KEYNES, F., *Women's Part in Cambridge's Progress*, 1938.

—— *Gathering up the Threads*, 1950.

KILGOUR, M. S., *Women as Members of Sanitary Authorities*, 1900.

KNOTT, E. A., *Women and Municipal Life*, 1902.

LANE-POOLE, S., 'Workhouse Infirmaries', *Macmillan's Magazine*, 1881, pp. 219–26.

LANSBURY, G., *My Life*, 1928.

LEACH, E., *Notes of a Three Months Tour in America*, 1883.

LEES, M., *An Account of the Women's Suffrage Movement*, 1912.

Leicester: Its Civic, Industrial, Institutional and Social Life, 1927.

LIDGETT, E., 'Poor Law Children and the Departmental Committee', *Contemporary Review*, Feb. 1897.

LINDSAY, A., *Christian Women as Citizens*, 1884.

LLANGATTOCK, Lady, 'The Primrose League: Why Women Should Support it', *The Lady's Realm*, June 1898, pp. 180–4.

LONSDALE, M., 'Platform Women', *Nineteenth Century*, Mar. 1884, pp. 409–15.

McCLAREN, E., *The Election of Women on Parish and District Councils*, 1894.

—— *The Duties and Opinions of Women with Reference to Parish and District Councils*, 1894.

MACDONALD, J. R., *Margaret Ethel MacDonald*, 1912.

MACDONALD, M., *Women and the Labour Party*, 1909.

M'ILQUHAM, H., *The Enfranchisement of Women*, 1892.

—— *The Rights and Duties of Women in Local Government*, 1892.

—— *The Duties of Women in Local Administration*, 1893.

—— *Women and the Parish Councils Act*, 1894.

—— *Local Government and its Limitations for Women*, 1895.

—— *Local Government in England and Wales*, 1896.

—— 'Women in Assemblies', *Nineteenth Century*, November 1896.

MACKAY, T., *History of the English Poor Laws*, 2 vols., 1903.

McMILLAN, M., *Life of Rachel McMillan*, 1927.

MALLET, L., *The Duties of Citizenship*, 1896.

MANSBRIDGE, A., *Margaret McMillan, Prophet and Pioneer*, 1932.

MARKHAM, V., *Return Passage: The Autobiography of Violet Markham*, 1953.

MARTINEAU, V., *Recollections of Miss Sophia Lonsdale*, 1936.

MARTYN, E., 'Women in Public Life', *Westminster Review*, 1889, pp. 278–85.

MASON, B., *The Story of the Women's Suffrage Movement*, 1912.

MASTERMAN, C. F. G., *The Heart of the Empire*, 1901.

METCALF, A. E., *Women's Effort: A Chronicle of British Woman's Fifty Years Struggle for Citizenship, 1865–1914*, 1917.

METCALFE, E., *Memoirs of Rosamund Davenport Hill*, 1904.

MIDDLETON, J., *Talking Him Over: An Election Dialogue*, 1908.

MITCHELL, H., *The Hard Way Up*, 1977 edn.

MORTEN, H., *Questions for Women*, 1899.

MULLER, H., *The Future of the Single Woman*, u.d.
—— 'Some Aspects of the London School Board', *Westminster Review*, Jan. 1888.
—— (?) 'The Work of Women as Poor Law Guardians', *Westminster Review*, Apr. 1885.

NIVEN, J., *Observations on Public Health Effort in Manchester*, 1923.

OAKESHOTT, J. F., *Humanizing the Poor Law*, 1894.
OAKLEY, C., 'Of Women in Assemblies', *Nineteenth Century*, Oct. 1896, pp. 559–66.
ORME, E., *Lady Fry of Darlington*, 1898.
OSTROGORSKI, M., 'Women's Suffrage in Local Self Government', *Political Science Quarterly*, 6.4, 1891–2, pp. 677–710.

PANKHURST. E., *My Own Story*, 1914.
PEABLES, F. W., *Great and Little Bolton Cooperative Society 1859–1909*, 1909.
—— *History of the Educational Department of Bolton Cooperative Society, 1861–1914*, 1914.
PEDDAR, D. C., *Where Men Decay: Pictures of English Rural Conditions*, 1908.
PHILPOTT, H. B., *London at School*, 1904.
PIGOTT, B., *Recollections of our Mother, Emma Pigott*, 1890.
PRESTON THOMAS, H., *The Work and Play of a Government Inspector*, 1909.
PRIESTMAN, A., *Recollections of H. B. Priestman*, u.d.
Quarterly Review, 'The Needs of Rural England', 1903, pp. 540–68.
—— 'Local Government', 1908, pp. 322–32.
—— 'The Past and Future of Rural England', 1913, pp. 490–512.

RAIT, R. S., ed., *Memorials of Albert Venn Dicey*, 1925.
RATHBONE, E., *William Rathbone: A Memoir*, 1905.
—— *Technical Education of Women and Girls in Liverpool*, 1910.
REDDISH, S., *Women and County and Borough Councils, a Claim for Eligibility*, 1903.
REDLICH, I., and HIRST, F., *Local Government in England*, 1903.
PEMBER REEVES, M., *Round About a Pound a Week*, 1913.
REID, Mrs H. G., 'Women Workers in the Liberal Cause', repr. *Westminster Review*, June 1887.
RIDDING, L., *Women on Education Authorities*, u.d.
ROGERS, F., *Labour, Life and Literature*, 1913.
ROGERS, Dr J., *Reminiscences of a Workhouse Medical Officer*, 1889.
ROSS, J., *Early Days Recalled*, 1891.
RUNCIMAN, J., *School Board Idylls*, 1885.

SHAEN, M., *Memorials of Two Sisters, Susanna and Catherine Winkworth*, 1908.

SHAW, E., 'The Workhouse from the Inside', *Contemporary Review*, Oct. 1899.

SIDDON, E., *Women's Work in the Administration of the Poor Law*, 1908.

SLACK, Mrs BAMFORD, *A Menace to Liberty*, 1903.

SMEDLEY, M., *Boarding Out and Pauper Schools: Mrs. Senior's Report*, 1875.

SNOWDON, E., *The Feminist Movement*, 1911.

SOUTTER, F., *Recollections of a Labour Pioneer*, 1923.

SPALDING, T., *The Work of the London School Board*, 1900.

STANLEY, M., *Clubs for Working Girls*, 1890.

STRACHEY, J., *The Part of Women in the State Organization of Taxation*, 1901.

STURGE, E., *Reminscences of My Life*, 1928.

SYKES, Dr J., 'The Role of the Medical Officer of Health', *Public Health*, 6, 1894, pp. 245–8.

SYNOTT, H., 'Little Paupers', *Contemporary Review*, 1894, pp. 954–72.

TANNER, S. J., *The Suffrage Movement in Bristol*, 1918.

TOWNSEND, F. M., *Election versus Cooption*, 1902.

TREVELYAN, C., 'Women in Local Government', repr. *Independent Review*, June 1904.

—— *Women on County Councils*, WLGS leaflet, 1904.

—— *The Life of Mrs. Humphrey Ward*, 1923.

TUCKWELL, G., *The State and its Children*, 1894.

TUCKWELL, Revd, 'Village Life and Politics', *Contemporary Review*, 1892, pp. 397–407.

TWINING, L., *Recollections of Workhouse Visiting and Management During Twenty-five Years*, 1880.

—— 'Women's Work, Official and Unofficial', repr. from the *National Review*, 1887.

—— 'Women as Public Servants', *Nineteenth Century*, Dec. 1890.

—— *Recollections of Her Life and Work*, 1893.

—— 'Women as Official Inspectors', *Nineteenth Century*, Mar. 1894.

—— *Workhouses and Pauperism*, 1898.

VARLEY, J., *Life in the Casual Ward*, u.d.

WARHURST, E., 'The Liverpool Women's Citizen Association', u.d.

WARWICK, FRANCES, COUNTESS OF, *Life's Ebb and Flow*, 1929.

—— *After Thoughts*, 1931.

WEBB, B., *Diaries*, vol. i, 1873–92; vol. ii, 1892–1905, ed. N. and J. Mackenzie, 1982–3.

—— *Our Partnership*, 1948.

WEBB, S., *London Education*, 1904.

A Word to Women Electors about the Borough Councils, Progressive leaflet No. 3, 1906.

SECONDARY SOURCES

vii. Theses

BILLINGTON, R., 'Women's Education and Suffrage Movements 1850–1914', Hull Ph.D. 1975.

DAWKINS, S., 'Perspectives on Childhood: Marianne Mason, Inspector of Boarded out Children', Bristol M.Ed. 1980.

DIGBY, A., 'The Working of the Poor Law in the Norfolk Economy in the 19th and 20th Century', Univ. of East Anglia Ph.D. 1971.

DOLTON, C., 'The Manchester School Board 1870–1902', Manchester M.Ed. 1969.

DOWNING, W. C., 'The Ladies Sanitary Association and the Origins of the Health Visiting Service', London M.A. 1963.

EVANS, A. J., 'A History of Education in Bradford, 1870–1904', Leeds M.A. 1947.

FOSTER, H. J., 'The Influence of Socio-economic, Spatial and Demographic Factors on the Development of Schooling in a Nineteenth Century Lancashire Residential Town' [Southport], Edge Hill M.Ed. 1976.

GANLEY, D. J., 'The Social and Political Composition of Oldham Town Council, 1888–1939', Lancaster M.A. 1976.

GARNER, L., 'The Feminism of Mainstream Women's Suffrage in Early 20th Century England: An Evaluation', Manchester Ph.D. 1981.

HARRIS, P. A., 'Class Conflict, Trades Unions and Working Class Politics in Bolton', Lancaster M.A. 1971.

JONES, A. W., 'The Work for Education of the Hon. E. Lyulph Stanley', Durham M.Ed. 1968–9.

JONES, D. W., 'Local Government Policy towards the Infant and Child with Particular Reference to the North West, 1890–1914', Manchester M.A. 1983.

KELLEY, S., 'A Sisterhood of Service: The First 25 Years of the National Union of Women Workers, 1895–1920', London M.A. 1985.

LEACH, C. E., 'The Feminist Movement in Manchester 1903–14', Manchester M.A. 1971.

MARSDEN, W. E., 'The Development of the Educational Facilities of Southport 1825–1944', Sheffield M.A. 1959.

MURPHY, L., 'School Board Elections in Bolton 1870–9', Bolton Institute of Technology B.A. 1979.

PENNYBACKER, S., 'The Labour Question and the L.C.C. 1889–1914', Cambridge Ph.D. 1984.

ROSS, E., 'Women in Poor Law Administration', London M.A. 1955.

ROSS, W. D., 'Bradford Politics 1880–1906', Bradford Ph.D. 1977.

SCOTT, J. C., 'Bradford Women in Organization 1867–1914', Bradford M.A. 1970.

SHAW, L., 'Aspects of Poor Relief in Norwich 1825–75', Univ. of East Anglia Ph.D. 1980.

SLOANE, W., 'The Cooperative Women's Guild and the Women's Suffrage Campaign 1893–1918', Loughborough Cooperative College B.A. 1978.

STEVENS, J. M., 'The L.C.C. under the Progressives', Univ. of Sussex M.A. 1966.

TAYLOR, A. E., 'The History of the Birmingham School Board 1870–1903', Birmingham M.A. 1955.

WALKER, L., 'The Women's Movement in England in the Late 19th and Early 20th Centuries', Manchester Ph.D. 1984.

WRIGHT, T., 'Poor Law Administration in Bradford, 1900–1914', Huddersfield M.A., u.d.

YEO, E. M., 'Social Science and Social Change: A Social History of Some Aspects of Social Science and Social Investigation in Britain 1830–1890', Univ. of Sussex Ph.D. 1972.

YOUNG, M. E., 'A Study in Public Administration: The Lancashire School Boards 1870–1902', Manchester Ph.D. 1977.

viii. *Modern Articles*

BLEWETT, N., 'The Franchise in the U.K. 1885–1918', *Past & Present*, 32, 1965.

BRISTOW, S. E., 'Women Councillors', *County Councils Gazette*, May, Nov., Dec. 1978.

—— 'Women Councillors: An Explanation of the Under-representation of Women in Local Government', *Local Government Studies*, May/June 1980.

—— 'More Women in County Government', *County Councils Gazette*, Feb. 1982.

CAHILL, M., & T. JOWITT, 'The New Philanthropy: The Emergence of the Bradford City Guild of Help', *Journal of Social Policy*, 1980.

CLARKE, F., 'The Progressive Movement', *Transactions of the Royal Historical Society*, 24, 1974.

CRUIKSHANK, M., 'Mary Dendy 1955–1933', *Journal of Educational Administration and History*, 8, 1976.

DAVIN, A., 'Imperialism and Motherhood', *History Workshop Journal*, 1978.

(Anon.), 'The Department of Social Science', *Social Science Society Bulletin* (Liverpool), July 1954.

DUNBABBIN, J., 'The Politics of the Establishment of County Councils', *Historical Journal*, 6.2, 1963.

DUNBABBIN (cont.), 'Expectations of the New County Councils and their Realization', *Historical Journal*, 8.3, 1965.

—— 'British Local Government Reform: The 19th Century and After', *EHR* Oct. 1977.

DYHOUSE, C., 'Working-class Mothers and Infant Mortality in England 1895–1914', *Journal of Social History*, 2, 1978.

FIDLER, G., 'Liverpool Socialists and the School Board', *History of Education*, 1980.

GOLDMAN, L., 'The Social Science Association 1857–1886: A Context for Mid-Victorian Liberalism', *EHR* Jan. 1986.

GOSDEN, P., 'The Origins of Co-option to Membership of Local Education Committees', *British Journal of Education Studies*, 1977.

GYFORD, J., 'Our Changing Local Councillors', *New Society*, 3 May 1984.

HARRISON, B. H., 'For Church, Queen and Family: The Girls Friendly Society 1874–1920', *Past & Present*, 61, 1973.

—— 'Women in a Men's House: The Women MPs, 1919–1945', *Historical Journal*, 1986.

HILLS, J., 'Women Local Councillors: A Reply to Bristow', *Local Government Studies*, Jan./Feb. 1982.

HUGHES, K., 'A Political Party and Education', *British Journal of Education Studies*, 1959–60.

KIDD, A., 'Charity Organization and the Unemployed in Manchester 1870–1914', *Social History*, Jan. 1984.

KITSON CLARK, G., 'Statesmen in Disguise', *Historical Journal*, 1959.

LEWIS, J., 'Parents, Children, School Fees and the London School Board 1870–1890', *History of Education*, 1982.

(Anon.), 'Liverpool Women's Citizen Association', *Liverpool Review*, Mar. 1932.

McCRONE, K., 'Feminism and Philanthropy in Victorian England: The Case of Louisa Twining', Canadian Historical Association, *Historical Papers*, 1976.

McDONAGH, O., 'The Nineteenth Century Revolution in Government', *Historical Journal*, 1958.

McKIBBON, R. I., 'Social Class and Social Observation in Edwardian England', *Transactions of the Royal Historical Society*, 28, 1978.

MOORE, M. J., 'Social Work and Social Welfare: The Organization of Philanthropic Resources in Britain 1900–1914', *Journal of British Studies*, 1977.

MUNSON, J. E. B., 'The L.S.B. Election of 1894', *British Journal of Education Studies*, 1975.

POPE, R., & M. BERBEKE, 'Ladies Educational Organizations in England 1865–85', *Paedogogica Historica*, 16, 1976.

Pugh, D., 'The Destruction of English School Boards', *Paedogogica Historica*, 12, 1972.

Pugh, M., 'A Note on School Board Elections', *History of Education*, 1977.

Rimmington, G., 'Leicester School Boards 1871–1903', *Leicester Archaeological and Historical Society*, 1976–7.

Roelofs, S., 'G.L.C. Women's Committees', *London Labour Briefing*, Apr. 1983.

Roper, N., 'Elections for Bradford School Board 1970–1903', *Bradford History of Education Bulletin*, 1968.

Rubinstein, D., 'Annie Besant and the L.S.B. Elections of 1888', *East London Papers*, 1970.

Sheppard, F., 'London and the Nation in the 19th Century', *Transactions of the Royal Historical Society*, 35, 1985.

Sheppard, M. G., 'The Effects of the Franchise Provisions on the Social and Sex Compositions of the Municipal Electorate 1882–1914', *Bulletin of the Society for the Study of Labour History*, Autumn 1982.

Showalter, E., 'Victorian Women and Insanity', *Victorian Studies*, Winter 1980.

Simey, M., 'Eleanor Rathbone', *Social Service Quarterly*, 40, 1966–7.

Simey, T. S., 'The School and Department of Social Science 1904–54', *University of Liverpool Recorder*, Jan. 1955.

Simon, B., 'The 1902 Education Act', *History of Education Society Bulletin*, 1977.

Spence, M., 'Hampshire School Boards', *History of Education Society Bulletin*, Spring 1978.

(Anon.), 'Takeover Town Hall', *Spare Rib*, Apr. 1983.

Taylor, T., 'The Cockerton Case Revisited: London Politics and Education 1898–1901', *British Journal of Education Studies*, 1982.

Thane, P., 'Women and the Poor Law in Victorian and Edwardian England', *History Workshop Journal*, 1978.

Toynbee, P., 'Going Public', *The Guardian*, 27 Oct. 1986.

Travis, M., 'The Work of the Leeds School Board 1870–1902', *Research & Studies*, 1953.

Van Arsdel, R., 'Victorian Periodicals Yield Their Secrets: Florence Fenwick Miller's Three Campaigns for the L.S.B.', *Warwick's Year Studies in English*, 1985.

(Anon.), 'The Victoria Settlement, Liverpool', *The Sphinx*, 24 Jan. 1906.

Wells, M. B., '6½ Years in Nethfield Rd.', *Liverpool Diocesan Review*, Sept. 1926.

WOODHOUSE, J., 'Eugenics and the Feebleminded: The Parliamentary Debates of 1912–14', *History of Education*, 1982.

WOODROOFE, K., 'The Royal Commission on the Poor Laws 1905–9', *International Review of Social History*, 22, 1977.

YOUNG, K., 'The Politics of London Government 1880–1899', *Public Administration*, 51, 1973.

ix. Books

ALEXANDER, A., *Borough Government and Politics: Reading 1835–1985*, 1985.

ARMYTAGE, W. H. G., *A. J. Mundella 1825–1897: The Liberal Background to the Labour Movement*, 1951.

—— *The American Influence on English Education*, 1967.

—— *Heavens Below: Utopian Experiments in England 1560–1960*, 1961.

ASHBY, M., *Joseph Ashby of Tysoe, 1854–1919*, 1961.

AUSUBEL, H., *In Hard Times: Victorian Reformers, 1867–1903*, 1960.

BANKS, O., *Faces of Feminism*, 1981.

—— *Becoming a Feminist: The Social Origins of 'First Wave' Feminism*, 1986.

BEHLMER, G., *Child Abuse and Moral Reform in England, 1870–1908*, 1982.

BELL, C. & R., *City Fathers: The Early History of Town Planning in Britain*, 1969.

BENNETT, W., *The History of Burnley from 1850*, 1951.

BENTLEY, M., and J. STEVENSON, eds., *High and Low Politics in Modern Britain*, 1983.

BIBBY, C., *T. H. Huxley on Education*, 1971.

BINGHAM, J. H., *The Period of the Sheffield School Board 1870–1903*, 1949.

BLUNDEN, M., *The Countess of Warwick*, 1967.

BOWMAN, E., *Stands there a School: Memories of Dame Frances Dove*, 1972.

BRETT, P., and F. S. FRANKLIN, *Thornham and its Story*, 1973.

BRIGGS, A., *The History of Birmingham*, 1951.

—— *Victorian Cities*, 1963.

BRISTOW, E. J., *Vice and Vigilance: Purity Movements in Britain since 1700*, 1977.

British Federation of University Women, *Participation in Public Life at Local Level*, 1985.

BROOKES, P., *Women at Westminster: An Account of Women in the British Parliament 1918–1966*, 1967.

BROWN, K. D., ed., *The First Labour Party 1906–14*, 1985.

BRUCE, M., *The Coming of the Welfare State*, 1966.

BURMAN, S., *Fit Work for Women*, 1979.

BURNETT, J., *A Social History of Housing*, 1978.

BURSTYN, J., *Victorian Education and the Ideal of Womanhood*, 1980.

BURTON, H., *Barbara Bodichon*, 1949.

BUTLER, A. S. G., *Portrait of Josephine Butler*, 1954.

BUTTON, S., *Women's Committees: A Study of Gender and Local Government Formulation*, School of Advanced Urban Studies working paper, 45, Bristol, 1984.

CAINE, B., *Destined to Be Wives: The Sisters of Beatrice Webb*, 1986.

CHAPMAN, S., ed., *The History of Working Class Housing*, 1971.

CHAPPLES, L., *The Last Resident* [Lady O'Hagan of Burnley], 1974.

CLARKE, P., *Liberals and Social Democrats*, 1978.

CRESSWELL, D'ARCY, *Margaret McMillan: A Memoir*, 1948.

CROWTHER, M. A., *The Workhouse System 1824–1929*, 1981.

CUTTLE, G., *The Legacy of the Rural Guardians*, 1934.

DAUNTON, H., *Coal Metropolis: Cardiff 1870–1914*, 1977.

DIGBY, A., *Pauper Palaces*, 1978.

—— & P. SEARBY, *Children, School and Society in 19th Century England*, 1981.

DONNISON, J., *Midwives and Medical Men*, 1977.

DUNBABBIN, J., *Rural Discontent in 19th Century Britain*, 1974.

EAGLESHAM, E., *From School Board to Local Authority*, 1956.

Educational Services Committee, ed., *Education in Bradford since 1870*, 1970.

EMY, H., *Liberals, Radicals and Social Politics 1892–1914*, 1973.

EVANS, R., *The Feminists: Women's Emancipation Movements in Europe, America and Australia, 1840–1920*, 1977.

FIDO, J., 'The C.O.S. and Social Casework in London 1869–1900', in A. Donajgrodski, ed., *Social Control in 19th Century Britain*, 1977.

FINNEGAN, F., *Poverty and Prostitution: A Study of Victorian Prostitution in York*, 1979.

FOLEY, A., *A Bolton Childhood*, 1973.

FOWLER, W. S., *A Study in Radicalism and Dissent: The Life and Times of H. J. Wilson 1833–1914*, 1961.

FRASER, D., ed., *The New Poor Law in the Nineteenth Century*, 1976.

—— *Power and Authority in the Victorian City*, 1979.

—— *Urban Politics in Victorian England*, 1976.

—— ed., *Municipal Reform and the Industrial City*, 1982.

—— & A. SUTCLIFFE, *The Pursuit of Urban History*, 1983.

FREEDEN, M., *The New Liberalism: An Ideology of Social Reform*, 1978.

FULFORD, R., *Votes for Women*, 1957.

GAFFIN, J., AND D. THOMS, *Caring and Sharing: The Centenary History of the Co-operative Women's Guild*, 1983.

GARDNER, P., *The Lost Elementary Schools of Victorian England*, 1984.

GARRARD, J., *Leadership and Power in Victorian Industrial Towns, 1830–80*, 1983.

GAULDIE, E., *Cruel Habitations: A History of Working Class Housing, 1780–1918*, 1974.

GIBBON, G., & R. BELL, *The History of the London County Council 1889–1939*, 1939.

GILL, A., 'The Leicester School Board', in B. Simon, ed., *Education in Leicestershire, 1540–1940*, 1968.

GORDON, P., *The Victorian School Manager: A Study in the Management of Education 1800–1902*, 1974.

GORHAM, D., 'Victorian Reform as a Family business: The Hill Family', in A. Wohl, ed., *The Victorian Family*, 1978.

GOSS, S., 'Women's Initiatives in Local Government', in M. Boddy & C. Fudge, eds., *Local Socialism*, 1984.

GRANT, I., *The National Council of Women: The First Sixty Years, 1895–1955*, 1955.

GRAVESON, R. & F. CRANE, eds., *1857–1957: A Century of Family Law*, 1958.

HAMILTON, M. A., *Margaret Bondfield*, 1924.

—— *Mary MacArthur*, 1925.

HAMMOND, J. L. & B., *James Stansfeld: A Victorian Champion of Sex Equality*, 1932.

HARRIS, J., *Unemployment and Politics: A Study in English Social Policy 1886–1914*, 1972.

HARRISON, B. M., *Drink and the Victorians: The Temperance Question in England 1815–1872*, 1971.

—— *Separate Spheres: The Opposition to Women's Suffrage in Britain*, 1978.

—— *Peaceable Kingdom: Stability and Change in Modern Britain*, 1982.

—— 'Women's Suffrage at Westminster 1866–1928', in M. Bentley & J. Stevenson, eds., *High and Low Politics in Modern Britain*, 1983.

HARROD, R. F., *The Life of John Maynard Keynes*, 1951.

HAYWOOD, J. H., *The History of the House of Haywood 1830–1930*, 1954.

HEALEY, E., *Lady Unknown: The Life of Angela Burdett-Coutts*, 1978.

HEASMAN, K., *Evangelicals in Action*, 1962.

HENNOCK, E. P., *Fit and Proper Persons*, 1973.

HOBSBAWM, E., and G. RUDÉ, *Captain Swing*, 1968.

HOLCOMBE, L., *Victorian Ladies at Work: Middle Class Working Women in England and Wales 1850–1914*, 1973.

—— *Wives and Property*, 1983.

HOLLIS, P., ed., *Pressure from Without*, 1974.

—— ed., *Women in Public: Documents of the Women's Movement 1850–1900*, 1979.

HOLROYD, M., *Lytton Strachey: A Critical Biography*, 1967.

HORN, P., *The Victorian Country Child*, 1974.

—— *Labouring Life in the Victorian Countryside*, 1976.

Howkins, A., *Poor Labouring Men: Rural Radicalism in Norfolk 1870–1923*, 1985.

Hume, L., *The National Union of Women's Suffrage Societies, 1897–1914*, 1982.

Humphries, S., *Hooligans or Rebels? An Oral History of Working-Class Childhood and Youth, 1889–1939*, 1981.

Hurt, J. S., *Elementary School and the Working Classes, 1860–1918*, 1979.

Hutchins, B., and B. Harrison, *A History of Factory Legislation*, 1903. 1903.

Huxley, L., *Thomas Henry Huxley*, 1920.

Jennings, H., *The Private Citizen in Public Social Work*, 1935.

Jephson, H., *The Sanitary Evolution of London*, 1907.

Johnson, J. H. & C. Pooley, *The Structure of 19th Century Cities*, 1982.

Jones, G. Stedman, *Outcast London: A Study of Relationships between the Classes in Victorian Society*, 1971.

Jones, G., *Borough Politics*, 1969.

Kamm, J., *Hope Deferred: Girls Education in English History*, 1965.

—— *Indicative Past: A Hundred Years of the Girls Public Day School Trust*, 1971.

—— *John Stuart Mill in Love*, 1977.

Kent, W., *John Burns, Labour's Lost Leader*, 1950.

Lambert, R., *Sir John Simon*, 1963.

Laski, H., ed., *A Century of Municipal Progress 1835–1935*, 1935.

Lee, J. M., *Social Leaders and Public Persons: A Study of County Government in Cheshire since 1888*, 1963.

Lewis, J., *The Politics of Motherhood: Child and Maternal Welfare in England, 1900–1939*, 1980.

—— *Women in England, 1870–1950*, 1984.

—— ed., *Labour and Love: Women's Experience of Home and Family 1850–1940*, 1986.

Liddington, J., *The Life and Times of a Respectable Rebel: Selina Cooper, 1864–1946*, 1984.

—— & J. Norris, *One Hand tied behind Us*, 1978.

Linklater, A., *An Unhusbanded Life: Charlotte Despard, Suffragette, Socialist and Sinn Feiner*, 1980.

Lipman, V. D., *Local Government Areas 1834–1945*, 1949.

Lord, M., *Margaret McMillan in Bradford with Reminiscences*, 1957.

Loughlin, M., M. D. Gelt, & K. Young, eds., *Half a Century of Municipal Decline 1935–1985*, 1985.

Lowndes, G., *Margaret McMillan, The Children's Champion*, 1960.

LOVENDUSKI, J., *Women and European Politics*, 1986.
—— & J. HILLS, *The Politics of the Second Electorate: Women and Public Participation*, 1981.
LUCAS, B. KEITH, *The English Local Government Franchise*, 1952.
McCANN, P., ed., *Popular Education and Socialization in the Nineteenth Century*, 1977.
McGREGOR, O., *Divorce in England*, 1959.
McHUGH, P., *Prostitution and Victorian Social Reform*, 1980.
McKENZIE, E. A., *Edith Simcox and George Eliot*, 1961.
MACKINTOSH, J. M., *Trends of Opinion about the Public Health 1901–1951*, 1953.
MCLACHAN, H., *Records of a Family, 1800–1933* [Mary Dendy], 1935.
MACLEOD, R., ed. D. Owen, *The Government of Victorian London 1855–1889*, 1982.
MACLURE, S., *One Hundred Years of London Education 1870–1970*, 1970.
MALMGREEN, G., ed., *Religion in the Lives of English Women 1760–1930*, 1986.
MANTON, J., *Elizabeth Garrett Anderson*, 1965.
—— *Mary Carpenter and the Children of the Streets*, 1976.
MARSH, J., ed., *Staleybridge Centenary Souvenir 1857–1957*, 1957.
MELLING, J., ed., *Housing, Social Policy and the State*, 1980.
MELLOR, H., ed., *Nottingham in the 1880s: A Study of Social Change*, 1971.
MERRETT, S., *State Housing in Britain*, 1979.
MIDDLETON, L., ed., *Women in the Labour Movement*, 1977.
MINGAY, G., *Rural Life in Victorian England*, 1977.
—— ed., *The Victorian Countryside*, 2 vols., 1981.
MITCHELL, J., & A. OAKLEY, eds., *The Rights and Wrongs of Women*, 1976.
—— eds., *What is Feminism?* 1986.
MOMMSEN, W. J., *The Emergence of the Welfare State in Britain and Germany*, 1981.
MORGAN, D., *Suffragists and Liberals*, 1975.
MOWAT, C. L., *The Charity Organization in Society, 1869–1913*, 1961.
NETHERCOTE, A. H., *The First Five Lives of Annie Besant*, 1961.
NEWSOME, S., *The Women's Freedom League, 1907–1957*, 1960.
NEWTON, J., M. RYAN, & J. WALKOWITZ, eds., *Sex and Class in Women's History*, 1983.
OFFNER, A., *Property and Politics 1870–1914*, 1981.
OLDHAM CENTENARY, *A History of Local Government 1849–1949*, 1949.
PANKHURST, S., *The Life of Emmeline Pankhurst*, 1935.
PELLING, H., *Popular Politics and Society in Late Victorian England*, 1968.
PHILLIPS, G., *The Diehards: Aristocratic Society and Politics in Edwardian England*, 1979.

POOLE, H., *The Council of Social Service, 1909–1959*, 1960.

PROCHASKA, F., *Women and Philanthropy in Nineteenth Century England*, 1980.

PUGH, M., *The Making of Modern British Politics, 1867–1939*, 1982.

—— *The Tories and the People 1880–1935*, 1986.

RANDALL, J., *The Origins of Modern Feminism, 1780–1960*, 1985.

RANDALL, V., *Women and Politics*, 1982.

REDFORD, A., *The History of Local Government in Manchester*, 3 vols., 1940.

REEDER, D., *Urban Education in the Nineteenth Century*, Proceedings of the 1976 Annual Conference of the History of Education Society of Gt. Britain, 1977.

RICHTER, M., *The Politics of Conscience: T. H. Green and his Age*, 1964.

ROBB, J., *The Primrose League 1883–1906*, 1942.

ROBERTS, C., *The Radical Countess: Rosalind, Countess of Carlisle*, 1962.

ROBERTS, E., *A Woman's Place: An Oral History of Working Class Women 1890–1940*, 1984.

ROEBUCK, J., *Urban Development in 19th Century London, Lambeth, Battersea and Wandsworth, 1838–1888*, 1979.

ROOFE, M., *A Hundred Years of Family Welfare, 1869–1969*, 1972.

ROSE, M., *The Relief of Poverty 1834–1914*, 1972.

ROSEN, A., *Rise up Women*, 1974.

ROVER, C., *Women's Suffrage and Party Politics, 1866–1914*, 1967.

RUBINSTEIN, D., *School Attendance in London, 1870–1904*, 1969.

SAMUEL, R., *Village Life and Labour*, 1975.

SANDERS, C., *The Strachey Family 1588–1932*, 1953.

SEARLE, G. R., *Eugenics and Politics in Britain, 1900–1914*, 1976.

SELLMAN, R. R., *Devon Village Schools in the 19th Century*, 1967.

SILTANEN, & M. STANWORTH, eds., *Women and the Public Sphere*, 1984.

SIMEY, M., *Charitable Effort in Liverpool in the 19th Century*, 1951.

—— *Charles Booth, Social Scientist*, 1960.

—— *Eleanor Rathbone 1872–1946, a Centenary Tribute*, 1974.

SIMON, S., *A Century of City Government 1838–1938*, 1938.

—— *Margaret Ashton and her Times*, 1949.

SKIDELSKY, R., *John Maynard Keynes*, vol. i, 1983.

SMITH, F. B., *The People's Health, 1830–1910*, 1979.

SMITH, W. H. RAWDON, *The George Mellys*, 1962.

SPRINGALL, L. M., *Labouring Life in Norfolk Villages 1834–1914*, 1936.

STEPHEN, B., *Emily Davies and Girton College*, 1927.

STOCKS, M., *Eleanor Rathbone*, 1949.

—— *My Commonplace Book*, 1970.

STRACHEY, R., *Millicent Garrett Fawcett*, 1931.

SUTCLIFFE, A., ed., *The Rise of Modern Urban Planning, 1800–1914*, 1980.

SUTHERLAND, G., *Policy Making and Elementary Education*, 1973.

THANE, P., ed., *The Origins of British Social Policy*, 1978.

THOMPSON, F. M. L., *English Landed Society in the 19th Century*, 1963.

THOMPSON, P., *Socialists, Liberals and Labour: The Struggle for London 1885–1914*, 1967.

TREBLE, J. R., *Urban Poverty in Britain, 1830–1914*, 1979.

TURNBULL, A., 'So Extremely Like Parliament: The Work of the Women Members of the London School Board, 1870–1904', in London Feminist History Group, *The Sexual Dynamics of History*, 1983, pp. 120–33.

VALLANCE, E., *Women in the House: A Study of Women M.P.s.*, 1979.

—— & E. DAVIES, *Women M.E.P.s and Equality Policy*, 1986.

VICINUS, M., ed., *Suffer and be Still*, 1972.

—— ed., *The Widening Sphere*, 1977.

—— *Independent Women: Work and Community for Single Women 1850–1920*, 1985.

WALKOWITZ, J., *Prostitution and Victorian Society*, 1980.

WALLER, P. J., *Democracy and Sectarianism: A Political and Social History of Liverpool 1868–1939*, 1981.

—— *Town, City and Nation: England 1850–1914*, 1983.

WALTON, R., *Women in Social Work*, 1975.

WALVIN, J., *A Child's World: A Social History of English Childhood, 1800–1914*, 1982.

WARDLE, D., *Education and Society in 19th Century Nottingham*, 1971.

WARDLEY, F., *The Founder: The Life of Dr. Mary Royce*, 1970.

WARREN, J. C., *A Biographical Catalogue of Portraits, High Pavement Chapel, Nottingham*, u.d. 1930–2.

WATSON, P., and V. ANAND, *Women and Politics*, 1984.

WEBB, S. & B., *The Parish and the County*, 1906.

WHYTE, P., *The Life of W. T. Stead*, 2 vols., 1925.

WIGHTWICK, H. W., *District and Parish Councils*, 1925.

WILLIAMS, G., *Mary Clifford*, 1920.

WILSON, R., *A History of the Sheffield Smelting Company Ltd., 1760–1960*, 1960.

WOHL, A., *The Eternal Slum*, 1977.

—— *Endangered Lives: Public Health in Victorian Britain*, 1983.

WOODROOFE, K., *From Charity to Social Work*, 1962.

WRIGHT, D. G., and J. A. JOWITT, eds., *Victorian Bradford*, 1981.

YOUNG, K., *Local Politics and the Rise of Party*, 1975.

Index